BETWEEN COPERNICUS AND GALILEO

BETWEEN COPERNICUS AND GALILEO

*Christoph Clavius
and the Collapse of
Ptolemaic Cosmology*

JAMES M. LATTIS

THE UNIVERSITY OF CHICAGO PRESS
Chicago and London

James M. Lattis is science director at the University of Wisconsin Space Place, and coeditor of the *Bibliography of the Italian Reformation*. He has taught astronomy and history of astronomy at the University of Wisconsin–Madison, and at Marquette University, and is the recipient of a 1994 Rome Prize, awarded by The American Academy in Rome.

The University of Chicago Press, Chicago 60637
The University of Chicago Press, Ltd., London
© 1994 by The University of Chicago
All rights reserved. Published 1994
Printed in the United States of America
03 02 01 00 99 98 97 96 95 94   5 4 3 2 1

ISBN (cloth): 0-226-46927-1
ISBN (paper): 0-226-46929-8

Library of Congress Cataloging-in-Publication Data

Lattis, James M.
    Between Copernicus and Galileo : Christoph Clavius and the collapse of
Ptolemaic cosmology / James M. Lattis.
        p.   cm.
    Includes bibliographical references and index.
    ISBN 0-226-46927-1. — ISBN 0-226-46929-8 (pbk.)
    1. Cosmology.   2. Astronomy, Medieval.   3. Clavius, Christoph,
1538–1612.   4. Ptolemy, 2nd cent.   I. Title.
QB981.L32   1994
523.1—dc20   94-8675
                                                                CIP

For Jen and Tony

Contents

Illustrations

Preface

One of the largest craters on the lunar surface is named for the subject of this book. The crater called Clavius, in the moon's southern hemisphere, measures some 232 miles in diameter. It achieved its only modern notariety in Arthur C. Clarke's *2001: A Space Odyssey* in which the crater was the location of the U.S. lunar base. (In the film version of Clarke's novel, the name was pronounced "CLAY-vius," in the characteristic manner of English. I prefer "CLAH-vius," in accord with the Latin origin of the name.) Most readers would have no trouble identifying the eponyms of such smaller craters as Copernicus, Tycho, Galileo, and the like but few will know who Clavius was. Thus Clavius's obscurity today is out of all proportion to the size of his monument on the lunar surface. This is due, in part, to the fact that Clavius's fellow Jesuits exercised considerable influence in the early days of lunar nomenclature. But even a non-Jesuit selenographer of the early seventeenth century would have had Clavius on his list of the more important astronomers of the time.

We encounter Clavius's name in other contexts, of course. He is routinely noted as one of the architects of Gregorian calendar reform and always occurs as a peripheral figure in studies of Galileo, Tycho, and Kepler, but Clavius has rarely been considered in his own right. Yet in his day, Clavius was considered to be as eminent and authoritative as any of those illustrious figures so widely studied today. The reason for the relative obscurity of Clavius's name is not difficult to fathom: Galileo, Tycho, Kepler, and Copernicus, too, represent new ideas, techniques, and methods in the history of science. We find in their work many of the most important roots of modern science. Clavius, in contrast, opposed many (though certainly not all) of the innovations represented by the others. The zeal to seek, identify, and exalt those in the history of science who were "right" and who seem to foreshadow our modern conceptions of ourselves has understandably (if unfortunately) allowed the study of figures who were "wrong," such as Clavius, to languish. The price of that neglect is the loss of appreciation and enjoyment of a fascinating part of the history of science.

Clavius was one of the foremost defenders of the ancient cosmology and mathematical astronomy epitomized by Ptolemy; and Ptolemy was one of the principal targets of Copernicus, Galileo, and Tycho. Furthermore, Clavius's work, both in its form and substance, was part of a tradition closely associated with the scholarship of the Middle Ages. His textbook, the *Commentary on the "Sphere" of Sacrobosco,* placed him firmly in a tradition that stretched back to the thirteenth century. Clavius's book also draws on the thirteenth-century *Theorica planetarum,* and on the fifteenth-century *Theoricae novae planetarum* (both of which are discussed briefly in Chapter 2). By the end of Clavius's life, both the venerable Ptolemaic theories and the established medieval and Renaissance textual traditions had become obsolete. In the view of some, perhaps, Clavius might have seemed an atavism, a lonely outpost of the ancient and medieval worlds in the midst of a new age. He was the last important Ptolemaic astronomer, and with him died that fifteen-hundred-year-old Ptolemaic tradition. (We must note, however, that this did not mark the end of geocentric astronomy, which continued yet a while in the guise usually identified as Tychonic and semi-Tychonic theories.)

Clavius would deserve to be studied even if he were significant only as a historical terminus. But we must step back and take a broader view. Who was Christoph Clavius? He was an early member of the Society of Jesus, whose system of schools, colleges, and universities grew rapidly in the late sixteenth and early seventeenth centuries. The Jesuit educational enterprise was international, in fact intercontinental, in scope and highly respected. Clavius himself was the founder of mathematical (and thus astronomical) studies at the center of the Jesuit academic network. He was the author of textbooks that were in use for a half a century nearly everywhere that Jesuits taught. As such, Clavius was one of the most authoritative and well-connected astronomers of his time. His opinions were very influential, and his support of or opposition to particular ideas carried great weight, not just among Jesuits or even Catholics but across Europe.

Novel ideas are, almost by definition, advocated by a few against the more accepted views of the many. It is the conservative establishment that evaluates and reacts to novelties, whether in literature, art, or science. Clavius, as Jesuit educator and spokesman for the established astronomy and cosmology, helped set the standards by which innovators, such as Copernicus and Galileo, would be judged. We cannot hope to appreciate fully the work and achievements of the innovators unless we understand

something of those who judged them. Furthermore, when we focus retrospectively on certain innovators and revolutionaries we risk seeing them
as isolated upstarts. But the defenders of orthodoxy had no more knowledge of their future than we do of ours. Thus Clavius, when he set out to
defend Ptolemaic cosmology against its rivals, had to address all of the
significant threats, not just those eventually vindicated by history. It turns
out that Copernicus was only one of a number of threats with whom
Clavius had to contend.

Clavius is also important to the history of astronomy because of the
wide publication of his textbooks, particularly his astronomy textbook.
Generations of university students learned what they knew of astronomy
and cosmology from Clavius's book. When we read Clavius's text, we
gain an understanding of what the average educated person of those days
knew and believed about the universe inhabited by humankind. We cannot learn this from the works of the innovators precisely because they
were innovators. The average student of the early seventeenth century
did not read Galileo, and certainly not Copernicus or Kepler. The dusty,
neglected textbooks of yesterday, not the celebrated controversial works,
give us a window into the minds of our real ancestors. As one of the
foremost explicators, defenders, and arbiters of late sixteenth- and early
seventeenth-century astronomy and cosmology, Clavius is a very important aperture to the past.

Clavius is important for another reason: his life spanned a very special
period in the history of astronomy. The first edition of Clavius's textbook
appeared in 1570, and his last revision appeared in 1611. This forty-one
year span included some of the most significant events of the "astronomical revolution": the nova of 1572, the comet of 1577, the novas of 1600
and 1604, the better part of Tycho Brahe's career, and Galileo's first
telescopic discoveries, to name the most important. His attempts to explain, interpret, and (sometimes) ignore the remarkable astronomical developments of his career are other ways in which Clavius and his book
offer us a window into early modern astronomy.

In the first chapter of this study I attempt to explain who Clavius was
and to summarize how he has been seen by historians. The second chapter
expands Clavius's context to include Jesuit education, the status of mathematical sciences in his day, and the textual traditions and content of medieval astronomy and cosmology. It includes a brief introduction to the
concepts and problems of geocentric astronomy in the Western tradition.
The fundamental source and primary focus of this whole study is Clavius's
famous and long-lived introductory astronomy textbook, and in the next
four chapters I explore that textbook. In chapter 3, I describe the "ortho-

dox'' cosmology that Clavius advocated and defended during his entire career. This is the universe as it was explained to his multitudes of readers. In chapters 4 and 5, I recount Clavius's battles with "rival" cosmologies—the alternatives (to Ptolemaic cosmology) that he noted and refuted. Chapter 5 is devoted to the Copernican rival. In chapter 6, I examine Clavius's responses to other challenges—both celestial and mundane— faced by traditional cosmology during the publishing history of his book. The study concludes with chapter 7, in which I recount the roles of Clavius and his fellow Jesuits in the reception and interpretation of Galileo's epochal telescopic discoveries.

My work on Clavius would never have begun without the advice and encouragement of Albert Van Helden, Owen Gingerich, and, most of all, David Lindberg. The scholarly work of each one has left its stamp on my own. From the very beginning Ugo Baldini unstintingly shared his time, experience, and his writings—both published and unpublished. His work on Clavius and the history of Jesuit science has made this book possible. Michael Shank's knowledge of the history of astronomy and his energies expended in critically reading my work are much appreciated as are Robert Smith's efforts on my behalf. I must also thank Steve Harris and Martha Baldwin for many helpful discussions and suggestions on the history of Jesuit science. And William Wallace has my sincere gratitude for his thorough reading of a draft of this work—the final version is much better, thanks to his efforts. Any errors in the work are, of course, my own.

I want to thank many people at Memorial Library, University of Wisconsin–Madison, for helping me locate important source material and navigate bibliographical thickets. The people of the Interlibrary Loan office and the Department of Special Collections were always ready to assist. The advice and expertise of John Neu and John Tedeschi have been particularly important. I also want to express my gratitude to the good fathers of the Jesuit archives in Rome (the Archivum Romanum Societatis Iesu), who amiably assisted me there.

Much of this work has been supported by the generosity of others. I thank the Pollock Award Committee of the Dudley Observatory whose award supported a crucial phase of my research and writing. Another vital stage was supported by a University Fellowship and a foreign travel grant from the University of Wisconsin–Madison Graduate School. Jeff Dorsey always saw to it that I had a place to sleep in Rome during my occasional research trips. I have received much support from family, especially Irene and John Sloan, Judy and Jim Sloan, and last but not least, Diana and Donald Lattis.

I wish to thank some of my good friends for things impossible to enumerate. Gary Sego, Peter Sobol, Pam Gossin, Jole Shackelford, Patty Harris, Jeff Percival, Virginia Green, and Herb Howe have unselfishly and in varying proportions provided help with translations, reading and criticism, stylistic expertise, good advice, constant encouragement, and friendship. Whatever failings this work has are fewer for their efforts. I must also acknowledge my friends Silvio and Elaine Senigallia. Silvio once attended the *liceo* that now occupies the Palazzo del Collegio Romano. Their interest in my work, insightful questions about its progress and significance, and "connection" to Clavius have inspired and shaped this work in ways they probably do not suspect.

And finally, I acknowledge more remote debts. I thank Joel Gwinn and John Kielkopf, who as teachers, friends, and colleagues encouraged my earliest ventures into history of astronomy. I must also thank my many friends and coworkers at the University of Wisconsin–Madison Space Astronomy Laboratory, especially Kathy Stittleburg, Bob Bless, and Art Code. This work might have been possible without the Space Astronomy Laboratory, but it would have been much less fun. Finally, I must express my deep gratitude to my wife, Jennifer, who helped bring this about, and to the two people who first taught me to love books and learning, my mother Evelyn Petersen and her mother Charlotte King.

Note on Editions, Quotations, Translations, and Names

Clavius's textbook on introductory astronomy, the *Commentary on the "Sphere" of Sacrobosco,* is the primary source of this study. It was printed in many revised editions over several decades, and the question of which one to quote and cite is not always an easy one. In general, I will quote from the version that was published in Clavius's collected works. That edition (cited as *Sphaera,* 1611), was published in the final year of his life and was the last printing over which he had control. Thus I have preferred it as representing his last word on the issues considered therein. (The first three volumes of his collected works, *Opera mathematica,* bear the date 1611, while the remaining two have 1612 on their title pages. The five-volume set has yet another title page stating 1612.) Other editions occasionally appear for special purposes, and citations to the original always accompany the quotations. I have taken certain liberties in transcribing Clavius's text, but they affect only the appearance of the text. For instance, I have altered capitalization of letters to reflect modern usage. I have also deleted superfluous punctuation, for example, eliminating the periods that follow numerals and commas that follow *et* when they serve no other purpose. All translations are my own unless they are otherwise noted.

Personal names of the sixteenth and seventeenth centuries often have multiple forms. I attempt to use a common form (though not necessarily in the vernacular) of any given name, generally (though not always) taking an authority such as the *Dictionary of Scientific Biography* or the *Dizionario Biografico degli Italiani* for my guide where possible. Arabic names, and the Latin forms of those names, present a particular problem. I have preserved the Latin forms in translations (e.g., Alpetragius) while using the more modern style (e.g., al-Bitrūjī) when the words are mine—except that I always use the more common Averroës in preference to Ibn Rushd. Preserving the Latin in translations avoids the possibly misleading suggestion that the sixteenth-century author shared our historical understanding and identification of the person behind the Arabic name.

ONE

Clavius's Astronomical Work and Life

I remember with particular pleasure having seen this demonstration while studying a most learned commentary on the Sphere *of Sacrobosco.*

—Sagredo in Galileo's *Two New Sciences*

The monuments of two powerful popes flank the chair of Saint Peter inside the Roman basilica that bears his name. The stony figures of Paul III (1534–49) and Urban VIII (1623–44) are but two among a forest of papal statuary, but their status within the central apse denotes extraordinary rank. Their monumental stations aptly reflect their historical roles in the tumultuous period during which, among other things, the world came to grips with the heliocentric planetary theory of Nicholas Copernicus. Copernicus dedicated *De revolutionibus orbium coelestium* (1543) to Paul III. That book set forth the theory that the earth is a planet, which orbits, like the other planets, about the stationary and central sun. Paul also blessed the formation of the Society of Jesus in 1540. The Jesuits would soon become one of the most potent intellectual forces within the church. Also during Paul's pontificate Christoph Clavius was born. After joining the Jesuits, Clavius would journey to Rome and help shape the church's reaction to the new astronomy. In 1632, Urban VIII presided over the climax of that reaction as the church reinforced its condemnation of Copernicus's theory by punishing one of its most famous champions, Galileo.

Clavius (fig. 1) has no statue, but if he did, the shadow cast by it would be a long one, touching many aspects of late sixteenth- and early seventeenth-century astronomy and mathematics. He still has much to teach us about the events and contexts of Renaissance and early modern astronomy. Yet he and his works are rarely studied in the history of the astronomical revolution of the late Renaissance and early modern era, so it is fair to ask why one should write, much less read, a book about him. Therefore, in the first part of this chapter I will discuss briefly Clavius's place and significance in the history of astronomy—in short, why he

Figure 1. Portrait of Clavius. From P. Freher, *Theatrum virorum eruditione clarorum* (Nuremberg, 1688). Photograph courtesy of the University of Wisconsin–Madison Memorial Library.

deserves our attention. Assuming success in that task, we must know in more detail how others have seen him, where he came from, and what kind of life and career he led. Thus this first chapter will hew the rough outlines, allowing a first glimpse of Clavius and preparing the way for the finer sculpting of later chapters.

Why Clavius Is Important

When Galileo had his first troubles with the church over the Copernican theory, in 1616, Christoph Clavius had been dead for nearly four years. But because he was an old friend and associate of both Galileo and Cardinal Robert Bellarmine, who led that early investigation into the theological nuances of the new astronomy, Clavius's influence was at work. The generation of scientists and theologians that would judge Galileo and debate the Copernican theory had learned its astronomy from Clavius's textbook, which he had been publishing and revising for decades before some of them were even born. For Jesuits through most of the seventeenth century he would be "our Clavius," and for them and nearly everyone else he was "the Euclid of his times." That epithet is not only an expression of great esteem, it is also a testimony to Clavius's important edition of Euclid's *Elements*, one of the earliest printed editions of this work. But the epithet has broader significance than that. As Euclid had been the geometrical tutor to many generations, so the mathematical scholars of the first half of the seventeenth century saw Clavius in an analogous role, not only because of his edition of the *Elements* but also because of the tutelage they had received from his many other textbooks on geometry, arithmetic, and, in particular, astronomy.

Clavius published his first edition of Euclid's *Elements* in 1574, followed by others in 1589, 1591, 1605, 1612, 1627, 1654, and even later.[1] Thus his Euclid spans one of the most important periods in the history of Western science, the so-called scientific revolution, during which great cosmological debates and discoveries transformed traditional concepts of the basic structure and physics of the universe. Clavius is a crucial figure of the scientific revolution because, among other reasons, of his felicitous place in the chronology of the history of astronomy. His first major work, the *Commentary on the "Sphere" of Sacrobosco,* was also published in many editions between 1570 and 1611, and through them we can trace the failure of the old cosmologies and witness the advent of the new. (Henceforth, *Sphaera* will denote Clavius's own commentary as distinct from Sacrobosco's original *Sphere* and its other commentaries.)

Clavius is commonly cited for his work as one of the members of the calendar reform commission established by Pope Gregory XIII. But I will not consider Clavius's important role in the establishment of the Gregorian calendar here. For one thing, the history of that unique process—with all of its social, religious, and scientific significance for the development of early modern Europe—deserves careful study in its own right.[2] But further, the Gregorian reform is not closely connected with the cosmological controversies of the sixteenth and early seventeenth centuries. The need for reform of the Julian calendar had been recognized and contemplated well before Copernicus published his *De revolutionibus* in 1543. And as it turned out, the reform itself involved a rather empirical process that depended little on novel astronomical theory. In fact Clavius, in his books on the reform, explained that the reformers deliberately avoided connecting the new calendar with specific astronomical hypotheses. This had the effect of insulating the reform scheme from scientific controversies over theoretical constructs.[3]

Here I concentrate on the work of Clavius the astronomer. Despite the relatively minimal attention given to it until now, Clavius's astronomical work is significant in its own right. As a theoretical astronomer, Clavius remained to the end of his life a follower of Ptolemy and (generally speaking) Aristotle. He argued always that Ptolemy's astronomy was consistent with the Aristotelian cosmos—an issue that had been debated since the early Middle Ages at least. In fact, Clavius was the last scion of a line with its roots in classical antiquity—the final serious defender of Ptolemy. Yet he was also an admirer of Copernicus and Galileo. Along with some other conservative astronomers of his time, most notably Galileo's rival Giovanni Magini, Clavius adapted some of Copernicus's innovations (his theory of the precession of the equinoxes, for instance) to the Ptolemaic framework, as I explain more fully in chapter 6, below.

Despite his conservatism in matters of theory, Clavius was, like Johannes Kepler, one of the earliest and most influential endorsers of Galileo's telescopic discoveries—though he stopped far short of agreeing with Galileo's interpretations of the observations. Clavius and Galileo had an amicable relationship dating back to the 1580s. Thirty years later, Clavius and his students formally (and later in print) endorsed Galileo's discoveries, an act that was very important in the initially favorable reception given to the Pisan professor in Roman ecclesiastical circles.

Clavius founded Jesuit astronomical studies. For thirty-seven years he taught at the capital institution of the Jesuit educational system. This school was created in Rome in 1551 by Ignatius Loyola, the founder of the Society of Jesus, to serve as a model for all the Society's other schools and to provide an advanced training center for its teachers. Clavius's reputation helped make the Collegio Romano (so called today to distinguish it from its modern Roman successor, the Gregorian University) an internationally famous center of astronomical activity. Its astronomers were able to make their own telescope at nearly the same time as Galileo, but apparently independent of him. This prior optical experience undoubtedly helped them give their prompt and fair evaluation of Galileo's discoveries after the announcement of his telescopic observations. Clavius's astronomical school also produced one of Galileo's foremost and worthiest antagonists, Orazio Grassi (of the comet controversies), and hosted another, Christoph Scheiner (of the sunspot debates), as a close associate.

The extent of Clavius's influence is nowhere greater than in his role as a textbook author. Even if his books had been read only in Jesuit colleges

Clavius would have had a considerable audience, and we know that students of the Jesuit schools such as René Descartes, Marin Mersenne, and Pierre Gassendi learned their mathematics and astronomy at least in part from Clavius's books.[4] But in fact, his books were read by students and teachers all over Europe for decades.[5] In the hands of Jesuit missionaries, such as Matteo Ricci, their influence spread to the Far East.[6]

In the quotation that opens this chapter, written nearly twenty-five years after Clavius's death, Galileo referred to Clavius's famous astronomy textbook, the *Commentary on the "Sphere" of Sacrobosco*.[7] Galileo drew heavily on Clavius's *Sphaera* in preparing his own *Trattato della Sfera* and other lectures at Pisa and Padua.[8] Moreover, William Wallace suggests that the Pisan used Clavius's 1589 edition of Euclid's *Elements* while lecturing on Euclid in 1590 and thus that Clavius is one of the most important links in a chain of Jesuit authors who connect medieval calculational science to Galileo.[9]

For these reasons Clavius's work is important to the history of science. Though lacking the retrospectively seductive aura of a great revolutionary, innovator, or martyr, Clavius was nonetheless a giant in his own time, and we neglect such a figure at the great peril of misunderstanding the history of which he was a part. The very course of his life itself, spanning a crucial period in the history of astronomy, allows us to witness the initiation and fruition of the astronomical revolution of the late sixteenth and early seventeenth centuries. Clavius was but five years old when Copernicus's *De revolutionibus* was published. As a twenty-two-year-old student in Coimbra he observed the total solar eclipse of 21 August 1560 that supposedly inspired the young Tycho Brahe (who would have seen it in Copenhagen as only a partial eclipse) to make a career of astronomy.[10] As a young professor at the Collegio Romano, Clavius observed and reported on the nova of 1572 and was one of those who concluded that it was a phenomenon of the supralunar realm. He may also have observed the great comet of 1577, which was so influential for Tycho's work. In 1600 and again in 1604, he and his students observed the novas of those years and concluded that they, too, were phenomena of the heavens. In 1610 he received the news of Galileo's telescopic discoveries with some skepticism, but by April of the next year Clavius and his astronomical colleagues pronounced Galileo's observations accurate. Despite the new evidence and the shifting cosmological allegiances even of his fellow Roman Jesuits, some of whom now preferred the Tychonic system, some perhaps even the Copernican, Clavius did not retreat from the Ptolemaic theories. His last published comments on the matter, in the 1611 edition of his *Sphaera,* make only the concession that Galileo's findings are of such merit that astronomers should find a way to incorporate them into the discipline.

When Clavius first published his full critique of the Copernican cosmology, in 1581, he was essentially free to evaluate rival cosmologies as he wished. The general of the Society of Jesus, Claudio Aquaviva would eventually order Jesuits to defend Aristotle in all matters, but that would not be until 1611. When Clavius died in February 1612, Galileo's first clash with the ecclesiastical authorities was four years in the future. Thus during Clavius's career, the church made no explicit endorsement or condemnation of the Copernican, Ptolemaic, or any other cosmological system. Later Jesuit authors, such as Christoph Scheiner and Giovanni Battista Riccioli, would write in the shadow of the Inquisition's 1616 pronouncement that limited the teaching of Copernicanism and placed Copernicus's *De revolutionibus* on the Index of Prohibited Books. As the founder and dominant early figure of Jesuit astronomy, Clavius's views had considerable influence on the very men, Cardinal Bellarmine for example, who would later deal directly with Galileo in the controversies over cosmological exegesis. But Clavius's published opinions on competing cosmological theories were formed decades before the bitter disputes and official prejudices of the early seventeenth century.[11]

Copernicus's publication of *De revolutionibus,* in 1543, presented the first truly worthy alternative to the Ptolemaic view, but there was no evidence to favor one theory over the other. It was Galileo's telescopic work of 1609 and 1610 that provided the first evidence capable of overthrowing Ptolemy. In the meantime, those remarkable celestial prodigies, the novas and comets, appeared just when European astronomers such as Clavius and Tycho were prepared to observe them and assess their cosmological significance. During his entire life, Clavius defended Ptolemaic astronomy against the mounting wave of contrary evidence and opinion. In Clavius's work we can see an important part of the process that Charles Schmitt described as the "shift in conceptual framework . . . by which the escape from Aristotelianism was accomplished."[12] Schmitt noted that the gradual abandonment of the Aristotelian worldview involved a countertendency: "As more and more Aristotelian doctrines were questioned, the convinced Aristotelians made slight internal modifications of the system to allow them to deal with objections within a quasi-Aristotelian framework. That is to say that Aristotelianism was in no way so static as is commonly believed, but serious . . . attempts were made to retain a basically Aristotelian system into which new scientific discoveries could be fitted."[13] One historian of the Collegio Romano characterizes Jesuit Aristotelianism in terms similar to Schmitt's. "It was not an extreme

Aristotelianism that the professors of the Collegio Romano defended, but an Aristotelianism purified and sufficiently broad to be capable of accepting and assimilating the discoveries of astronomy and of modern physics."[14] Because Ptolemaic astronomy had deep links to the Aristotelian framework, the story of the decline of Ptolemaic astronomy is a vital part of the broader history in which a major conceptual shift took place. In the late Renaissance and early modern periods a transition occurred in which assumptions, preferences, methods, and sources dating back to antiquity were replaced by new ones. Clavius played an important part in that story.

Clavius and His Reputation

Clavius has today achieved a degree of obscurity that his contemporaries would not have expected from "the Euclid of his century." His contemporaries, both admirers and detractors, saw Clavius as an important figure. Even a glance at the names of his correspondents is revealing: Galileo, Tycho, Giovanni Magini, Guidobaldo del Monte, François Viète, Adrian van Roomen, and many others. In addition to scientists and mathematicians, he corresponded with such wealthy patrons of science and learning as the German banking heirs Mark Welser and Georg Fugger, along with a sprinkling of royalty.

Leaving aside the effusive praises sung by Clavius's Jesuit contemporaries, he still received many published expressions of respect. In 1614, for instance, Jean Taxil praised both Clavius and Galileo and said of Clavius that he "has spoken more clearly of astrology and all other parts of mathematics than any author of past centuries."[15] Seventy years after Clavius's death, the encyclopedic biographer Isaac Bullart counted him among the world's great mathematicians, printed his portrait, and accorded him a biographical *éloge*.[16]

Clavius's views were so important to some—John Wilkins, Alexander Ross, and Libert Froidmond—that they indulged in a debate long after Clavius was dead about whether he would have endorsed the Copernican system had he sufficiently understood Galileo's evidence.[17] Galileo, who knew Clavius personally, judged the senior mathematician "worthy of immortal fame" despite the Jesuit's failure to agree with him on the meaning of lunar surface features. Galileo generously excused Clavius for this failing on the grounds of the latter's advanced age![18] *(~72)*

1538 -1612
or 1537
(see p. 12)

Not everyone saw Clavius in a positive light. Perhaps one of the harshest and, at the same time, more perceptive opinions of Clavius was penned by the English poet and preacher John Donne (1573–1631), whose early tutors were Jesuits. In his satirical and anti-Jesuit *Ignatius His Conclave*

(1611), Donne depicts Ignatius Loyola in hell. Ignatius argues before Satan that Copernicus does not deserve a place in the nether realm because he has not done enough to obfuscate the minds of men and thus deliver them further from the truth—in fact, he may have done just the opposite. One proves worthy of a place in hell by defending falsehoods, not truths. Addressing himself to the candidate Copernicus, Ignatius continues,

> If therefore any man have honour or title to this place in this matter, it belongs wholly to our *Clavius,* who opposed himselfe opportunely against you, and the truth, which at that time was creeping into every mans minde. Hee only can be called the Author of all contentions, and schoole-combats in this cause; and no greater profit can bee hoped for heerein, but that for such brabbles, more necessarie matters bee neglected. And yet not onely for this is our *Clavius* to bee honoured, but for the great paines also which hee tooke in the *Gregorian Calender,* by which both the peace of the Church, & Civill businesses have beene egregiously troubled: nor hath heaven it selfe escaped his violence, but hath ever since obeied his apointments: so that *S. Stephen, John Baptist,* & all the rest, which have bin commanded to worke miracles at certain appointed daies, where their Reliques are preserved, do not now attend till the day come, as they were accustomed, but are awaked ten daies sooner, and constrained by him to come downe from heaven to do that businesse.[19]

When Donne wrote and published *Ignatius,* Clavius was still alive, which explains why Clavius does not appear in hell to confront his rival claimant Copernicus. Donne's infernal comedy shows us that even during Clavius's lifetime, he was well known for his textbooks ("schoole-combats"), his anti-Copernican stance, and the unwelcome (in Protestant England) Gregorian calendar. And note that Donne's account stresses the importance of Clavius's anti-Copernican views more strongly than his calendar reform work. Donne, of course, was writing for a Protestant audience and admired Protestant astronomers, in particular the Copernican Kepler. Nevertheless, Donne's view of Clavius could also be in part a satirically inverted reflection of the esteem shown for the astronomer by the English Jesuit teachers of Donne's early years.

Modern estimations of Clavius have been just as uneven as those of his contemporaries. Jesuits today will proudly claim that Clavius is one of only two members of their Society (the other is St. Ignatius Loyola himself) depicted in St. Peter's Basilica in Rome. There, on a bas-relief on the tomb of Pope Gregory XIII, the sculptor Camillo Rusconi depicted

Figure 2. Bas-relief on Camillo Rusconi's monument to Pope Gregory XIII in St. Peter's Basilica, Vatican City. Jesuit tradition says that Clavius is depicted in the center presenting the reformed calendar to the pope. Photograph courtesy Scala/Art Resource, New York.

a priest, said to be Clavius, presenting the reformed calendar to the pope (fig. 2). Most histories of the Society of Jesus identify Clavius as one of their first and most eminent scientists and a precursor to such worthies as Christoph Scheiner, Athanasius Kircher, and Rudjer Bošković. Some Jesuit scholars, Riccardo Villoslada, Juan Casanovas, and Frederick Homann, have produced excellent work treating aspects of Clavius's significance. Villoslada's unique history of the Collegio Romano presents Clavius in his proper role and context as a founding teacher of that institution. Casanovas has published studies on certain aspects of the history of the Collegio Romano, while Homann has made careful, focused studies of Clavius's mathematical work.[20] Another generally useful modern source by a Jesuit historian is the little book by Pasquale D'Elia, *Galileo in Cina,* which, despite its title, summarizes handily some of the early astronomy in the Collegio Romano.[21] It is especially useful as a guide to some of the Jesuit archival materials. On the other hand, an example of a clearly apologetical work is that by Bellino Carrara in which he goes to some trouble to point out the many niceties and formal expressions of esteem (which are, in fact, almost purely formulaic) in the surviving correspondence between Galileo and Clavius as if to provide evidence that they were not enemies.[22]

However, the desire to rescue Clavius from the perceived ignominy of his "failure" to embrace Galileo's views has produced a less innocuous tendency to see him as, if not favorable to Galileo, at least impartial in the Copernican controversies. The worst examples of this interpretation are in the unfortunate articles of Charles Naux, in which he inexplicably concludes that Clavius "did not take a position" in the debate over alternative world systems.[23]

Of comparably low quality but at the other end of the apologetical spectrum from Naux are the judgments of Peter Aufgebauer who claims, apparently through a serious misreading of the texts, that Clavius could not or would not confirm the existence of the moons of Jupiter. He then goes on to read the 1616 prohibition of Copernicus's *De revolutionibus* back into Clavius's work of the early 1580s, saying that Clavius "as a true son of the Church" could not approve the Copernican doctrines.[24] Clavius's vocation as a Jesuit is certainly an important factor in any historical interpretation of his work, but the historian must not allow that fact alone to color the whole enterprise, especially when the texts and historical circumstances themselves contradict polemical presuppositions about how Clavius's religious committments affected his thinking.

It was, or course, Galileo who reshaped astronomy within Clavius's lifetime, and their relationship is a central aspect of Clavius's significance for the history of astronomy. The simplistic idea of friend or enemy of Galileo is apparent in the approaches of Naux and Aufgebauer, but much good work has been done seeing Clavius as a predecessor, teacher, and colleague of Galileo. Adriano Carugo, Alistair Crombie, and William Wallace have all written of Clavius's considerable influence on Galileo both in broad philosophical and methodological issues and in specific textual dependence. They have made a strong case for Jesuit influence on Galileo's attitudes toward the role of mathematics in natural philosophy and his appreciation of the power of certain types of reasoning in physical demonstrations.[25]

Authors of some recent work have begun to consider Clavius in his own right. Peter Dear, William Donahue, and Nicholas Jardine have treated Clavius on his intellectual merits as a significant source in early modern education and cosmological theories.[26] Edward Grant chose Clavius as an important source of scholastic arguments for the earth's centrality and immobility.[27] Frederick Homann has studied certain aspects of Clavius's mathematics, and Eberhard Knobloch has published several good studies.[28] Ugo Baldini, however, has produced the most important investigations to date of Clavius and Jesuit astronomical activity at Rome. Baldini's wide-ranging work, in addition to many articles, includes a volume of studies largely concerned with Clavius and his students in the Collegio Romano (including part of Clavius's recently discovered *Theorica planetarum* text). Most recently, Baldini and P. D. Napolitani published an edition of all of Clavius's known surviving correspondence, which makes available for the first time Clavius's letters, heretofore accessible only in archives and scattered sources; it adds an invaluable biographical supplement and bibliography.[29]

Clavius has received less than he is due from historians of astronomy who dismiss him as primarily a compiler or as a defender of methods and views that were becoming obsolete even as he wrote them down. A useful historical judgment can only come from understanding the proper context in which the history itself was played out. As Mark Twain sarcastically put the point, history "must be judged by the standard of the time and place, not by ours or heaven's—which are much the same no doubt."[30] Twain was talking about political revolutions, not astronomical ones, but the point is suited to both. In this spirit, we must bear in mind the fact that before 1610 there was no observational evidence for preferring Copernican cosmology over Ptolemaic. At the same time, the accepted Aristotelian physics and its close coupling with Ptolemaic cosmology rendered the hypothesis of the earth's motion completely implausible to all but the most daring imaginations. As a result, before Galileo's *Sidereus nuncius* appeared in 1610, only a handful of astronomers accepted Copernicus's arguments that the earth actually orbits the sun and not vice versa.[31]

Nevertheless, Clavius has, now and then, been belittled by shortsighted evaluations. Jean-Baptiste Joseph Delambre was especially severe, remarking, "His commentaries themselves show that if Clavius was more of a scholar and mathematician than the English monk [Sacrobosco], he was not really more of an astronomer—he performed no observations nor theoretical research."[32] Clavius, while not a dedicated observer like Tycho, not only made and published influential observations, he was also an active worker in theoretical, albeit Ptolemaic, astronomy. Nevertheless, a proper evaluation of his importance as an astronomer must go beyond a consideration of the technical details of his work. The opponents of the new astronomy were just as important as the innovators because the authoritative opponents, like Clavius, were the ones who determined the criteria and the tests by which the new theories and observations would be judged. In other words, in order to appreciate the eventual vindication of Copernicus and Galileo, we must grasp the standards to which their contemporaries would hold them. Clavius played a major part in setting the terms of cosmological debates for an entire generation of scholars.

Clavius's astronomical observations are not the most important aspect of his work, but they have, even recently, found a place in modern astronomical investigations. His observations of the solar eclipse of 9 April 1567 were called upon in our own century to support the claim that the sun may be subject to long term variations in its diameter.[33] While this kind of use of old astronomical data is a very questionable procedure, no doubt the modern interest in his efforts would have pleased the modest Clavius—and surprised Delambre![34]

Early Life

Up to the point when Clavius first appears in the records of the Society of Jesus at Rome in 1555, we know almost nothing about him except that he was born on 25 March 1538 in the German town of Bamberg, the see of the prince-bishop of Franconia.[35] Records of the Society's Roman archives (the Archivum Romanum Societatis Iesu [ARSI]) give the year of Clavius's birth, though the date is otherwise attested only in the short biographical sketch by the Italian mathematician and historian Bernardino Baldi, a younger contemporary of Clavius.[36] A number of modern authorities give 1537 as Clavius's birth year, thus giving rise to an ironic chronological problem for the formulator of the Western world's calendar. But Eberhard Knobloch has made the sensible suggestion that the discrepancy comes about from differences in defining the beginning of the new year. In 1538, Easter fell on 21 April. If, in early sixteenth-century Franconia, the new year was considered to begin with the celebration of Easter, then Clavius, born on 25 March, could be said by that reckoning to have been born in 1537.[37]

That he was from Bamberg (fig. 3) can never be doubted. In almost every book he published he proudly styled himself "Christophorus Clavius Bambergensis." He dedicated successive editions of his *Sphaera* to the archbishops of Bamberg and writes fondly of the town in his dedicatory letters. A slightly more subtle hint appears on the title page of Clavius's collected works, entitled *Opera mathematica*. The illustrations include, near the top, the figures of Saint Heinrich (the eleventh-century Holy Roman Emperor Heinrich II) and his consort Saint Kunigunda. Both are prominently entombed in the magnificent thirteenth-century Bamberg cathedral, which must have been a familiar and impressive place for the young Clavius.

Figure 3. The town of Bamberg. The cathedral complex is at left, just above center. From J. H. Lochner, *Geographische Bilder Lust von Deutschland* (Nuremberg, 1750). Photograph courtesy of the University of Wisconsin–Madison Memorial Library.

The *Allgemeine Deutsche Biographie*[38] states that Clavius's original German name was Schlüssel, apparently reasoning that he took the Latin form as a play on the Latin word "clavis" (key). Other standard sources have repeated this inference about Clavius's original name. However, an analogous argument would just as well get us back to the German "Nagel," or even better "Klaue," if the pun were on the Latin "clavus" (nail). The fact that all three names (Schlüssel, Nagel, and Klaue) actually occur in the region of Bamberg does not help.[39] If there were a complete absence of evidence either way, it would seem best to forget linguistic word play in favor of phonetic similarity and therefore to guess, along with the Jesuit bibliographer Carlos Sommervogel, that Clavius's original German name was Klau or something similar. But there is one hint in the other direction: Upon taking his simple vows, Clavius signed his name "Christophorus Clavis Bambergensis," which Baldini sees as favoring Schlüssel.[40]

Early Influences

Of Clavius's early life in Bamberg we know nothing. His family background, early education, and first contact with the Society of Jesus are a blank. Though Bamberg was the site of a prominent Jesuit school, it could not have been Clavius's connection to the order because the school was not founded until 1611.[41] Clavius's religious vocation, however it occurred, took place during the remarkable initial growth of the Society of Jesus, which numbered ten men in 1540 but about a thousand by 1556.[42] In the period leading up to the Peace of Augsburg (1555), which gave official recognition to the Lutheran presence in the Holy Roman Empire, Catholic efforts at consolidation of their traditional strongholds such as Franconia would have been vigorous. The Jesuit preacher, diplomat, and saint Peter Canisius traveled widely in Germany and Bohemia in these years and personally attracted many recruits to his Society. The young Clavius may have been caught up in the spirit of this Catholic retrenchment, which the early Jesuits, like Canisius, helped advance by furnishing teachers and founding many schools in the Germanic principalities.[43]

By February 1555, a month before his seventeenth birthday, Clavius found himself in Rome, where, on 12 April of that year, he was received into the Society of Jesus by Ignatius Loyola himself.[44] The years from 1555 to 1557 were particularly difficult for the young Society, particularly because of the hostility, created by the accession of Pope Paul IV, between the papacy and Spain. The Jesuits, almost destitute, could not afford to maintain all their young recruits in Rome, and for this reason many were dispersed to other Jesuit colleges.[45] Thus Clavius was sent off to study at the University of Coimbra, in Portugal, where he was enrolled in 1556.

The University of Coimbra was already a venerable institution in the mid-sixteenth century, and the young Society of Jesus was rapidly gaining influence there. As early as 1544 there was a community of forty-five Jesuits established at Coimbra,[46] and by October 1555, they were so numerous as to be able to open their own college and offer instruction.[47] The Jesuits remained a strong influence at Coimbra until their expulsion from Portugal in 1759.[48]

Clavius entered the newly instituted Jesuit College when he arrived in Coimbra. We have only a few details of Clavius's experience in Coimbra. Many years later, Baldi wrote that Clavius studied the humanities under two famous Jesuit scholars, both Spaniards, Pedro Perpinyá and Cipriano Soares.[49] In philosophy he heard the lectures of the famous Portuguese Jesuit Pedro Fonseca, a prolific dialectician and philosopher.[50] Baldi's words are partly confirmed in the ARSI records, which include the Jesuit catalogs of the Portuguese province. The beginning of his studies in 1556 is noted, and the catalog of 1558/59 counts Clavius among the second-year philosophy students (''inter auditores 2di anni'') and mentions as one of the professors Petrus Alfonsequa (i.e., Fonseca). It is surprising that mathematicians and astronomers are absent from the list, in particular the Portuguese mathematician and cosmographer Pedro Nuñez, who is known to have taught at Coimbra at about the time Clavius was there.

The same catalogs that help confirm Baldi's account also offer some of the very few personal glimpses we have of Clavius. We know that by May 1557 he had recovered from a mysterious and grave illness, which had left him very thin.[51] The effects seem to have lingered, as the catalog from 1559 records Clavius's presence and observes that "he is talented and in good health, but very skinny."[52]

We know that Clavius was still in Coimbra in August 1560, because in his *Sphaera* he records a total eclipse of the sun that he observed there at that time.[53] He may have left Coimbra soon thereafter, however, because he is reported in that same year to have made a pious visit to Monserrat, in Spain, a site important in the spiritual progress of Saint Ignatius.[54] As Montserrat is not out of the way of an overland route from Portugal to Italy, he probably worked this pilgrimage into the journey that would return him to Rome, where he spent most of the rest of his life.

By May 1561 he was numbered among the students studying physics and metaphysics at the Collegio Romano, and in 1562 we find Clavius enrolled as a theology student. Because a "fully formed" Jesuit had to be a priest and was expected to do advanced work in theology, it is not surprising to find Clavius, whose fame is founded in other areas, pursuing advanced theological studies. Because the pope or the Society might at any time call on him for any duty (e.g., missionary or diplomatic assignments, a post as a confessor or advisor, etc.), even a Jesuit mathematician had to be a credible theologian. In 1566 Clavius is still listed among those studying theology at the Collegio Romano, though in 1564 he had been ordained. However, the process of full formation was a long one. It was only in September 1575, when he was thirty-seven years old, that he is recorded as having professed the solemn vows, including the fourth vow, peculiar to Jesuits, of obedience to the pope, and thus became a fully formed member of the Society.

Clavius is first listed as a professor in the Collegio Romano in 1567, when he is recorded as teaching mathematics. However, he actually began teaching mathematics as early as 1563, which is consistent with the statement recorded in 1584 that Clavius had been teaching that subject for twenty years.[55] If Clavius studied little or no mathematics at Coimbra and had only arrived in Rome sometime in 1561 yet was teaching mathematics by 1563, where did he acquire his mathematical training?

Given the early age at which Clavius left Bamberg, it seems unlikely that he acquired much mathematical training there. However, a clue from

Clavius's later writings may reveal Bamberg as an early influence. The *Bamberger Rechenbuch* (Bamberg, 1483), a manual of practical arithmetic for business calculations, contains a very early representation of common fractions in something resembling the modern form, that is, the two terms of the fraction written as half-sized numerals one above the other, but without the horizontal fraction bar of modern notation.[56] A very similar notation appears in Clavius's *Epitome arithmeticae practicae* (Rome, 1583) in the discussion of the concept of fractional fractions. H. L. L. Busard refers to it as "a distinct notation" whose peculiar characteristic is the omission of the fraction bar in one of the fractions of the pair.[57] If Clavius's use of this notation is a reflection of notational practices at Bamberg, then the calculators of Bamberg's commercial sector may have been an important early influence, perhaps through a practical arithmetic text, on Clavius's later mathematical interests.

Another suggestion of the activity of Bamberg's mathematical community comes from Ernst Zinner's survey of works on instruments in the early sixteenth century. Of the mere handful of books on astrolabes published in German-speaking areas in that century, he notes that one of them appeared at Bamberg in 1525.[58] Zinner also found that Bamberg was a minor center for publication of astronomical yearbooks and a few other astronomical and mathematical texts from 1481 until 1525 or so.[59] So although Clavius left Bamberg at the age of sixteen and, as far as we know, never returned, there is enough evidence of mathematical and scientific activity at Bamberg to suggest that it was an early influence on his professional inclinations.

It would seem natural to assume that Clavius was strongly influenced by Pedro Nuñez, who was teaching at Coimbra while Clavius was a student there. Nuñez (d. 1578) was the author of several astronomical works including a commentary on Sacrobosco's *Sphere*, a short epitome of spherical astronomy, some notes on Peurbach's *Theoricae novae planetarum*, and several other standard mathematical texts. Aside, however, from some references to him in Clavius's later published works, it is difficult to identify any real influence of Nuñez on Clavius. Baldi's biographical sketch reports that Clavius claimed Nuñez, along with Girolamo Cardano, Federico Commandino, and others, among the famous mathematicians he knew.[60] But Nuñez is not mentioned in Baldi's list of Clavius's professors at Coimbra.[61] And the only likely opportunity for Clavius to have encountered Nuñez was in Coimbra in the late 1550s.

Baldi also reports that Clavius claimed to be self-taught in mathematics. According to Baldi's account, Clavius first became interested in mathematics through his study of Aristotle's *Posterior Analytics,* though it is unclear

Figure 4. The city of Rome. Buildings associated with the Jesuits are shown enlarged. The building marked *B* identifies the Collegio Romano, and *A* is the Society's principal church, SS. Nome di Gesù. From S. Rose, *St. Ignatius Loyola and the Early Jesuits* (London, 1891). Photograph courtesy of the University of Wisconsin–Madison Memorial Library.

exactly when he studied this text.[62] "Desiring then to understand [mathematics] well, he set himself to exhaust the subject on his own without any teacher's help, so that in this discipline he affirms himself to be, as the Greeks say, an autodidact."[63] By the time of Baldi's account, Clavius was an established and respected scholar who would have had no reason to conceal his debts, if any, to someone like Nuñez. So it seems reasonable to accept his word that he had not studied, at least formally, with Nuñez.

If, in fact, Clavius did not study mathematics (with or without Nuñez) at Coimbra, then it may well be that he was self-taught or nearly so, for in Rome (fig. 4) there were no chairs of mathematics or even regular lectures on mathematics—either at Rome's La Sapienza, the city's university, or at the Collegio Romano.[64] He did, however, study with the early Jesuit teacher Jerónimo Torres for a time, and some of that work could have been mathematical in nature.[65] But given the short interval between Clavius's arrival in Rome (probably 1561) and the beginning of his mathematics teaching at the Collegio Romano (1563), not to mention that in those three years he was studying theology and becoming a priest, it seems likely that he must have given considerable attention to mathematics while still in Portugal, whether self-taught or not.

Early Career

In the same year that he is first officially recorded as a professor, Clavius observed the solar eclipse that darkened Rome on 9 April 1567. Clavius's account of this event suggests that it was an annular eclipse (so called because at mideclipse the moon does not completely obscure the sun but leaves a bright ring, or annulus, around the dark lunar disk). But this contradicts accounts of other observers as well as the accepted opinion of the day that annular eclipses never happen. Kepler, in particular, drew attention to Clavius's account, thus ensuring that this eclipse would generate controversy for many years, lasting even to our own day.[66] Three years later, in 1570, he published the first edition of the *Commentary on the "Sphere" of Sacrobosco,* and then, in 1574, his edition of Euclid's *Elements,* both of which he would revise and republish for the rest of his life.

Clavius's edition of the *Elements* was part of a broad movement among European mathematicians of the sixteenth century to recover the mathematical texts of antiquity.[67] In the years just before Clavius's publication of the *Elements,* at least two other major editions of Euclid appeared, namely, Henry Billingsley's English translation of the *Elements* in 1570 (with John Dee's well-known preface) and the Latin edition of 1572 by Commandino, whom Clavius, as noted above, counted among his personal acquaintances.[68] Nor was Clavius's edition his only effort in this movement to recover and publish ancient mathematical texts, for in 1586 he published an edition of the *Spherics* of Theodosius.[69]

Clavius's edition of Euclid's *Elements* is not so much a translation of Euclid's geometrical texts as a paraphrase and commentary. His goal was not merely to present Euclid, but to make him accessible. To this end, Clavius departed substantially from the original, often using, for instance, demonstrations different from those of Euclid. Later editions even introduced some topics foreign to Euclid, such as issues raised by other commentators, and even some original contributions, such as Clavius's own attempt to prove Euclid's fifth postulate and to solve the quadrature of the circle.[70] Like his *Sphaera,* and indeed like Clavius himself, his *Euclid* had a long life. He revised and republished it at least five times, his last revision appearing in his collected works in 1611, the year before he died.

As we have seen, the significance of Nuñez for Clavius's mathematical roots is unclear, but the same cannot be said about Francesco Maurolico (1494–1575). Maurolico was the descendant of Greeks who had fled the fall of Constantinople in 1453 and settled at Messina, on the island of Sicily. He had a deep knowledge of Greek and an inclination for editing classical mathematical works. Maurolico had already taught for many years in Messina before Ignatius Loyola established his first school there in 1548 and seems to have had a close connection with Balthazar Torres, the Jesuit from Palermo who became the Collegio Romano's first mathematics professor and Clavius's immediate predecessor.[71] Maurolico apparently taught mathematics for the Jesuits at their Messina college and was well respected enough that in his later years they asked him to produce some textbooks to be used in their schools. Thus in 1569, Maurolico wrote to Francis Borgia, the general of the Society of Jesus, asking him to send the young Clavius to Messina to help put the desired texts into publishable shape.[72] Since this request predates Clavius's first publication, the fact that Maurolico knew of him indicates that Clavius was already acquiring something of a reputation, at least within the Society. Perhaps Torres was the original link between the elder mathematician and Clavius. For some reason, Maurolico's original request for assistance was put off. So it was not until 1573 that then General Everard Mercurian gave Clavius permission to travel to Messina.

Clavius seems to have arrived in Messina in April 1574 and departed in September. During that time he taught a course on the fifth and sixth books of Euclid.[73] His own *Euclid* first appeared that very year. Clavius's reward for his efforts in Messina was considerable. He brought back several of Maurolico's scholia and proofs, which Clavius incorporated into editions of his *Euclid* and his 1581 *Gnomonices libri octo*.[74] It was probably during that stay in Messina that Clavius acquired Maurolico's observations and treatise on the nova of 1572, much of which Clavius later published (giving its author due credit) in his *Sphaera*. Maurolico's treatise on the nova was a valuable source for Clavius, whose opinion on the location of the nova in the heavens was very influential. Clavius also received from Maurolico a considerable number of manuscript treatises on various mathematical topics, including a set of optical treatises (Clavius's edition of which was finally published in Maurolico's name in 1613) as well as minor works on sundials, the five regular solids, music, the ecclesiastical calendar, and mathematical instruments.[75] In short, from Maurolico Clavius received a small library of texts on subjects that coincide with many of Clavius's later mathematical publications. With the exception of the treatise on the nova of 1572, none of the material directly concerns pure astronomy.

Calendar Reform

Another major event in Clavius's career occurred in the mid-1570s, this one as significant as his expedition to Messina, for it remains a principal source of his fame. That event was the institution by Pope Gregory XIII (1572–85) of the commission to reform the calendar. It is surprising that we know relatively little about the body of scholars convened by the pope for the express purpose of literally changing Western history, but, in fact, its date of origin is unknown and must be inferred from much later statements of the participants. It seems to have operated on and off for about ten years before the actual promulgation of the papal bull of 1582 announcing the reform. A reasonable guess is that the commission commenced its work sometime between 1572 and 1575.[76] This conclusion leads to another puzzle, because these dates are still early in Clavius's career, and it seems odd that such an important post as one of the two technical advisors on a panel including distinguished cardinals and bishops should be given to a young Jesuit scholar. The answer probably lies in several circumstances. First, as we saw earlier, aside from the Collegio Romano, there was little or no institutional basis for mathematical scholarship—such as an established chair of mathematics—in Rome of the 1560s and early 1570s. Thus there would have been no other obvious local source of technical authority for a pontifical commission seeking advice. Coincident with this partial academic vacuum, the growing academic reputation of the Jesuits and their Roman college, as well as the growing influence of individual Jesuits in the church hierarchy, made them a credible source for technical expertise. Further, Pope Gregory was favorably disposed toward the Jesuits and was one of their more generous benefactors. Finally, if Clavius's career to that point had been relatively short, it was not undistinguished, for he had, by 1574, two major publications to his credit, his *Sphaera* (1570) and Euclid's *Elements* (1574).

Clavius's primary contribution to the calendar reform commission was to review and explain the alternatives that had been suggested in the years before the commission began its work. The commission eventually adopted a variation of the scheme invented by Luigi Giglio (Aloisius Lilius).[77] However, the bull of 1582 announcing the reform was not at all the end of Clavius's work, for he became the principal expositor of the reform itself by writing books explaining the bases and applications of the new calendar and defending it against its many detractors, among whom

Figure 5. Palazzo del Collegio Romano, designed by B. Ammannati. It is located in Piazza del Collegio Romano. Built during Clavius's lifetime, the Collegio Romano was the central university of the Jesuit educational system for nearly 300 years. Photograph courtesy Alinari/Art Resource, New York.

were the famous humanist Joseph Scaliger and the astronomer Michael Maestlin. Clavius published seven works explaining and defending the Gregorian calendar between 1588 and 1612.[78]

A Distinguished Professor

In 1584, the Jesuits of the Collegio Romano moved into a new building (fig. 5), which still stands today in the Piazza del Collegio Romano (though it now serves as a public high school). It was created specifically for the Jesuits and their thriving university largely through the munificence of their great papal patron, Gregory XIII. (His patronage is honored in the later name Gregorian University, borne by the Collegio Romano and its successor.) The unsatisfactory nature of their earlier circumstances is the point of one of the rare personal anecdotes that survive about Clavius. It dates from the time of Clavius's service on the pope's calendar reform commission. According to the story, the pope asked Clavius whether he had good living quarters, comfortable and suited to his studies. "Good? The best!" responded the mathematician. "All I have to do is move my bed from one room to another when it rains at night so that the water doesn't fall on my head."[79]

Other anecdotes suggest Clavius's modesty and stolid piety. When the people of Bamberg invited him to visit the town on the occasion of the dedication of a monument celebrating their famous mathematician, he declined the honor saying that his only desire was to remain at his post and continue his work. Yet another story tells that Clavius was so piously attentive to his priestly vows that when he chanced to see a woman on the street outside one of his chamber's windows he had the aperture barred shut so as to prevent such an intrusion ever happening again.[80] The authenticity of such anecdotes is questionable, of course, but they do tell us at least something of how Clavius was remembered.

Clavius was fifty years old in 1588, when his first exposition and defense of the Gregorian calendar was published. Then he was at the peak of his powers and prominence, was still teaching actively, and was beginning to train the generation of mathematical scholars who would succeed him at the Collegio Romano. The biographer Bernardino Baldi, who knew Clavius at this time, has left us the only contemporary sketch of the Jesuit mathematician.

> He is a man untiring in his studies and is of a constitution so robust that he can endure comfortably the long evenings and efforts of scholarship. In stature he is well proportioned and strong. He has an agreeable face with a masculine blush, and his hair is mixed in black and white. He speaks Italian very well, speaks Latin elegantly, and understands Greek. But as important as all these things, his disposition is such that he is pleasant with all those who converse with him. At the time of this writing he is in his 50th year, and we should pray that that life is prolonged so that the world may continue to receive those fruits that intellects as cultured and fertile as his are accustomed to produce.[81]

Of other human factors in Clavius's life we know very little. A small number of surviving musical compositions unveil a completely neglected aspect of his life interests.[82] We also know that, like many Jesuits of his generation, Clavius's life was hardly sedentary. Baldi reports that Clavius traveled a great deal in Spain, Germany, and Italy so that he could meet the principal scholars of the humanities, philosophy, and mathematics. Yet the extant records indicate only a number of journeys in and around Italy (Sicily, as mentioned earlier, Naples, and several tours of duty in Campania). Perhaps the references to Spain and Germany correspond to the voyages of his youth (from Bamberg to Rome, to Coimbra, and back to Rome), but as a Jesuit novice he would hardly have been meeting with famous scholars.

There is some discussion in Clavius's correspondence of a trip to Portugal at the behest of King Philip II in about 1593. Edward C. Phillips concluded that Clavius's voyage actually took place and cited as confirmation an inscription, quoted by Gottlieb von Murr, on a monument to Clavius in Bamberg.[83] But the Bamberg inscription actually says only that Clavius replied to the rulers of the Spanish peninsula, not that he delivered the response in person—and in fact the voyage never took place. Letters in the Society's archives show that in 1592 King Philip did request that Clavius's superiors transfer him to Lisbon to supervise the publication of a geography covering the lands subject to Portuguese rule. But Clavius was reluctant to go and managed to arrange a deferral for a few years, after which the king's request apparently was not renewed.[84]

There need be no such doubts about Clavius's sojourn in Naples for a little over a year beginning in autumn 1595, at least partly for reasons of his health. A letter from his student and colleague Christoph Grienberger suggests that Clavius traveled first to Frascati (about ten miles southeast of Rome and at that time a customary retreat for Roman Jesuits) and then on to Naples, where he arrived sometime in October 1595, "scarcely alive."[85] Whatever his malady, the correspondence also records his rapid recovery, apparently by the end of October. The purpose of his journey to Naples and the nature of his activities there are unknown, but it is quite likely that he taught or otherwise served at the Jesuit college in Naples. His correspondence indicates that by the beginning of 1597, Clavius had returned to Rome.

Clavius seems to have taught the "public" mathematics curriculum (which was the elementary sequence, including astronomy, required by the *Ratio studiorum*) from 1563 until 1571, then again in 1575–76, and finally, after a one-year pause, in 1577–78.[86] But documentation of Collegio Romano courses during Clavius's career is fragmentary, and he could certainly have offered his courses in other academic years after 1578. The longer break, between 1571 and 1575, is understandable as it included the publication of his *Euclid* (1574), the sojourn in Messina, and coincides roughly with the period leading up to his profession of solemn vows in September 1575. However, the required introductory mathematics course represents only one side of Clavius's teaching enterprise. He led a kind of seminar in advanced mathematical studies, sometimes known as the "academy of Clavius," from his earliest days at the Collegio Romano until approximately 1610. The activities of this circle of teachers and advanced students included not only instructional work but also the development and refinement of textbooks and the pursuit of research.[87] The research of the members of the academy was not only theoretical, but encompassed astronomical observations and even experimental optics (as shown by Paolo Lembo's experiments building telescopes, discussed in chap. 7 below). In addition to leading his academy, Clavius in his later years continued his personal writing and research, being listed from time to time in the college catalogues as a *scriptor*.[88] His status as a *scriptor*, freed from teaching duties as one of the Society's privileged writers, indicates the importance and prestige that the Society of Jesus associated with Clavius's name and work.

Though he did not teach in his later years, his correspondence, publications, and other records make it clear that Clavius remained active in the intellectual life of the Collegio Romano up until a few months before his death in February 1612. Clavius's fame made him part of the circuit toured by important visitors to Rome. Typical is the visit by the Spanish explorer Fernandez de Quirós, who was also a noted inventor of navigational instruments. In the reference to the meeting, Clavius alone is mentioned by name as one of the mathematicians and geographers who were impressed by Quirós.[89] The visit that Galileo would have with Clavius in the wake of the publication of *Sidereus nuncius* in 1610 also fits the pattern of the triumphant visitor paying his respects to Rome's elite.

During the rest of Clavius's life, the mathematics chair at the Collegio Romano was alternately held by Grienberger (1595–98, 1602–5, 1612–16, 1624–25, 1628–33); Gaspar Alperio (1598–99, 1600–1602); Angelo Giustiniani (1599–1600); Vincenzo Filliucci (1610–11); and Odo van Maelcote (1605–10). The first and last on this list, Grienberger and Maelcote, were Clavius's longtime students and colleagues, who dominated the activity of the mathematics faculty in Clavius's later years. The others are relatively minor figures about whom little is known.[90]

From Clavius's later years at the Collegio Romano there exists a very interesting portrait of him done by Francesco Villamena in 1606.[91] Vil-

CHRISTOPHORVS · CLAVIVS
.E · de Boulonie fec

Figure 6. Portrait of Clavius in about 1606. By E. de Boulonois after Francesco Villamena. From I. Bullart, *Académie des Sciences* (Amsterdam, 1682). Photograph courtesy of the University of Wisconsin–Madison Memorial Library.

lamena was known for portraits that depicted his subjects in their natural, unromanticized settings surrounded by the tools or symbols of their lives.[92] This portrait of Clavius (see fig. 6) shows the aged professor, now sixty-nine years old, seated at his desk and wearing the habit of a Jesuit priest. On the wall in the background hang an astrolabe and a quadrant—basic tools of the pretelescopic astronomer and themselves the subjects of some of Clavius's books. On the desk rests an armillary sphere, not the kind used for actual observations, but rather the kind that a teacher would use in explaining the fundamentals of spherical astronomy (the title page of

Clavius's *Sphaera* contains such a sphere). Also on his desk, in addition to inkwell and penknife, are a pen and a straightedge, and his hand grasps a compass—all the indispensable tools of the Euclidean geometer. And, of course, Clavius has spread before him many books, some stacked, some open, as if the artist had captured him in a working moment.

Clavius had many working moments in the years after his retirement from the classroom. After 1597 he published nine titles (one of which was the long-delayed volume of optical works by Maurolico), a couple of minor tracts, and eight revisions of his earlier books, and he oversaw the publication of the volumes of his collected works. He also continued actively in the research work of the Collegio Romano astronomers and mathematicians. Documents of the Collegio Romano record his participation in the discussions over the observations and the nature of the nova of 1604, and he was present during the observations of a lunar eclipse that took place in January of his seventy-first year, 1609, the very year in which Galileo learned to use the telescope as an astronomical instrument.[93]

After the publication of Galileo's *Sidereus nuncius* (March 1610), Cardinal Bellarmine wrote to Clavius asking that the Collegio Romano astronomers send their opinion of Galileo's observations and confirm them if they could. Apparently Clavius took an active role alongside his younger colleagues in the difficult process of teaching themselves, first, how to build and use an astronomical telescope and then how to interpret the observations. This is clear from correspondence between Clavius and Galileo in which the former complains that he and his fellow astronomers have been unable to see the Jovian satellites and Galileo responds with a discussion of the subtleties of telescope construction and the necessity of a stable mounting. (A fuller account of these events is given in chap. 7.)

In March and early April 1611, Clavius was still an active participant in the life of the Collegio Romano, for Galileo reported having long discussions with him during a trip to Rome that also included a ceremony at the Jesuit college celebrating the Pisan's discoveries. However, soon after this it would appear that Clavius's health began to fail. A letter of 14 June 1611 from Giuseppe Biancani, a Jesuit mathematician and astronomer in Parma who had studied briefly at the Collegio Romano, to Grienberger at the Collegio Romano indicates that Clavius had fallen ill and remained so long enough for word to spread.[94]

The ARSI records indicate that Clavius died on 6 February 1612 in the Collegio Romano itself. This is confirmed by a Vatican Library manuscript in an entry dated 8 February (which was a Wednesday). "On Monday Father Christoph Clavius of Bamberg died. [He was] a Jesuit, a famous mathematician, and 74 years old, of which he had passed 56 in his Order, and into which he had been received and invested by the blessed Ignatius Loyola himself, founder of the Society of Jesus.''[95] Some sources recount an odd story about Clavius's death, namely, that he was killed by a bull, sometimes a wild bull, while visiting the seven churches of Rome. It seems that no sources close to Clavius confirm, or even hint at, this bizarre end, nor do they suggest that he died anywhere but in his room at the Collegio Romano.[96]

At the time of his death, Clavius's reputation was sufficiently authoritative to generate another deathbed story, less peculiar but even less plausible than the attack by a raging bull. It is reminiscent of the tale about Copernicus receiving the first copy of *De revolutionibus* in his last moments. In Clavius's case the story was told by John Wilkins in his influential defense of Copernicus and Galileo entitled *Discourse Concerning a New Planet* (1640). The story focuses on Clavius's reaction to Galileo's discoveries and their significance for the Ptolemaic astronomy.

> 'Tis reported of *Clavius,* that when lying upon his Death-bed, he heard the first news of those Discoveries which were made by *Gallilaeus* his Glasse, he brake forth into these words: *Videre Astronomos, quo pacto constituendi sunt orbes Coelestes, ut haec Phaenomena salvari possint:* That it did behoove Astronomers, to consider of some other *Hypothesis,* beside that of Ptolomy, whereby they might salve all those new appearances. Intimating that this old one, which formerly he had defended, would not now serve the turne: and doubtlesse, if it had been informed how congruous all these might have been unto the opinion of *Copernicus,* he would quickly have turned on that side.[97]

Lest anyone be deceived by this story we must note that Clavius did not die until nearly two years after Galileo's discoveries, and he was aware of them as soon as Galileo had made them public, if not before.

Wilkins's book was very influential in England in the second half of the seventeenth century, and throughout the text he refers frequently and respectfully to Clavius, treating him less as an adversary of Copernicus than as a source of reputable opinions on the Ptolemaic side.[98] Wilkins, a convinced Copernican, may have known this deathbed story to be false but repeated it to dramatize the power of Galileo's discoveries, that is, to portray them as so influential that they forced even Clavius, the Jesuit, to surrender his mistaken beliefs.

The quotation of Clavius's words is lifted from the final edition of the *Sphaera.* There Clavius reported and discussed Galileo's observations and concluded, ''Videant Astronomi, quo pacto orbes coelestes constituendi sint, ut haec phaenomena possint salvari.''[99] Wilkins, in his own book, occasionally cites Clavius's *Sphaera* on other matters and so must have known the work itself. But perhaps he did not have the last edition, because he seems to have taken the quotation itself from an intermediate source, Johannes Kepler's *Epitome of Copernican Astronomy*. In the dedicatory letter to that work, Kepler relates that Clavius, "near death," had learned of Galileo's discoveries and issued the famous directive.[100] Kepler paraphrased Clavius's published words, and Wilkins copied Kepler's paraphrase verbatim. Kepler, however, was not so brash as to conclude or even speculate on what Clavius might have thought had he been "informed."

Wilkins was not the first to suggest that Clavius intended that his words close the book on Ptolemaic astronomy. For instance, Galileo, in his *Letter to the Grand Duchess Christina,* observes, "I might also name other mathematicians who, moved by my latest discoveries, have confessed it necessary to alter the previously accepted system of the world, as this is simply unable to subsist any longer." Stillman Drake writes that a marginal note in Galileo's text identifies Clavius as one of these mathematicians.[101]

Wilkins's Copernican sympathies, like Galileo's, colored his interpretation of Clavius's words. Despite Wilkins's suggestion to the contrary, there is no doubt that Clavius was well aware of "how congruous all these might have been unto the opinion of *Copernicus.*" The Jesuit astronomer knew the arguments of the *Sidereus nuncius* and had discussed the matter fully with Galileo himself. And that was before the final edition of the *Sphaera* was published. Clearly, Wilkins was not on firm ground with his assertions, and Alexander Ross, in his diatribe against Copernicanism and Galileo's teachings, responded directly to Wilkins's speculations about what the long-dead Clavius would have thought. "Now to wish Aristotle alive, or to think that he or Clavius would ever be of your opinion, are meere dreames and phansies. And although Clavius had found that Ptolomies Hypotheses had not beene so exact as should be; yet he would not have beene so mad as to beleeve the Earths motion, and the Suns rest.''[102] Ross showed in this a clearer insight than Wilkins into Clavius's deeply held cosmological beliefs.

Like Ross, Libert Froidmond felt called upon to defend Clavius against becoming a posthumous tool of the Copernicans. In 1631 Froidmond published his *Ant-Aristarchus sive orbis-terrae immobilis,* which is a defense of the church's 1616 ruling against teaching or holding the Copernican theory to be true. He quoted Kepler's paraphrase of Clavius's supposed dying words and responded that "truly no one who knows Clavius would believe that as he was dying it would have occurred to him to agree that the earth moves, for it is one thing to have judged that Ptolemy's eccentrics and epicycles have been thrown into some confusion, but quite another to think that the whole ancient system is completely wrecked and to decide to give up to Copernicus.''[103] Wilkins's fable may be considerably less vicious than Donne's earlier satire, but the point for us is the same: Clavius's reputation among his contemporaries and immediate successors rested, for good or ill, on his opposition to Copernicus. As a principal voice of the opposition, Clavius was still influential enough nearly thirty years after his death that his opinion remained a serious concern for the Copernican camp.

Thus we have the rough outlines of Clavius's long-overdue monument. He was a product of the explosive early growth of the Society of Jesus and founder of its traditions in the mathematical sciences; a venerable and authoritative educator, who taught and wrote among the Roman Jesuits for nearly half a century; an internationally known astronomer and mathematician whose textbooks became standards; the expert to whom ecclesiastical authorities looked as they formed their earliest opinions on Galileo's challenges to the established natural philosophy; and finally, the defender, and almost the very embodiment, of the traditional amalgam of Aristotelian cosmology and Ptolemaic astronomy. One other preparatory task remains, and that is to ensure a platform suitable for the more detailed sculpting. In chapter 2, I discuss the situation of mathematics and the mathematical sciences in the educational schemes of sixteenth-century Jesuits followed by a summary of Ptolemaic astronomy as they taught it.

TWO

Jesuit Mathematics and Ptolemaic Astronomy

*In astronomy our spirits are enraptured and lifted above the con-
cerns of this terrestrial world, which never endures, to a world
not subject to corruption at all. From the despicable lands of this
narrow speck the spirit may wander happily through the vast air
and among the golden suns, the changeable silvery moons, and
the brilliant stars.*

—Clavius, *Sphaera* (1596)

Our primary source for studying Clavius's thought is his astronomical
textbook, the *Sphaera*. But we cannot attempt to understand the contents
of a textbook of any era without some background to help us understand
for whom the book was written, why it took its particular form, and what
it expected to teach its readers. In this chapter I will build that background
by sketching Jesuit education and Clavius's place in it, the place of mathematical sciences in the Jesuit curriculum, and the astronomical traditions
from which sprang Sacrobosco's *Sphere,* and Clavius's commentary on
it. The final sections should be particularly helpful for those readers who
desire an introduction to or brief review of the ideas that dominated Western astronomy from antiquity to the Renaissance.

Pragmatic Values in Jesuit Education

Clavius received a humanistic education broadly construed but strongly
influenced by the philosophy and theology of Thomas Aquinas, whom the
Jesuits favored as their authoritative interpreter. He learned Latin and
Greek from classical authors, especially Cicero, with a strong emphasis
on rhetorical practice, and he studied Aristotelian natural philosophy from
the texts of Aristotle himself, interpreted, where necessary, according to
Aquinas. These were standard requirements because Ignatius prescribed
them in his *Constitutions.* Clavius's minimal formal training in mathemat-

ics seems to have been common, for he would eventually join the movement deploring the low place of mathematical studies in the standard humanistic curriculum and advocating the improvement of their status. Clavius was also a humanist in his appreciation of the importance of recovering, understanding, and propagating the fundamental texts of classical antiquity, as he did with his editions of Euclid and Theodosius.[1]

The Jesuit formation through education placed its own peculiar stamp on the humanist ideals. That is to say, identity as a Jesuit implied certain intellectual orientations, preferences, and options. Early Jesuit science was rooted in the same ideology that gave rise to the Jesuit apostolates—the primary roles undertaken by the Society. Thus Jesuit scientific traditions were generally those activities that proved useful in those apostolates: education, service among the nobility, and foreign missionary work.[2] Clavius's writings eventually became source material for the development and application of the educational apostolate both because of their early date relative to Jesuit traditions and also because of the success of mathematical science during the seventeenth and eighteenth centuries.

Clavius's extensive interests in applied mathematics and astronomy (for instance calendar theory, timekeeping, instrument construction, optics, arithmetic, mensuration, etc.) are consistent with the generally pragmatic orientation among Jesuits. This practical disposition resulted in part from their "active engagement with the world." As Steven Harris puts it, "Apostolic spirituality engendered within the Society an intensely goal-oriented, purposeful attitude . . . whatever could contribute to the achievement of practical results was highly valued."[3] Esteem for utility was not peculiar to Jesuit education. It was a common concern for sixteenth-century humanists, who were reacting against what they saw as the sterility of the scholastics. The influential humanist pedagogue Peter Ramus, for example, often emphasized the utilitarian aspects of education.[4] While Clavius's practical orientation might have its roots in the Bamberg commercial calculators, his pragmatism could only have been reinforced by the Jesuit cultivation of applied sciences. His how-to books on instruments and timekeeping, as well as some of the ones on calendrical matters, were intended to serve in the educational apostolate of the Society.

Clavius also had an impact on the apostolate of the foreign missions, though it was probably unforeseen by him. His books went to China in the hands of Jesuit missionaries like Matteo Ricci and Johann Adam Schall, who were very effective in impressing the Chinese aristocratic classes with Western learning. Ricci translated the first six books of Clavius's Euclid into Chinese in 1607 and followed with Clavius's *Sphaera* and *Gnomonices*.[5] He reported great success among the Chinese with his ability to construct instruments, such as the astrolabe, celestial globe, and quadrant, which he had studied under Clavius at the Collegio Romano.[6]

Mathematical Sciences in the Jesuit Curriculum

Clavius spent most of his life as a professor of mathematics, the role chosen for him by the Jesuit hierarchy. Well before the Society's *Ratio studiorum* established detailed rules, the fundamental duties of the mathematics professor were spelled out in the guidelines promulgated for the Collegio Romano in 1566. "Concerning mathematics, the mathematician shall teach, in this order, the [first] six books of Euclid, arithmetic, the sphere [e.g., of Sacrobosco], cosmography, astronomy, the theory of the planets, the Alphonsine Tables, optics, and timekeeping. Only the second year philosophy students shall hear his lecture, but sometimes, with permission, also the students of dialectic."[7] These subjects read almost like a list of the titles of Clavius's lifetime publishing output. That fact begins to tell us what many later documents confirm, that Clavius's teaching and writing careers were primarily and consciously aimed at developing a complete mathematical curriculum for the Jesuit educational system. Not satisfied merely to promulgate, as he occasionally did, detailed schemes of what topics should be taught in what order, Clavius undertook and largely completed the enormous task of writing lengthy textbooks to occupy every niche in the rich curriculum he engineered.[8]

It was not a trivial task to establish a respectable place for mathematics in the Jesuit educational curriculum. Ignatius Loyola had sketched his pedagogical goals in rather broad strokes, leaving vague the precise place of mathematics. "Logic, physics, metaphysics, and moral philosophy should be treated, and also mathematics in the measure appropriate to secure the end which is being sought."[9] The vast room here for interpretation of Ignatius's intent (including even "the end" of the whole process) makes it clear why the detailed plans of the preliminary 1586 *Ratio studiorum,* and the final version of 1599, were required for the regulation of his international system of colleges.

We get some idea of the early status of mathematical studies in the Jesuit colleges from a curriculum document written by Clavius at an unknown date but probably before the 1586 *Ratio studiorum.* In the document, entitled "A Method for Promoting Mathematical Studies in the Schools of the Society," Clavius suggested that the mathematics professor, like other professors, should be invited to the solemn ceremonies of public disputations and the conferring of degrees, and that he should take part in the disputations.[10] Apparently the mathematics professor was not generally accorded the same dignities as the others in these matters.

Clavius wrote that the mathematics teacher should be one who has an inclination for that discipline and that he should not be assigned many other duties in addition to teaching mathematics. The implication is clear: mathematics teachers were sometimes unenthusiastic, or even inept, and overworked. It would also be good, he added, if occasionally, in front of all the other students, some of the mathematics students received a commendation for their skill in mathematics and, at the same time, gave presentations of some geometrical or astronomical problems. Such student assemblies and awards were standard practice in other disciplines.[11]

Finally, he observed how unfortunate it was that the philosophy teachers belittled mathematics as something less than a true science, as lacking in true demonstrations, and as abstracting from being and goodness (which suggests that it lacks virtue and relevance to the real world). In the Jesuit schools of Clavius's early days, neither the mathematics professors nor their discipline were treated as peers of the other faculties. This prejudice against mathematics was not peculiar to Jesuit schools but was consistent with a broader atttitude on the part of sixteenth-century academics that mathematics was a peripheral subject and somehow less worthy than disciplines such as theology and philosophy.[12]

The ambiguous status of the professor of mathematics is one aspect of a dispute among early Jesuit educators over the role that mathematics should have in the curriculum of their schools. That dispute was, in turn, a reflection of the ambiguous status of mathematics itself in the academic settings of the sixteenth century. The sixteenth century's renewed emphasis on mathematical studies and the recovery of their ancient sources raised the issue of whether methods of geometrical proof could be reconciled with Aristotle's standards for a demonstrative science.[13]

The syllogism was a crucial problem. Aristotle held the syllogism to be the ideal and most powerful reasoning tool. For many, a discipline that employed this mode of reasoning was more important, more reliable, and thus of higher status than one that did not. The detractors of mathematics as a science maintained that mathematical demonstrations are not comparable to the syllogisms of natural philosophy, because in a true demonstration, according to Aristotle, the premises are the proper causes of the conclusion.[14] The critics argued that this is not the case with a geometrical demonstration.

Take an example. We might (if we were medieval philosophers) explain the blotchy appearance of the moon with this syllogism: the celestial and terrestrial regions are characterized respectively by clarity and obscurity; the moon is intermediate between the celestial and terrestrial regions; therefore the moon shares in both clarity and obscurity.[15] The premises of a syllogism, in traditional Aristotelian logic, must be causally related to the conclusion if it is to yield the surest possible argument. A syllogism that meets this criterion, among others, constitutes the most powerful form of demonstration, or what was called a *demonstratio potissima*. In our example, we would have to agree that the premises (the clarity and obscurity of celestial and terrestrial matter and the moon's intermediate position between the celestial and terrestrial regions) are causally related to the conclusion if this syllogism were to be a *demonstratio potissima*. That being the case, we would be assured of the reliability of our explanation.

But mathematics, in contrast, treats quantity in the abstract and therefore alienated from physical reality and causal factors. For example, in a typical exercise of mathematical astronomy we might calculate the distance between the earth and moon by measuring a trigonometric parallax. The procedures involved would be justified by demonstrations based on Euclidean geometry. But such demonstrations depend only on the properties of lines and angles, which have nothing in particular to do with the earth and moon. Moreover, the procedures would apply to any similar geometrical construction whether celestial, terrestrial, or completely imaginary. The critics viewed the demonstrations of Euclidean geometry as arbitrary constructions bearing no necessary connections to the conclusions and failing to meet the higher standards of natural philosophy.[16]

A good example of this problem is the dispute in the 1550s between the Paduan professors Alessandro Piccolomini and Francesco Barozzi. Piccolomini held that Aristotle's *demonstratio potissima* could not be found in mathematics, pure or applied.[17] This would mean that mathematics could not be considered a science comparable with physics, for example. An important follower of Piccolomini was the Jesuit philosopher Benedict Pereira, a close contemporary of Clavius and one of his colleagues in the Collegio Romano. In his influential text on natural philosophy, Pereira stated clearly, "I believe that the most powerful demonstration, described by Aristotle in *Posterior Analytics* I, is rarely found in the mathematical sciences, if at all."[18]

Pereira and other Jesuits of like mind carried the conflict over methods of proof in natural philosophy and mathematics into the midst of the Jesuit scholars at the Collegio Romano itself. The resulting prejudice against mathematics, the discipline less clearly anointed by Aristotle, explains Clavius's efforts to raise its status. One strategem, not unique to Clavius, for enhancing the status of mathematics was to try to apply syllogistic reasoning to mathematics, that is, to show its conclusions to be justifiable through the *demonstratio potissima*. Clavius tried this halfheartedly. In his solution to the first problem in Euclid (to construct an equilateral triangle on a given line segment) Clavius gave a laborious syllogistic proof

of the solution and then made a grand claim for its generality even as he
dismissed the method as cumbersome. "All other propositions, both Euclid's and those of all other mathematicians, can be solved in this way."
But "mathematicians disregard this solution [by syllogism] in their demonstrations because proofs are quicker and easier without it, as can be
seen from the preceding example."[19] If taken seriously, his claim for the
generality of the syllogistic method in mathematical reasoning would be
an effective response to the likes of Piccolomini and Pereira. That is
doubtless how Clavius intended it. However, the failure of Clavius or
anyone else to carry through the entire program and back up the claim of
generality (by reworking all of Euclid into syllogistic form, for example)
sapped the argument of its power. In the end, the mathematicians won
but not by proving that astronomy and the other mathematical sciences
(optics, mechanics, and the like) could meet the standards of the extremists
among the natural philosophers. Rather, the early seventeenth-century successes in mathematical sciences by the likes of Galileo and Kepler allowed
the whole arid argument to dry up and blow away.[20]

The difficulty of supporting the claim that mathematics could be expressed syllogistically drove Clavius to defend mathematics in other ways.
He sought to enhance the status of mathematical studies by exalting the
dignity of the discipline, pointing out the certainty achieved by its methods, appealing to its utility in the other disciplines and in the Jesuit curriculum, and, finally, praising it as pleasurable and edifying to the mind. He
made his most complete and concise case in the prolegomena to volume
1 of his *Opera mathematica,* published in 1611.

Clavius argued that the dignity of mathematics can be seen by the
unanimity of opinion of mathematicians in comparison to philosophers
and natural philosophers, for in those other sciences one frequently finds
a great multitude of judgments, viewpoints, and uncertainties. But

> the theorems of Euclid and the rest of the mathematicians, still today
> as for many years past, retain in the schools their true purity, their
> real certitude, and their strong and firm demonstrations. . . . And
> thus so much do the mathematical disciplines desire, esteem, and
> foster the truth that they reject not only whatever is false, but even
> anything merely probable, and they admit nothing that does not lend
> support and corroboration to the most certain demonstrations. So
> there can be no doubt but that the first place among the other sciences
> should be conceded to mathematics.[21]

The high standards of mathematical reasoning are, for Clavius, not
only a source of the dignity of the mathematical disciplines but are also

the reason for the certainty of its conclusions. That certainty, in his opinion, qualified mathematics as a discipline by the Aristotelian definition, "[Mathematical demonstrations] always proceed from some foreknown principles to demonstrated conclusions, which is the proper function and purpose of a doctrine or discipline, as Aristotle states in *Posterior Analytics* I; nor do mathematicians accept anything which is not proven."[22]

But the admirable ironclad certainty of mathematical reasoning is not enough to answer the criticisms of Piccolomini and Pereira, because they denied that mathematics, regardless of its internal integrity, had any fundamental relevance to the physical world and natural science. Clavius countered this by first drawing distinctions between the mathematical disciplines and then appealing to the great utility of those disciplines in the natural world and the world of learning. He borrowed the disciplinary distinction from his fellow defender of mathematics, Barozzi, who got it from Proclus. The distinction admits that some branches of mathematics do abstract from matter but others do not. Clavius tells us that "some of the mathematical sciences are directed toward the intellect insofar as they are separated from all matter, but others pertain to the senses insofar as they pertain to passive sensible matter."[23] Of the former kind there are two: arithmetic and geometry. Of the latter there are six: astronomy (which he calls *astrologia*), perspective, geodesy, music, practical arithmetic (*supputatrix*), and mechanics. Clavius then indulges in a substantial discussion of the various subdivisions of each of these, especially astronomy and perspective, emphasizing the many practical applications they have—timekeeping, surveying, accounting, machine design, and so forth.[24]

Clavius also pointed out the utility of mathematics to the other academic disciplines, for which mathematics is "not only useful, but in fact necessary." Theologians must know arithmetic, music, geography, astronomy, and geometry if they are to understand the meaning of the Scriptures, and he cites a number of church fathers in support of this. Further, natural philosophy, morals, dialectic, and the rest are greatly enhanced by a knowledge of mathematical sciences. He cited Plato as his primary authority on the importance of mathematics to philosophy.

Clavius concluded his exposition of the utility of mathematical studies by pointing out the spiritual pleasure and edification that they provide. "Of all these benefits [of mathematical studies], perhaps the greatest is the entertainment and pleasure that fills the soul as a result of the cultivation and exercise of those arts. Indeed they, among the seven liberal arts, are special because they strongly move toward great glory and delightful freedom the souls not only of sincere youths, but also of noblemen, princes, kings, and emperors."[25]

By the early years of the seventeenth century, the mathematical disciplines in the Collegio Romano had achieved something like the status Clavius had envisioned for them. Courses in mathematics were offered as part of the regular course rotation. Mathematics professors participated in disputations, and their students offered demonstrations of their skills at formal assemblies. More important, the Jesuit schools were producing a growing number of skilled mathematicians whose publications were of the highest caliber. Under Clavius's leadership, mathematics and astronomy became sources of pride and more than a little fame for the Jesuits.

In the Latin Middle Ages, astronomy held a distinct place in the framework of the seven liberal arts. Astronomy was a member of the quadrivium, which (as Clavius reminds us) comprised the mathematical disciplines: arithmetic, geometry, astronomy, and music. Sixteenth-century doubts about the standing of mathematics among the sciences and within the educational curriculum would logically affect the status and support of all the quadrivial subjects, including astronomy. Thus Clavius's success in securing the place of mathematics in the Jesuit colleges raised the profile of astronomy as well. But the character of astronomy was changing dramatically during Clavius's lifetime. Traditional classifications of subject matter into the trivium and quadrivium came to be seen as antiquated and associated with unfashionable medieval Scholasticism. And astronomy itself grew beyond its place as one of the four quadrivial subjects. While remaining mathematical, it became ever more linked to questions of natural philosophy. Clavius's long-lived commentary on Sacrobosco, his *Sphaera,* follows many of the medieval formats and conventions for astronomy texts, but it also shows the signs of the changing nature of astronomy in the midst of the scientific revolution.

Both his own contemporaries and modern historians have judged Clavius's *Sphaera* to be the greatest of all *Sphere* commentaries. Clavius's Paduan contemporary and correspondent, Francesco Barozzi, wrote in his own astronomical textbook that Clavius's *Sphaera* was to be preferred above all others, while Kepler mentioned Clavius first among the many "very learned and ample" *Sphere* commentaries he praised.[26] Jean-Charles Houzeau and Albert Lancaster, in their bibliography of astronomy, also evaluated it as the best of its kind.[27] Francis R. Johnson, in his study of Renaissance textbooks, stated that Clavius's *Sphaera,* first printed in 1570, was the most successful *Sphere* commentary of all, going on to become "the principal advanced text-book of astronomy throughout Western Europe for the next three-quarters of a century."[28]

It is precisely the long publication history of Clavius's *Sphaera* that makes it such a useful source in the history of astronomy of this period.

Clavius attempted to deal with all of the controversial issues of astronomy and cosmology of his day, at least insofar as they could be addressed in an introductory textbook. The continual revisions and expansions of his *Sphaera* allow us to trace his responses and shifts of opinion as the world of astronomy changed around him. Clavius continually strove to interpret these developments for his readers. To understand fully the significance of Clavius's text, we need to examine the *Sphaera* itself, consider how it differs from other *Sphere* commentaries, what subjects it emphasizes, and what approaches it takes. We will begin by looking briefly at the history of astronomy texts and putting them in the context of a book tradition dating back to the Middle Ages.

Textual Traditions in Medieval Astronomy

The *Sphere* of John of Sacrobosco, together with the many commentaries on it, was the most popular textbook of elementary astronomy in Europe from the Middle Ages to the early modern era. It was written and first came into use in early thirteenth-century Paris, and it finally began to pass out of use in the schools of the early seventeenth century.[29] Galileo himself, in his early teaching days, chose to write a brief sphere commentary to accompany his astronomy lessons. The title ''Sphere'' signified the central pedagogical theme of elementary astronomy in the early European universities. The sphere was a universal concept underlying all the particulars that constituted the subjects of the science: the heavens and the earth are spherical, the heavenly bodies are spherical, their motions are circular paths on the surface of spheres, and celestial geometry is the study of circles on the surface of a sphere.

In medieval and Renaissance astronomy the term ''sphere'' had a range of meanings beyond the strict geometrical definition. The correct meaning often becomes clear only in context. It was, furthermore, often used interchangeably with ''orb'' (*orbis*), ''heaven'' (*caelum* or *coelum*), and even ''globe'' (*globus*). (For an example commingling all these terms see the Latin extract from Clavius's *Sphaera* of 1611, in chap. 5, n. 24, below.) In one sense, a sphere is a region defined by the qualities of the matter it contains; thus the elemental spheres of earth, water, air, and fire are often enumerated among the spheres constituting the universe. In another sense, a sphere is the region of space bounded by two concentric geometric spheres—the radius of the inner corresponding to a planet's minimum distance from the earth and the radius of the outer to the maximum distance. Such a spherical shell is the meaning usually intended in such a phrase as ''the sphere of Saturn,'' for example. These are also generally

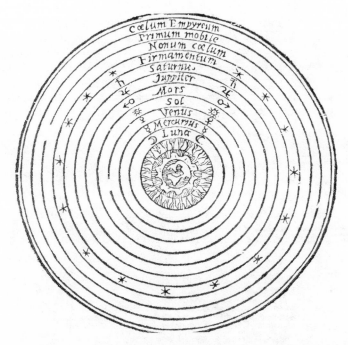

Figure 7. Clavius's depiction of the eleven-sphere cosmos. Earth and the elements are at the center, and the empyrean heaven is on the circumference. Later editions showed three spheres between the firmament and the empyrean heaven. From Clavius, *Sphaera* (Venice, 1596), 72. Photograph courtesy of the University of Wisconsin–Madison Memorial Library.

the spheres depicted in diagrams of the entire cosmos (e.g., fig. 7). Finally, "sphere" can refer to a component—an eccentric or epicycle, for example—of the mechanism that moves a given planet. Clavius sometimes calls these "partial spheres." The simple rotation of each component sphere combines with the rotations of the other components to produce the more complex observed motion of the planet itself. All the components, or partial spheres, for a given planet, taken together, constitute the sphere of the planet in the second sense, above.

Clavius also gave "sphere" a characteristically practical interpretation. The "sphere" of *Sphere* commentaries was, for Clavius, that most useful of astronomical teaching tools, the *sphaera materialis,* or what we might call a kind of armillary sphere. In its pedagogical applications it was roughly the early modern version of a celestial globe. Because the *sphaera materialis* represents all of the primary celestial circles—zodiac, equator,

Figure 8. Title page of Clavius's *Sphaera* (Venice, 1596) showing the *sphaera materialis*. Photograph courtesy of the University of Wisconsin–Madison Memorial Library.

tropics, and the like—it is almost indispensable in visualizing the concepts of spherical astronomy. Clavius chose an engraving of a *sphaera materialis* as the title page illustration for many editions of his *Sphaera* and frequently advised the reader that some concept or other would be made clearer by studying one of these implements (see fig. 8).

Whether the sphere itself was abstract or concrete, Sacrobosco's text on it was not the only such treatise,[30] but it was the only one that fathered a long line of progeny in the form of commentaries—and there were many commentaries on the *Sphere* of Sacrobosco. Lynn Thorndike discusses no less than twenty-six manuscript versions that he dates earlier than the fifteenth century.[31] Francis Johnson estimates the number of printed editions before 1501 to be at least thirty, while those printed between 1501 and 1600 number at least two hundred distinct editions.[32]

Sacrobosco's *Sphere* contains four chapters. The first presents the fundamental cosmological tenets of ancient and medieval European astronomy: the sphericity of the heavens and Earth, the basically circular motions of the heavens, the centrality and immobility of the earth, and the negligible size of the earth with respect to the heavens. The second chapter explains the basic geometrical concepts of the celestial sphere: the celestial great circles (such as the celestial equator, ecliptic, and meridian), the poles of the equator and ecliptic, the tropics, and the various implications of the word "sign" (of the zodiac). Sacrobosco's third chapter discusses a variety of temporal phenomena and how they vary (when they do) with the latitude of the observer: the rising and setting of celestial bodies, right ascensions, inequalities of days, and the motion of the sun along the ecliptic. He also explains the seven climates in this chapter. The fourth and final chapter sketches—with breathtaking brevity—the Ptolemaic movements of the sun, moon, and planets. It then considers the meaning of retrograde motion and the causes of solar and lunar eclipses.

Sacrobosco's little book, only twenty-four pages in the modern translation and edition, contains only those bare essentials of geocentric astronomy. Such a humble introduction as the *Sphere* could not serve as the sole astronomical textbook of the medieval and early modern period, thus in the vigorous intellectual atmosphere of the thirteenth century the need to supplement it was urgent. Because the *Sphere* gave only a rudimentary account of planetary motions, there appeared texts covering more advanced planetary theory and calculations. Most notable among them were the *Theorica planetarum,* explaining basic Ptolemaic constructions, and the *Canones* (or rules) accompanying the Toledan and Alphonsine Tables (tabulations of planetary motion to aid in calculations). These three texts, *Sphere, Theorica planetarum,* and *Canones,* formed the core of astronomical teaching texts in manuscripts of the Latin Middle Ages.[33] In addition, there were treatises on the construction and use of instruments such as quadrants, sundials, and astrolabes. As common as the *Sphere* itself was the *Computus,* which treated calendrical conundrums, the most important of which was the determination of the date of Easter. Sacrobosco himself wrote one of the most popular of these.

Even in teaching elementary astronomy from Sacrobosco's *Sphere,* teachers found it necessary to go beyond the original scope of the work. Some, like Michael Scot, were concerned to address issues of natural philosophy that Sacrobosco did not raise. While Sacrobosco dealt primarily with mathematical questions, the medieval natural philosopher inquired after such issues as the nature of celestial matter, how it is moved, how celestial bodies interact with each other and with terrestrial bodies, and so on. Scot's commentary on Sacrobosco's *Sphere* includes scholastic discussions of some fifty-three questions on natural philosophy accompanied by many citations of Aristotle. On the other hand, the commentaries by Robertus Anglicus and Cecco d'Ascoli discuss far fewer questions, cite Aristotle less, and cite other authors more. They go beyond basic exposition of the text to supplement Sacrobosco with additional information, most of it astrological.[34]

It is intriguing that the commentary on Sacrobosco's *Sphere* was so very long-lived. There are published editions as late as the middle of the seventeenth century.[35] The first British Astronomer Royal, John Flamsteed (1646–1719), was, by his own account, first introduced to astronomy through a study of Sacrobosco's venerable *Sphere.*[36] What sustained this remarkable popularity?

The practice of writing a commentary on some authoritative text, typically Aristotelian or scriptural, dates from classical antiquity. But why a commentary rather than an original treatise? We might suppose that commentators chose their format out of reverence for the ancient texts. Medieval Renaissance commentaries on Aristotle, for instance, could be seen as an acknowledgment of the intrinsic superiority or importance of the pristine, primary text, to which the commentator adds his ruminations. The importance of the older work justifies the commentator's preservation of the form and content of the primary text. In the case of texts that carried unquestioned authority in the Middle Ages and Renaissance—Scripture and the church fathers, for example—this interpretation is entirely plausible.

But many commentaries are extremely critical of the original author even when that source is of great antiquity and authority, as was Aristotle. In such a case, it makes more sense to see the original text as defining an area (and sometimes even methods) of discourse that the commentator accepts without necessarily judging the original author to be intrinsically superior. There are perfectly practical reasons why a writer would choose to comment on Sacrobosco rather than make a fresh start. In medieval European universities one did not necessarily discard a textbook simply because of its lack of depth. Instead of writing a completely new text, the ambitious textbook author could write a commentary on the popular and familiar *Sphere,* thereby conveniently allowing himself to retain the traditional order of presentation and the core of the subject matter. At the same time, the commentator could offer clarifications, supplements, and digressions within the basic framework of the original text. Thus the commentary was a very flexible format that could embody the status and recognition of the original text and simultaneously allow the commentator considerable freedom to apply his own emphasis and direction.

The popularity of *Sphere* commentaries was not constant but seems to have waxed and waned. In the early sixteenth century, among writers of astronomical textbooks, there was a tendency to distance themselves from Sacrobosco's text and to compose instead original manuals treating the traditional subjects. In his study of sixteenth-century astronomical textbooks, Johnson noted two basic types of original astronomical texts.[37] One is the popular encyclopedia, typified by Gregor Reisch's *Margarita philosophica* (published very early, sometime before 1500, and reissued many times in the early sixteenth century). In his encyclopedia Reisch devoted one book to astronomy as one of the liberal arts and gave a summary of the basic content of Sacrobosco. The other common type of text was the cosmographical book. Perhaps the most popular of these was Peter Apian's *Cosmographicus liber* (originally published in 1524 but frequently revised, translated, and reprinted). The cosmographical text generally departed from Sacrobosco's format and presented introductory astronomical concepts as part of a general description of the terrestrial and celestial worlds. Cosmographies also tended to be brief manuals rather than comprehensive treatises (though this is not true of Apian's book). Popular cosmographies were written by Oronce Finé, Gemma Frisius, Francesco Maurolico, Francesco Barozzi, and many others.[38] There were many other popular introductory texts in the mid-sixteenth century that fall outside of both the Sacrobosco commentary category and the cosmographical category but which still share the tendency toward brevity characteristic of both.[39]

In the last third of the sixteenth century, the cosmographical and encyclopedic astronomical textbooks seem to have fallen out of fashion in favor of a new generation of Sacrobosco commentaries. These new-wave *Sphere* commentaries differed considerably from their late fifteenth- and early sixteenth-century ancestors. In them the commentary text has become so vast and wordy that it dilutes Sacrobosco's original twenty-four or so pages almost beyond recognition. Johnson identifies the earliest of these lengthy expansions as Erasmus Schreckenfuchs's *Commentaria in sphaeram Joannis de Sacrobusto* (1569), which fills a folio volume of over three hundred pages. Of comparable bulk was the compendium of Francesco Giuntini, *Commentaria in sphaeram Ioannis de Sacro Bosco* (1577), which occupied two thick octavo volumes.[40] But the grandest of these Sacrobosco revivals was the five-hundred-page tome published in 1570 by Christoph Clavius, *In sphaeram Ioannis de Sacrobosco commentarius.* Clavius issued at least seven different revisions of his *Sphaera,* which eventually had more than sixteen printings between 1570 and 1618.[41] It was obviously a very successful book.

The corpulence of this new generation of *Sphere* commentaries is reminiscent of encyclopedic works like Reisch's *Margarita philosophica* (which was still in print as late as 1583).[42] These chubby *Sphere* commentaries attempted to be encyclopedias of astronomy. They gathered and summarized opinions and information from both ancient and modern sources. The very concept of astronomy as a discipline is broader and more ambitious in these books than in earlier *Sphere* commentaries, which often focused closely, though not exclusively, on spherical astronomy. In particular, there was more discussion of problems and questions from the nonmathematical branches of philosophy—especially physics. Such problems as the nature of celestial matter, the problem of change in the heavens, the possibility of extramundane worlds and the like were traditionally considered not in mathematical texts such as the *Sphere* but in texts on natural philosophy. But textbook authors after the middle of the sixteenth century became much more willing than their predecessors to introduce (albeit sometimes hesitantly) questions of physics into astronomy. The genre acquired added bulk from, for example, discussions of instruments and observations, reference tables such as star catalogs, and discursive expositions.

Perhaps the appearance and success of this next generation of *Sphere* commentaries reveals a dissatisfaction arising from the earlier generation of compact manuals and narrow treatises. Both fashions (handbook and encyclopedia) make sense as different kinds of reactions by sixteenth-century humanists to the medieval tradition of the commentary on Sacrobosco. The manuals, including the cosmographical genre, seem to be deliberate rejections of the medieval commentary tradition and its scholastic associations from which many humanist educators sought to distance themselves. The later megacommentaries of Clavius and others seem to blend the undisputed pedagogical utility of Sacrobosco's *Sphere* with the encyclopedic inclination and comprehensive attention to ancient and modern opinions characteristic of early humanists. Moreover, the humanist esteem for utility often manifests itself in the great number of tables, procedures, and examples provided for the student in these textbooks.

Specialized mathematical treatises were a major ingredient in the growth of the successive editions of Clavius's *Sphaera*, which eventually encompassed theoretical digressions on isoperimetric figures, combinatorics, and Archimedes' *Sand-Reckoner*. In addition, he wrote extended segments on more applied topics such as calculating right ascension and declination of a given point on the sky, calculating the times of twilights, and so on. While fitting the encyclopedic propensities of the late *Sphere* commentaries, these mathematical essays probably also played a role in the debates over the status of mathematics, for they are strong expressions of Clavius's position that mathematics is a proper and highly useful science. In a related way, his discussions of parallax (a fundamentally mathematical technique) in connection with the order of the planets and celestial phenomena strongly suggest that mathematical sciences are suitable, even essential, to the cosmological discussions customarily reserved to the realm of philosophy.

Clavius's *Sphaera* is thus the product of a tradition more than three hundred years long. Over that time Sacrobosco's little manuscript flourished in the new world of printed books, outlived scholasticism to find a place on the bookshelves of Renaissance humanists, and persisted through the sixteenth century as one of the most popular introductory astronomical texts in Europe. This remarkable survival was largely possible because of the work of generations of skilled commentators such as Clavius, who could remake the text to meet the needs of new circumstances, whether astronomical or institutional. But not even Clavius could salvage Sacrobosco in the new world wrought by the likes of Copernicus and Galileo.

The Planetary Astronomy of Sacrobosco and Clavius

Sacrobosco's *Sphere*, though a beginner's text, offered its reader a more thorough understanding of astronomical appearances than that possessed by today's average college graduate—unless, of course, the hypothetical modern student has actually studied astronomy. The *Sphere* becomes truly obsolete only when presenting cosmological information or theoretical matters. We still teach, as did Sacrobosco, that the stars, sun, and moon rise daily in the east, cross the sky, and set in the west. Sacrobosco attributes this to a true daily westward rotation of the entire sky around the motionless earth, while today's teacher will explain that the appearance results from an eastward rotation of the earth. This diurnal motion of the sky is one of the most fundamental concepts of ancient and medieval astronomy, because from it came the idea that the sky was, in some sense, a single body moving as a whole. Supporting this conception was the fact

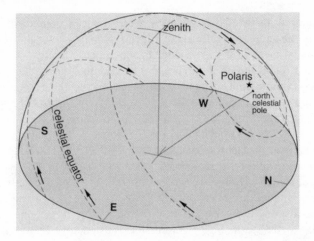

Figure 9. Celestial geometry and some diurnal paths as seen from a temperate northern latitude. Cartography Laboratory, University of Wisconsin–Madison.

that the patterns of stars, the constellations, appear to be unchanging from generation to generation. Instead of shifting patterns within a group, like the changing positions of individual birds in flight within a flock, the stars, as they rotate about the earth, seem more like nails in the rim of a wagon wheel—the nails are fixed in relation to each other even though the wheel as a whole may move. Thus they were commonly called the "fixed" stars. The body to which they were fixed, like the rim of the wheel, was generally called the "firmament" or, more plainly, the sphere of the fixed stars. Not all ancient people necessarily pictured the heavens this way, but Greek philosophers had developed the idea explicitly, at least by the time of Plato.[43]

The firmament, or stellar sphere, makes its daily rotation about an axis that defines north and south. In the north, the axis points approximately toward the star called Polaris, or North Star. The exact points in the sky where the axis meets the firmament are called the "north" and "south celestial" poles. Around the middle of the stellar sphere is the "celestial equator." From some place in the mid-northern hemisphere, say, Rome, imagine looking south on a clear night (see fig. 9). Behind us, about halfway between the northern horizon and the point directly overhead (called the "zenith"), stands Polaris. Very near it (though invisible, being imaginary) is the north celestial pole. In front of us, stretching from the eastern point of the horizon to the western, and culminating about halfway between the zenith and the southern horizon, is the (also imaginary) celestial equator—every point of which is ninety degrees from the celestial poles. Another imaginary line, the local meridian (not shown), arches overhead between north and south and passes through both the zenith and the north celestial pole.

On their daily paths, stars rise from the eastern horizon, reach their culmination on the meridian some hours later, then descend into the west and set. The diurnal path of any given star is parallel to the celestial equator. From our position in the northern hemisphere, the more southerly stars have shorter paths, and are thus above the horizon for less time, than stars farther north. Stars in the general vicinity of the north celestial pole, circumpolar stars, have diurnal paths that do not cut the horizon at all, so they never rise or set but are perpetually above the horizon as they complete their daily march around the pole. Because our principal interest here is planetary theory, this discussion need not pursue celestial geometry in much detail, but Sacrobosco explains all such fundamental concepts, including the various risings and settings of stars, the seasons, climates, twilights, and the like.

Even prehistoric people were aware that celestial motions involve more than the basic diurnal motion. England's Stonehenge, the basic constructions of which are more than four thousand years old, demonstrates that its builders followed closely the motions of the sun and moon. Fairly casual observation of the two luminaries, as well as the five bright "stars" we call planets, reveals that their motion is not identical to the diurnal motion of the stars. Though also rising in the east and setting in the west, these seven objects seem continually to fall a bit behind the stars, rising somewhat later each day in comparison with the stars. If we ignore the diurnal motion and think of the fixed stars as a stationary backdrop (as we will do in the rest of this discussion), then the motion of the sun, moon, and planets is generally eastward.

This eastward motion carries the sun in a complete circuit about the sky in one year. The moon moves much faster, completing its revolution in about a month. Many aspects of astronomy would be much simpler than they are if the sun simply slid eastward along the celestial equator. (That simplicity would come, however, at the cost of the seasons, for there would be none.) In fact, the sun's path through the stars, called the "ecliptic," is inclined some twenty-three-and-one-half degrees to the celestial equator and crosses it at two points on opposite sides of the sky. For this reason the sun spends half of the year to the north of the equator and half to the south. The two crossing points are called the "equinoxes." The sun reaches one equinox around 21 March each year, and at that time it crosses from south of the celestial equator to the north. At the other equinox the sun crosses the equator, from north to south, around 21 September. Our historical prejudice for the viewpoint of the northern hemisphere leads us to call the March equinox the "vernal equinox," while the other one, of course, is the "autumnal equinox." The sun is moving northward when it crosses the celestial equator at the vernal equinox, and it continues to move farther north of the equator until it reaches its maximum distance from the equator around 21 June. This point is called the "summer solstice" (again in the northern hemisphere), and the opposite point, where the sun is farthest south of the celestial equator, is called the "winter solstice," which occurs around 21 December.

Though we cannot usually see the stars and the sun at the same time, we can visualize the sun's yearly course along the ecliptic as it moves in front of (but we usually say "through") different groups of stars. Astronomers of the ancient near east laboriously determined which groups of stars the sun's path crosses and identified the twelve constellations that we call the zodiac. Sacrobosco taught his reader to memorize the names of the zodiacal constellations in the order in which the sun traverses them: Aries, Taurus, Gemini, Cancer, Leo, Virgo, Libra, Scorpio, Sagittarius, Capricornus, Aquarius, and Pisces. As luck (or the laws of nature) would have it, the paths of the moon and planets, while they do not adhere strictly to the ecliptic, also never deviate far from it. So the zodiac, for thousands of years, has served as the background against which astronomers have measured the motions of the sun, moon, and planets.

Like the sun and moon, each of the five planets (no others were known before the eighteenth century) circumnavigates the zodiac at its own unique pace. The motion peculiar to a planet is called its "proper motion." Saturn, Jupiter, and Mars complete a zodiacal circuit in about thirty, twelve, and a bit less than two years, respectively. Venus and Mercury accompany the sun around the zodiac on its annual round. The first three planets can, at any given time, be in the same direction as the sun (in "conjunction"), directly opposite to it (at "opposition"), or at any point in between. But Mercury and Venus are never found very far from the sun. Indeed they appear to wander endlessly east and west of the sun, like dogs on leashes, while they follow the sun's perpetual eastward motion. Mercury has the shorter leash and stays so close to its master that it is almost always difficult to see in the midst of the solar glare. Brilliant Venus, while straying considerably farther away than Mercury, always sets soon after the sun or rises soon before it—thus earning the names "evening star" and "morning star." Mercury takes about 116 days to complete one perambulation from its eastern extremity to the western and back, while Venus requires more than a year and a half.[44] Their motions

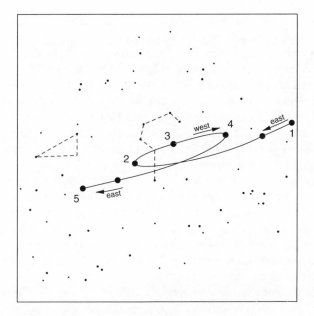

Figure 10. An example of retrograde motion of a planet. This shows the motion of Mars against the background of the constellation Leo between October 1994 and July 1995. Opposition (point 3) occurs on 11 February 1995. Cartography Laboratory, University of Wisconsin–Madison.

divide the five planets into two categories: Those that must remain near the sun (Venus and Mercury) and those that need not (Saturn, Jupiter, and Mars). The first two became known as the "inferior" planets and the rest as the "superior" planets from the general belief that the former were lower (closer to the earth) than the sun and the others higher. (In chap. 3 I consider more fully the problem of the order of the planets.)

Just as the inferior planets do not always move eastward, the superior planets have their own perversity (see fig. 10). Some months before reaching opposition, a superior planet's normal eastward motion begins to slow. Eventually it slows to a stop (point 2), then reverses and accelerates toward the west. While moving westward it will pass through opposition (point 3) and then begin to slow. After some time it comes to another stop (point 4), then reverses again and accelerates eastward resuming its normal motion. The whole process is slow, requiring some months. As it goes through these motions the planet also becomes dramatically brighter, reaching a peak just at opposition, and then dimming. The anomalous westward motion is called "retrograde" motion. Finding a way to explain the motions and brightness changes and predict their reoccurence was a central problem for ancient Greek astronomers.[45]

Tradition has it that the Greek philosopher Plato (427–348 B.C.) introduced the idea that celestial bodies move only in perfect circles in some sense, and that their motion along the circles is uniform. Remarkable though it may seem, his notion of uniform circular motion would form the basis of all astronomical theorizing until Kepler discarded it in the early seventeenth century. The centrality and stability of the earth, supported by both intuitive predisposition and rational argument, were the other major assumptions of astronomical theory. Philosophers and astronomers from antiquity to early modern times often considered the possibility of the earth's motion, but it is difficult (though not impossible) to find any who took it seriously before Copernicus in the early sixteenth century. So the task of the ancient astronomer was to explain the motions of the celestial bodies on the assumptions that the earth was motionless at the center of the universe and that everything else moved in some kind of uniform circular motion around it.

At the time of Plato and Aristotle (384–322 B.C.), explaining the proper motions of the sun, moon, and planets must have seemed to be a matter of extending the idea of the firmament to the planets. That is, a celestial body moves because it is attached to a sphere, just as the stars are fixed to the firmament. Unlike the stars, the proper motion of each planet is different from the others and from the stars, so each planet needs its own sphere (or spheres). A planet's sphere must move slightly slower than the firmament and on a different axis. Thus if the sun had its own sphere, tilted twenty-three-and-a-half degrees from the north-south axis, and if that sphere's rotation were to lag behind the stellar sphere just enough, then the sun would traverse the zodiac as we observe it to do—on the average. That is the essence of the idea. In reality, the day-to-day motion of the sun is not uniform, and a single, uniformly moving sphere cannot yield acceptable predictions of such events as the solstices and equinoxes. The retrogradations of the planets posed an even greater problem for astronomical theory.

Homocentric Spheres in Planetary Theories
Combinations of rotating spheres form the basis of the earliest known mathematical model for the motion of the planets. Each sphere, like the firmament, rotates uniformly about a diametrical axis whose midpoint is the earth. Earth is thus the common center of all the spheres. Such a system of spheres, often called "homocentric" (having the same center), combines several uniform circular motions into a compound motion that

is nonuniform. The motions of the spheres are cumulative, that is, each sphere receives the sum of the motions of the spheres above it. While not a part of Sacrobosco's astronomy, homocentric spheres were nonetheless part of the traditions of mathematical astronomy and appeared prominently—if not favorably—in Clavius's *Sphaera*.

The ancient Greek astronomer Eudoxus (fl. 370 B.C.) showed that four homocentric spheres could provide a fair approximation to the motions of some of the planets. (Eudoxus's own descriptions of his planetary theories were lost in antiquity, and only accounts by later commentators survive. Thus modern descriptions of his theories are necessarily reconstructions.)[46] In the models of Eudoxus, the first and outermost sphere rotates from east to west once daily, carrying the three lower spheres with it. This sphere rotates about an axis passing through the celestial poles, which means that its equator is equivalent to the celestial equator (see fig. 11). The motion of this sphere makes the planet rise and set daily. (Eudoxus seems to have treated each planet as a separate case, without regard for the motions of any other planet, which means, among other things, that each model has its own sphere providing the diurnal motion). The second sphere rotates from west to east about an axis tipped twenty-three-and-a-half degrees from the celestial poles, thus its equator represents the ecliptic. This sphere rotates with a period of motion equal to the observed zodiacal period of the planet. In the case of Jupiter, the second sphere would rotate once in about twelve years, thus producing the planet's average progress through the zodiac. To bring about retrograde motion, Eudoxus added the third and fourth spheres. The axis of the third sphere is embedded in the equator of the second, which is to say in the plane of the ecliptic. The planet itself is found on the equator of the fourth sphere, the axis of which is inclined at a small angle to the axis of the third. The third and fourth spheres rotate with the same period but in opposite senses, and their combined motion causes the planet to execute a figure-eight curve, which is called a "hippopede" (see fig. 12a). The hippopede represents the regular retrogradations of the planet. The second sphere carries the hippopede, in effect, through the zodiac. The resulting path of the planet against the fixed stars would look something like that shown in figure 12b.

The homocentric spheres of Eudoxus bear the significant distinction of being the earliest known mathematical theories of planetary motion. But they also have serious deficiencies as representations of the observable planetary phenomena. While judicious selection of theoretical parameters can yield fair approximations to the retrogradations of Saturn, Jupiter, and Mercury, the approach Eudoxus takes is incapable of doing the same for Venus and Mars.[47] Furthermore, the retrograde path produced by the

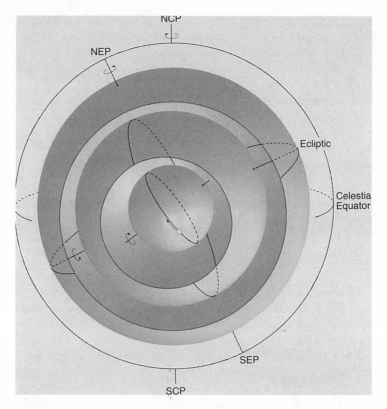

Figure 11. An example of a simple system of homocentric spheres. The north and south celestial poles are labeled *NCP* and *SCP*, respectively; *NEP* and *SEP* are the north and south ecliptic poles. The size of the gaps between the spheres is arbitrary. Cartography Laboratory, University of Wisconsin–Madison.

hippopede of a given planet is identical at every passage of the planet through opposition, while in reality planetary retrogradations exhibit a variety of shapes. Finally, because the motion of the planet is produced by the rotations of spheres that are all, by definition, centered on the earth, the planet will, at every moment, be the same distance from the earth. Yet the real planets vary considerably in brightness, as mentioned earlier, and intuition strongly suggests that the cause of this phenomenon is variation in distance between the earth and the planet. In addition, the apparent diameter of the moon, and thus its distance, was known to vary, so it could not be moved simply on homocentric spheres. Theories using homocentric spheres are thus very unsatisfactory in many respects from the point of view of the astronomer.

Despite the shortcomings of homocentric theories, there were no competent alternatives for Greek philosophers of the fourth century B.C. Aristotle endorsed a version of the theory, and his approval helped the general idea take hold. He proposed a scheme by which all the spheres of all the planets could be nested within each other to constitute a cosmos—a single unified system of the heavens. Following Aristotle, the universe came to be pictured as a series of concentric spherical regions, each region containing the spheres required to move its celestial body, with the earth at the center and the firmament around the outer edge. (Note again the ambiguity of "sphere," meaning sometimes the region bounded by the lowest and highest spheres of a given planet, but sometimes denoting any of the component moving spheres of a given planet.) The exact order of the planetary spheres (in the larger sense), especially those of the sun, Mercury, and Venus, provoked some debate. Their relative rates of progress around the zodiac suggested that Saturn was the outermost, followed by Jupiter, then Mars, then the problem area of Sun, Mercury, Venus, and

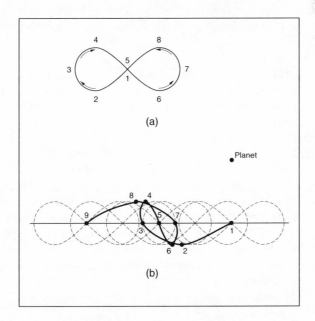

Figure 12. Example of retrograde motion generated by homocentric spheres. Part (*a*) shows the "hippopede" generated by the motion of the two innermost spheres, and (*b*) shows the path generated as the hippopede is translated eastward by the next higher (third) sphere. After Hargreave, "Reconstructing the Planetary Motions of the Eudoxean System" (1970). Cartography Laboratory, University of Wisconsin–Madison.

finally the moon, which was undoubtedly the closest. The entire assembly was visualized as eight spheres (moon, sun, five planets, firmament), and the entire assembly rotated about the stationary earth once a day (see fig. 7).

Now this is a remarkable conception, and we are entitled to ask questions about it, such as, What keeps the spheres moving, and of what are they made? Aristotle did not shirk his duty to answer, but that does not mean that he was absolutely clear. He said that the spheres do not move themselves, but all are moved by some other mover or movers. Aristotle spoke sometimes as if the firmament were the mover of all the lower spheres, but at other times he made statements that appear inconsistent with that idea. His inconsistencies (for this was far from the only one) provided Aristotelian philosphers with centuries of busy disputation about just what he did mean. As to what material constitutes the heavenly bodies, Aristotle taught that it is not one of the four terrestrial elements (earth, air, fire, and water), but some fifth element, called by medieval philosophers the "quintessence," or fifth essence. This remarkable stuff moves naturally and eternally in perfect circles—it is the material of the celestial spheres. It is, he taught, changeless—except for position. That is to say that unlike our experience on earth, where things grow, develop, corrode, corrupt, and so on, in the heavens the celestial bodies are forever in the same state. For Aristotle, things that seemed to come and go in the heavens, such as meteors or comets, were actually phenomena of the upper atmosphere, where change is possible. This concept, the incorruptibility of the heavens, became the subject of much disputation among natural philosophers.

A couple of centuries after Aristotle, the astronomer Hipparchus made good use of the general technique of compounding spheres to explain complex celestial motions. Hipparchus noticed that star positions, which he measured with respect to the vernal equinox, seemed to have shifted as compared to measurements made many years before. It seemed to him as if all the stars had shifted very slightly eastward with respect to the ecliptic. We call this motion the "precession of the equinoxes." The displacement is very slow—even Hipparchus realized that a complete revolution of the stars around the ecliptic would require tens of thousands of years. He chose to explain this motion by positing a ninth sphere, this one beyond the firmament. The ninth sphere carries the firmament and the rest of the spheres in their westward daily motion, but the firmament, or eighth sphere, lags slightly behind the ninth (similar to the way the planetary spheres lag the firmament and each other). The discrepancy between the speeds of the eighth and ninth spheres explains the precession. In the

Middle Ages, some Arabic astronomers became convinced that the rate of precession was not constant. This motion, sometimes called "variable precession" but better known as "trepidation," provoked a number of schemes to account for it, including some that added yet more spheres. Regardless of the astronomical shortcomings of homocentric spheres, they had great appeal because of the authority of Aristotle, the strength of his ideas about celestial physics, and perhaps the success of additional spheres in explaining precession and trepidation. Astronomers and philosophers of antiquity, the Middle Ages, and the Renaissance were continually tempted to make homocentrics work, despite the great success of a rival system.

Eccentrics and Epicycles in Planetary Theories

The failure of the homocentric spheres to explain planetary motions led Greek astronomers to explore other theoretical approaches. The most successful approach was to abandon the earth as the precise center of the planetary circles. Instead, combinations of several different types of uniform circular motions were used to produce reasonably good explanations of planetary motion. The development of these models took centuries, beginning with Apollonius of Perga around 200 B.C. The most successful practitioner was Ptolemy (fl. 150 A.D.), who compiled his planetary theories in the book we know as the *Almagest*. The three basic mechanisms are called the "eccentric circle" (or simply "eccentric"), the "epicycle," and the "equant." The planetary theories that Ptolemy built from these basics were so successful that they remained the standard models for fifteen hundred years. Sacrobosco mentioned Ptolemy's three tools only very briefly, but Clavius considered the first two at some length.[48]

The eccentric is probably the easiest to understand. The celestial body moves uniformly (that is to say, at a constant speed along its path) in a circle. The center of the circle is not the earth, which is why the circle is called eccentric (out of center) because its center is away from the center of the universe—the earth. In figure 13, *T* represents the earth, *C* is the center of the eccentric circle, and *e* is the planet. (Disregard the other elements of the figure for now.) In the case of the eccentric, the larger solid circle is the actual path of the planet. The planet's closest approach to Earth *T*, will be at *y*, which is called "perigee." Its greatest distance from Earth will be at *x*, called "apogee." (Sacrobosco and Clavius use the term "aux" for apogee and "opposite aux" for perigee. These medieval words came to be seen as archaic and were replaced by the modern terms in the early seventeenth century.) Since these two points are known as the "apses," the line *xy* is naturally known as the "line of apses."

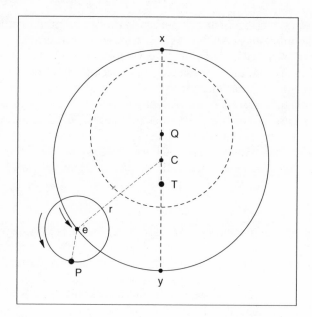

Figure 13. Three basic Ptolemaic planetary mechanisms: eccentric circle, epicycle and deferent, equant. Cartography Laboratory, University of Wisconsin–Madison.

Despite the fact that the true speed of the planet is constant along its path, the eccentricity of its circle produces an apparent nonuniformity in its speed as viewed from Earth, because as it passes near perigee it will seem to move faster than average, and as it moves through apogee it will seem to move slower than average. To an observer at *C,* the center of the eccentric circle (which is just a point in space), the motion of the planet would appear perfectly uniform. The degree of nonuniformity seen from Earth can be adjusted in the theory by choosing *C* to be closer to or farther from *T.* Ptolemy used the eccentric without any of his other tricks to construct a rather successful explanation for the motion of the sun.

The epicycle is a small circle whose center moves along the eccentric (fig. 13). Now let *e* represent not the planet, as earlier, but the epicycle center, which is merely a point in space. Because it carries the epicycle center along, the eccentric circle is often called the "deferent," from the Latin *deferre,* to carry or convey. (The deferent need not be eccentric in principle, but in Ptolemy's models it always was.) The planet, represented by *P,* is carried along the circumference of the epicycle. The two constructions, eccentric deferent and epicycle, produce two uniform circular motions that combine to produce an apparent nonuniform motion of the planet. For instance, most of the time the motion of the epicycle adds to the counterclockwise motion of the deferent, and the planet moves counterclockwise as viewed from Earth T. But at some point, as the planet swings toward r, the motion of the epicycle cancels, then overcomes the motion of the deferent as viewed from T. The result as seen from Earth is that the planet slows, stops, then reverses its motion, just as happens during the retrograde motion of a superior planet. Note that in both the eccentric and epicycle mechanisms, the precise center of the planetary motions is not Earth. These devices are therefore in technical violation of Plato's dictum that celestial motions should be geocentric. On the other hand, they preserve the principle that celestial motions must be composed of uniform circular motions.

Ptolemy invented neither the eccentric circle nor the epicycle-deferent techniques. He did introduce the equant, which is a somewhat more subtle concept than the others. It violates even Plato's dictum on uniformity of circular motion in the heavens, as we shall see. The equant, which Ptolemaic astronomers (including Sacrobosco and Clavius) often defined as a circle, is better represented by its center, Q in figure 13. (It is somewhat misleading to define the equant as a circle, but because it was traditional to do so, I have included a dashed circle to represent an equant circle. The actual size of the circle is unimportant.) The equant was introduced to provide a more flexible rule for governing the motion of the epicycle center, e, along the deferent. The idea is that the motion of the epicycle center should be uniform *as viewed from Q*, the equant point. (Another way of saying this is that the intersection of radius Qe with the equant circle should move in uniform circular motion along the equant circle. The equant circle is not unique because any circle centered on Q would serve the purpose.) If Q is some distance from C, then the epicycle center must move faster than average along the deferent when near y, for example, and slower than average when near x, for its motion to appear uniform as viewed from Q. The resulting nonuniform motion would be even more dramatic as viewed from Earth T.

By introducing the equant, Ptolemy introduced into the theory another controllable source of nonuniform motion. His judicious mixing and matching of the parameters at his disposal (the size of the epicycle, the speeds of the deferent and epicycle, position of the equant, etc.) secured the very long-lived success of his planetary models. But the price of the equant was the surrender of the philosophical principle of uniform circular motion, because now the motion along the deferent was uniform only in the sense that it appeared that way as viewed from Q, a mere point in empty space. Many later astronomers, among them Copernicus, would criticize the equant because of this philosophical shortcoming.

A Note on Parallax

Determining the relative order and distances of the sun, moon, and planets was a slippery problem (I discuss the arguments in more detail in chap. 3). Finding their actual distances, however, was even more difficult. It depended ultimately on the determination of one distance, usually the moon's, from which all the other distances could be derived.[49] The basic method of directly measuring celestial distances (and indeed the only one until the advent of ranging methods such as radar) is called trigonometric parallax, which is an application of basic geometry. It is basically triangulation. An object is observed from two vantage points, and the differences between its observed positions from those two points, combined with the relationship between the two observing sites, allows the observer to calculate the actual distance to the object. This could be accomplished by two widely separated observers making essentially simultaneous observations and later comparing their findings—Clavius would do just this in his discussions of the location of the nova of 1572. Alternatively, a single observer could measure the position of a celestial body when it is rising, for example, and then measure it again some time later, such as when it is setting. This kind of measurement is called diurnal parallax, because it depends on the diurnal motion of the heavens, or of the earth if you are a Copernican—the geometry is equivalent in both cases. (Note that diurnal parallax is different in practice, though not in principle, from "annual" parallax, in which the motion of the earth in its orbit about the sun moves the observer from one observing station to the other. Its existence is a natural consequence of the Copernican theory. Stellar parallax, though used today to obtain distances to the nearer stars, is a very small effect and was undetected before the nineteenth century. For an astronomer such as Clavius, who believed the earth to be completely stationary, there would be little point in looking for stellar parallax, except to show by its absence the earth's lack of motion. Clavius, however, did not make this argument.)

The technique of diurnal parallax is no mystery (see fig. 14). The observer at B measures the position of the moon at C as it ascends in the eastern sky. Our observer can note the lunar position either with respect to the stars or to the zenith, Z, overhead. Some hours later, the observer notes the moon's position at D as it descends through the western sky. As the diagram makes clear, the angle between these two observations

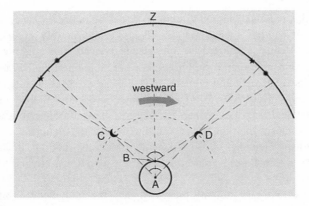

Figure 14. The fundamental geometry of diurnal parallax. Cartography Laboratory, University of Wisconsin–Madison.

would be different had they been recorded by a hypothetical observer at the center of the earth, *A*. Note also that parallax will cause an apparent change in the position of the moon with respect to the stars as seen by the observer at *B*: at *C* the moon appears east of the pair of stars, but at *D* it appears west of the stars, while its "true" position (as seen from the center of the earth) is between the stars. The angle *ACB* (or *ADB*) is called the moon's diurnal parallax. We can find it because it is one-half of the difference between the angle *CBD*, measured by the observer, and the angle *CAD*. We know angle *CAD* by the time difference between the observations (in the figure, angle *CAD* is ninety degrees, which means the observations were six hours apart). If we know the distance *AB*, which is the radius of the earth, we can solve the triangle and find *AC* (same as *AD*), the Earth-Moon distance. Of course, the moon's proper motion will also change its position in the time between the two observations, but we can correct for that, in theory. In practice, the entire procedure was tricky and error prone.

A more distant object will have a smaller parallax, and a closer object a larger parallax. More precisely, the parallax of an astronomically distant object is inversely proportional to the distance between the object and the observer. The moon's diurnal parallax was measurable even without telescopic instruments (nearly one degree on the sky, which is twice the width of the full moon) and could yield a reasonably good value for the moon's distance relative to the radius of the earth. Clavius (following Maurolico and others) gives the moon's average distance as sixty earth radii, which is very close to the modern figure.

Diurnal parallax was also applied to celestial bodies other than the moon, but without a telescope the effect is unmeasurably small even for Mars and Venus at their closest approaches to earth. (This did not prevent Tycho Brahe and other acute observers from believing they had measured parallaxes of planets and comets.) Yet diurnal parallax would be a key factor in the discussions of novas and comets, because if their measured parallaxes had been larger than the moon's, it would have followed that they were closer than the moon and thus within the changeable realm of the terrestrial elements. A smaller parallax than the moon's, including an undetectably small one, would imply that the object was beyond the moon and truly in the supposedly changeless celestial region.

In the *Almagest,* Ptolemy presented the planetary models, built with the three basic mechanisms, as little more than mathematical theories. But Ptolemy himself and later astronomers strove for ways to make his mathematical theories part of Aristotle's cosmology of concentric spheres. They did so by means of "materialized" eccentrics and epicycles, which is to say, theories of how celestial spheres could be arranged so as to be equivalent to Ptolemy's mathematical eccentrics and epicycles. Such schemes, though dating back to Ptolemy and having been known to Arabic astronomers during the Middle Ages, were new to Latin Europe when Sacrobosco wrote his *Sphere.* If he knew of them he gave no indication of it. But near the middle of the fifteenth century, a complete treatise on materialized Ptolemaic constructions had been produced by Georg Peurbach (1423–61) in his *Theoricae novae planetarum.* Clavius's *Sphaera* brings together Ptolemy's mathematics, Aristotle's cosmology, Peurbach's spheres, and Sacrobosco's text in a comprehensive presentation of the long astronomical tradition to which they had all contributed. With the framework of Jesuit education, Jesuit mathematics, and Ptolemaic astronomy now established, I proceed, in the next chapter, to reveal in much finer detail the cosmological messages embedded in the mass of Clavius's *Sphaera.*

THREE

The Defense of Ptolemaic Cosmology

In science as in war, history is written by the victors. Those who first embraced a new science are styled as precursors of the latest orthodoxy. Those who stubbornly clung to the old are featured as historical curiosities.

—James R. Moore, *Post-Darwinian Controversies*

The astronomy of Ptolemy and the physics of Aristotle were major pillars of the "modern" structure of astronomy at the time Clavius was writing his *Sphaera*. The legacies of the two ancient philosophers, far from being historical curiosities, were themselves the scientific orthodoxy, and the burden of proof rested on any ideas that challenged the orthodoxy. Sacrobosco's thirteenth-century *Sphere* presented that orthodox worldview in the confident tone of an exhortation to the converted. The same confidence suffuses most of the medieval commentaries on Sacrobosco's *Sphere*. Certain issues and questions continued to provoke disputes in medieval commentaries on Sacrobosco, but generally these were disagreements over details more than fundamentals. There were even some internal inconsistencies, but these were typically glossed over and do not seem to have threatened the conventional worldview. This is not to suggest that the issues disputed in *Sphere* commentaries were always trivial, or that beyond the genre of *Sphere* commentaries there was any lack of criticism of the conventional cosmology. But such cosmological examination usually appeared in commentaries and questions on Aristotle's *De caelo* or *Physics*, not in *Sphere* commentaries.

In every edition of Clavius's commentary on Sacrobosco there is a tone in the cosmological exposition that is distinctly different from the medieval *Sphere* commentaries—less confident and more conscious of a world of alternatives. His suspicion of the alternatives is evident in his language. Those who disagree with "us" are "the adversaries," and they come in camps, like the followers of Averroës, whom he calls the "Averroistas."[1] This defensive stance is also evident in his approach. Unlike Sacrobosco's guileless exposition of how things are, Clavius itemizes and destroys competing ideas wherever they occur, whether on the number and order of celestial spheres, the nature of their motions, the nature of celestial substance, or the causes of comets and novae, and so on. Some ideas, such as the single fluid heaven, which had been calmly considered in earlier *Sphere* commentaries, become, in Clavius, rivals to be critically evaluated and rejected. Thus Clavius's astronomy sometimes seems to be an astronomy under siege. We might attribute this mentality in part to Clavius the Jesuit—the chief mathematician of an elite order whose members consciously saw themselves in the vanguard of the great army of the Counter Reformation. But Clavius's defensive stance must also be understood in the context of the diversity of cosmological opinions circulating and competing in the late sixteenth century—a diversity that we learn about from Clavius himself. The Copernican cosmological alternative was only one of several.[2]

Did an Orthodox Cosmology Exist?

Before we ask, What is the orthodoxy that Clavius dedicated himself to defending? there will be, for many readers, a prior question: Was there an orthodoxy for Clavius to defend? That there existed an acceptable orthodoxy commanding consensus in medieval and Renaissance astronomy is a proposition that has not always been clear in historical interpretations of that period. The fault lies in historical accounts that focus on the scientific revolution as a beginning, as the foundation of modern cosmological thought—a perspective perhaps natural enough for a post-Newtonian historiography but one to be used cautiously. By searching history retrospectively for the roots of modern successes we run the great risk of misunderstanding and losing the intellectual property that an earlier era valued highly. We should take a (perhaps unintended) hint from an eminent historian of science, who wrote "Geniuses are more apt to indicate the line of march into the future than to reflect the consensus accepted in their own time."[3] This could be taken as a Whiggish prescription for recognizing geniuses. But, rather than worry about that problem, let us accept it as advice that if we want to understand the consensus of a given period of history, we should beware of focusing on those figures whom historians retrospectively label "geniuses."

How has this tendency to focus on geniuses, in this case Copernicus, distorted the Ptolemaic worldview's status as an orthodoxy? Historians of astronomy often point out, for instance, that the Copernican theory bases the arrangement of Earth, Sun, and planets on a structural rationale, a mathematical harmony that makes the order of the planets a necessary consequence of the observed motion of the planets. In contrast, they say, the Ptolemaic ordering of the planets is ad hoc and could save the phenomena just as well with any given order. That statement, while technically true, is historically misleading. To give a concrete example, Thomas Kuhn's widely used modern textbook puts it this way:

> In the Ptolemaic system the deferent and epicycle of any one planet can be shrunk or expanded at will. . . . The order of the orbits *may be* determined by assuming a relation between size of orbit and orbital period. In addition, the relative dimensions of the orbits *may be* worked out with the aid of the further assumption . . . that the minimum distance of one planet from the earth is just equal to the maximum distance between the earth and the next interior planet. But though both of these seem natural assumptions, neither is necessary. . . . There is no similar freedom in the Copernican system. If all the planets revolve in approximately circular orbits around the sun, then both the order and the relative sizes of the orbits can be determined directly from observation without additional assumptions.[4]

This account is literally true, but it is misleading in that it portrays the contrast between the Copernican cosmology and its Ptolemaic alternative as if it were a question of logical coherence.[5] It implies that (in the absence of Galileo's famous observations) a Ptolemaic cosmology could not have had an internal coherence simply because it lacked the mathematical harmonies of the Copernican theory. A different kind of internal coherence, founded on other principles than the Copernican, could and did render the Ptolemaic view as satisfactory to its followers as the Copernican was to its small group of partisans. Thus the distortion arises in the quotation above because Kuhn's admission that in the Ptolemaic scheme certain things *may be* determined by certain ''natural assumptions'' ignores the fact that those assumptions were, in fact, made. And once those natural assumptions are granted, the Ptolemaic scheme could be—as in fact it was until the invention of the telescope—as coherent and compelling as the Copernican.

There is another common theme that unfairly detracts from the significance of the Ptolemaic worldview and, in fact, comes close to denying its existence. That is the assertion that there was some considerable discord and dissatisfaction with late Ptolemaic astronomy among its practitioners. The following is a typical example, again from Kuhn: ''By the thirteenth century Alfonso X could proclaim that if God had consulted him when creating the universe, he would have received good advice. In the sixteenth century, Copernicus's co-worker, Domenico da Novara, held that no system so cumbersome and inaccurate as the Ptolemaic had become could possibly be true of nature. And Copernicus himself wrote in the Preface to the *De revolutionibus* that the astronomical tradition he inherited had finally created only a monster."[6] Even granting, for the sake of argument, the appropriateness and accuracy of these examples, they are hardly representative and, in fact, are quite isolated from the general opinion of European astronomers from the thirteenth to seventeenth centuries.[7] There is no reason to think that during the late Middle Ages and Renaissance, astronomically informed people, or even the majority of trained astronomers, were particularly dissatisfied with the Ptolemaic scheme—and Clavius's account gives us every reason to think otherwise. There were, of course, dissenters from the Ptolemaic cosmological orthodoxy—I do not suggest that it was unanimously accepted. But Clavius presents a cosmology that is, as far as he is concerned, traditional, generally accepted, well established, and, of course, true. In a word, orthodox. We may now ask again, What is the orthodoxy that Clavius dedicated his *Sphaera* to defending?

An Orthodox Cosmology

The outlines of Clavius's cosmological teachings were largely in place even in the first edition of his textbook, and they retained the same form, often even the same wording, in all subsequent editions. The biggest cosmological enhancement to the text appeared in the 1581 edition (*nunc iterum recognita*)[8] when he added the lengthy *disputatio* on eccentrics and epicycles, which is an important source for my discussion in this chapter and the next.[9] Other changes are generally much smaller additions that expand or clarify some point already present in previous editions, although some recognize new astronomical developments and will thus be important in this discussion. However, the basic content of Clavius's cosmological teachings remained constant after 1581, so I will not be concerned to note unimportant variations of wording that occur in successive editions.

Ptolemaic cosmology, as Clavius presents it in his *Sphaera,* is an amalgam of various doctrines all very familiar in the Middle Ages.[10] Although the cosmology embodied centuries of work by astronomers who came between Ptolemy and Clavius, it was still Ptolemaic in the same way that we could say that Kepler's cosmology was Copernican. That is to say, the cosmology still derived its inspiration, essential rationale, and a measure of authority, from the eponymic founder, who, despite its refinements, would still recognize his own work.

Clavius describes and defends this late Ptolemaic cosmology, but he does not label or encapsulate it in a neat package. Rather, it emerges slowly in the course of the exposition, accreting layer on layer around a nucleus of fundamental beliefs. That nucleus comprises ideas that a sixteenth-century European reader could reasonably be expected to accept with little or no argumentation or debate. That is to say, Clavius's presentation rests on assumptions and prejudices with which his audience is already comfortable. I will not try to enumerate all the concepts constituting this nucleus but must be content to offer some obvious examples. To begin with, one did not have to be an Aristotelian philosopher to believe in the stability, even immobility, of the earth. Its centrality, once the question was raised, would also require minimal argument. Clavius, in the tradition of *Sphere* commentaries (not to mention Ptolemy), does give arguments for the centrality and immobility of the earth, but the treatment is not adversarial and seems more a matter of establishing formally what is already believed intuitively.[11] Similarly, one did not have to be a theologian to acknowledge the authority of scriptural writings and the doctrine of divine creation in time. If the reader were Christian, then patristic writings also have great authority. Clavius also frequently depends on the reader's acceptance of the authority of some, though not all, ancient philosophical sources, especially Aristotle, Ptolemy, and occasionally Plato. Finally there is the principle of parsimony, which can be interpreted in many ways but generally means that God created nothing in the world in vain. A major consequence of this for many commentators, including Clavius, is that there can be no nonfunctional spaces or unnecessary mechanisms in the universe.

parsimony

There are four closely related astronomical or cosmological elements and two supplementary elements that rest on this nucleus of beliefs and that logically compose late Ptolemaic cosmology as presented by Clavius. They are

1. Acceptance of the planetary constructions of Ptolemy's *Almagest* (though occasionally with minor variations) as an explanatory system for planetary motions
2. Acceptance of the materialization of the Ptolemaic constructions as formulated by Peurbach
3. Acceptance of the order of the planetary spheres according to Ptolemy

4. Consensus on the dimensions of the celestial spheres and the overall size of the cosmos
5. Reconciliation of the Ptolemaic planetary system with Aristotelian physics
6. A Christian vision of the cosmos

Each of these points will be explained in the sections that follow.

Ptolemy's Constructions

The first element is acceptance of the planetary constructions of the *Almagest,* more commonly summarized in the medieval *Theorica planetarum,* as a complete explanatory system for planetary motions. Though the philosophical literature of the Middle Ages and Renaissance, especially commentaries on Aristotle's *Physics* and *De caelo,* continued to question the propriety of Ptolemy's eccentrics and epicycles as any more than calculational devices, the astronomical literature always employed them to some extent. Even when the immediate goal was not calculational, as was usually the case in *Sphere* commentaries, eccentrics and epicycles remained the preferred mechanisms. The question of their reality did arise in *Sphere* commentaries from time to time. Sacrobosco himself did not consider the issue of their reality. Some of his later commentators, for instance, Capuanus de Manfredonia and Jacques Lefèvre d'Étaples, took a cautious approach by not endorsing the reality of the mechanisms yet nevertheless presenting eccentrics and epicycles as explanatory devices. In his *Sphaera,* Clavius admits to no uncertainty about the reality of eccentrics and epicycles in the celestial spheres, except in the minds of certain skeptical philosophers, the refutation of whom we shall soon see.[12]

Peurbach's Materializations

The second element is acceptance of the materialization of the Ptolemaic planetary mechanisms in the tradition of Ptolemy's *Planetary Hypotheses* but known in Clavius's time through George Peurbach's *Theoricae novae planetarum.*[13] Despite the proximity and importance of Peurbach as a source for Clavius, the Jesuit prefers to cite Ptolemy by name as the source of these material mechanisms. Clavius shows no indication that he knew Ptolemy's *Planetary Hypotheses,* and his explanations and illustrations of the materialized spheres are straight out of Peurbach. So his penchant for citing Ptolemy on this point must be a conscious decision to cite the older, and thus more authoritative, source. He certainly knew Peurbach's *Theoricae* but seems to have used Regiomontanus's and Erasmus Reinhold's commentaries on Ptolemy and Peurbach, respectively, as his usual reference works.

In fact, Clavius had relatively little need to refer to Peurbach, because a detailed account of Peurbach's planetary models is far beyond the scope of Clavius's textbook. Peurbach had provided detailed mechanical equivalents for each mathematical planetary model described by Ptolemy. But such advanced planetary theory is not the business of a *Sphere* commentary, and Clavius did no more in his *Sphaera* than promise that they would appear in a later work. What Clavius took from Peurbach are the general principles and simplest examples of the materialized planetary concepts. So Peurbach's influence is there, and it is significant, but it is not overt.

For Clavius, a logical prerequisite to the materialized planetary models is the general idea of a rigid body, a true celestial sphere that bears the luminous celestial bodies we see. In the case of the planets, we never see these spheres but merely the planets themselves. Only in the case of the sphere of the fixed stars do we see anything suggesting a continuous spherical vault. Clavius does not make an explicit case for the solidity of the sphere of the fixed stars until he feels compelled to address a theory (discussed in chap. 4) that the heavens are not rigid but fluid, and that the stars are all pursuing independent paths through this fluid. He offers three arguments to demonstrate that the stars must be fixed in a rigid body that defines their relative positions. First he observes that one can find many examples of three stars that form a straight line (i.e., a great circle) on the celestial sphere. At any time, Clavius points out, these stars will fall on a great circle, no matter where they are—near the horizon or toward the meridian. Since this has been true throughout history, the stars must be fixed in place. Second, he argues that the diffuse and irregular bright patches of the Milky Way always appear the same and follow the regular rotation of the rest of the stars. How could this be if the Milky Way were not part of the same celestial body as the stars? Finally, "a certain learned and religious man passing his life in the province of Peru," where he can see the south celestial pole, has written that near that pole there appear dark places in the night sky as if the heaven had been perforated.[14] These features are moved regularly along with the stars and so must be part of the same sphere that bears the stars.

These arguments emphasize the rigidity, the dimensional stability, of the celestial spheres. Taken together with metaphors that Clavius will use to describe the situation of a planet within its sphere (e.g., as a nail in the rim of the wheel) the clear impression conveyed to the reader is that the "solid" spheres are not merely solid in the geometrical sense, but are hard and rigid bodies.[15] The lack of an explicit discussion of the material nature of the celestial spheres in Clavius's text is probably because this subject was generally seen to be in the realm of physics, not astronomy.

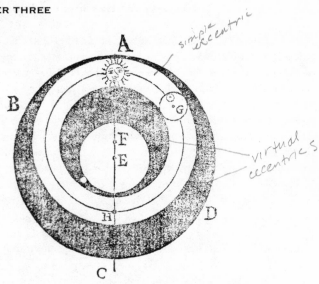

Figure 15. Materialized eccentric and epicycle constructions as adapted by Clavius from Peurbach. From Clavius, *Sphaera* (Venice, 1596), 433. Photograph courtesy of the University of Wisconsin–Madison Memorial Library.

But that by no means implies that the astronomers were agnostic on the question, and they sometimes let slip their thoughts on the matter, as when Christoph Scheiner, in his dispute with Galileo over sunspots, said that "the common opinion of astronomers [on] the hardness and constitution of the heavens cannot stand" thus confirming that, for him at least, the concept of solid sphere included the physical property of hardness.[16]

Physical visualization of the Ptolemaic mechanisms is more difficult than conceiving of a basic celestial sphere. Materialized eccentrics and epicycles (such as Peurbach's) are physical equivalents to the constructions that, in Ptolemy's *Almagest,* are purely geometrical devices. The basic idea, which Roger Bacon in the thirteenth century called "a certain conception of the moderns,"[17] is, in fact, traceable to Ptolemy himself, though this was not known until recently.[18]

Clavius illustrates the idea with what was, because of its simplicity, the most common example, the theory of the sun (see fig. 15, which is a cross-section of a spherical construction through the plane of the ecliptic). The earth, at the center of the universe, is point *E*. The sun's entire construction (generally called the sphere or heaven, *coelum*) is contained between two surfaces—the outer, labeled *ABD,* and the inner, the smallest circle centered on *E*. Note that these inner and outer surfaces are both centered on *E* so that, as a whole, the sun's heaven is concentric with the cosmos. The solar heaven, or sphere, comprises three component orbs, two of nonuniform thickness (shown dark) and one of uniform thickness between the other two. The uniform orb is the *eccentricus simpliciter,* or "eccentric in the simple sense," and it corresponds to the eccentric circle in Ptolemy's geometrical constructs. This eccentric, which defines the path of the sun itself, is centered on *F*. The two nonuniform orbs that complement the simple eccentric are called *eccentrici secundum quid,* that is, "eccentrics in a certain sense." They must be stationary with respect to each other and oriented so that the thickest part of one always corresponds to the thinnest part of its sister sphere, separated only by the thickness of the simple eccentric. From here on we will translate *eccentricus simpliciter* and *eccentricus secundum quid* by the terms "simple eccentric" and "virtual eccentric," respectively.[19]

Clavius uses the same diagram to show how a materialized epicycle may be constructed, though this example is too simple to correspond to any real planet. The epicycle is simply a smaller sphere, centered at *G,* whose diameter corresponds exactly to the thickness of the simple eccentric, which now serves the purpose of an eccentric deferent. The hypothetical planet, located on the circumference of the epicyclic sphere, is carried around *G* by the rotation of the epicycle, and the epicycle, in turn, is carried around *F* by the eccentric deferent. Thus the functions of the Ptolemaic eccentric and epicycle can be carried out by a mechanism that is, taken as a whole, concentric with the earth and the cosmos. Clavius makes little mention of the Ptolemaic equant, which is harder to explain and visualize than the eccentric and epicycle. He considers it a subject more suitable to the advanced treatment in the *Theorica planetarum* and promises the reader that it will be explained there.

Clavius undoubtedly had in mind here his own *Theorica planetarum,* which never appeared in print despite his frequent promises to produce it. It is a difficult problem to envision a mechanical realization of a Ptolemaic equant, and it would be interesting to know whether by postponing the discussion Clavius was evading a difficult issue, or if he actually was satisfied that it could be done but thought it too complicated for his present purposes. He may even have passed over the equant simply because that is what Sacrobosco does. The question of whether a Ptolemaic equant can be successfully materialized in a mechanical arrangement of celestial spheres is almost entirely a question of what mechanical and dynamical properties one is prepared to grant to those spheres. Without the more advanced discussion Clavius would presumably have provided in his *Theorica planetarum,* we cannot know how he conceived of a true mechanical equant.

It seems, nevertheless, that concerning the equant Clavius was prepared to follow the tradition of materialized constructions from Peurbach and, ultimately, Ptolemy's *Planetary Hypotheses*. In them the equant does not correspond to a particular sphere, thus the uniform angular motion of the planet's deferent does not seem to have a strictly mechanical origin. His only clue on the problem appears in the tabular "theory of the planets" (appended to all but the earliest editions of the *Sphaera* but not to be confused with a theoretical exposition in the tradition of the medieval *Theorica planetarum*). Noting that there are four component spheres in the theories of the superior planets, Clavius adds that one could count five by including the equant, "which is only a circle."[20] Unfortunately this is not nearly enough to understand what kind of connection he envisioned between the equant circle and the motion of the deferent sphere it must regulate.

In any case, Clavius accepts these orbs, at least the various eccentric and epicyclic ones, as actually existing. In fact, he believes they are direct consequences of the phenomena, and he argues this point energetically in his refutation of the Averroist skeptics. In his view, the explanatory power of the spheres argues forcefully for their actual existence, and he supports this by showing that even the simplest model—a first approximation, we might say—explains many things. After briefly noting some details of the solar and lunar mechanisms, he says, "So already [after this elementary explanation] we come close to explaining those phenomena that so strongly drove astronomers to discover the eccentric and epicyclic orbs in the celestial spheres."[21]

There are two immediate advantages if we accept, as Clavius does, the reality of the materialized constructions. The first is that we can now understand how eccentrics and epicycles can be real mechanical things (not merely abstract mathematical constructions), which can have a well-defined relationship to each other while moving the planets around. The second is that all the planetary heavens are, if we take each as the aggregate of its parts, concentric with the earth and the cosmos. This is consistent with one of Aristotle's central cosmological doctrines, namely, that the celestial spheres should be concentric with the earth and the cosmos. Though Clavius does not say so, it seems that for him the concentricity of the whole sphere lends a bit of much-needed legitimacy to the eccentric motion of the planets themselves—legitimation needed because, as we will see, the eccentric planetary motion will be a major point of contention in his debate with the skeptical adversaries.

This concentric structure is important for another reason as well, but to understand it we have to appreciate the large-scale architecture of the traditional geocentric universe. In an Aristotelian cosmos, the diurnal motions of all the celestial bodies are ultimately traceable to the primum mobile, or first mover, which was often identified with the outermost sphere. But the successive orbs, from the outermost inward, do not all move at the same rates or in the same directions. The planetary orbs move eastward, contrary to the westward diurnal motion, and rotate about axes that are closer to the poles of the ecliptic than to the terrestrial poles (or poles of the world). These eastward motions with respect to the background of fixed stars are the proper motions of the planets, which can take days or weeks to produce a noticeable change of position. The planetary spheres move at differing speeds—Saturn completes a revolution every thirty years or so, Jupiter about every twelve, and so on with the others. Despite these slow eastward motions, the westward diurnal "locomotive virtue" stretches from above the firmament all the way down to the sphere of the moon and causes all the spheres to rotate about the poles of the world once each day. This power that produces a diurnal rotation in all the spheres was called the *motus raptus*, a transporting impulse, so to speak. (The exact nature of this motive power and how it was transmitted to the lower spheres was grist for many discussions from antiquity to the seventeenth century. Clavius, reflecting the traditional disciplinary distinctions, saw this question as a matter for physics, not mathematics, and thus does not consider it in the *Sphaera*.)

Aristotle was not aware of the phenomenon of precession of the equinoxes and thus had no need for the stars to do anything but complete their diurnal rotation. Later astronomers, following Hipparchus, found that the stars collectively shift slowly with respect to the equinoxes and concluded that the primum mobile, provider of the *motus raptus*, must be a ninth sphere above the eighth. This ninth sphere was starless and its presence was manifested only by its diurnal impulse. The precession of the equinoxes was thus explained by a uniform eastward rotation of the eighth sphere with respect to the ninth—a complete rotation taking tens of thousands of years.[22]

Still later astronomers concluded that the motion of the eighth sphere was not uniform.[23] Instead it seemed to vary—increasing and decreasing in speed periodically every few thousand years. This trepidation, as it became known, was explained as an oscillation of the eighth sphere superimposed on the uniform precessional motion. The theory now required another invisible sphere above the firmament. The primum mobile became the tenth sphere, still providing the diurnal motion to all the lower spheres. The ninth sphere by its motion moved the eighth and lower spheres with the very slow and uniform precessional motion. The eighth sphere, bearing the stars, now exhibited the peculiar motion of trepidation: the "apparent" equinoxes of the eighth sphere were understood to be moving in small circles around the "true" equinoxes of the ninth sphere, thus producing the desired oscillatory motion superimposed on the precessional and diurnal. There are thus ten mobile spheres plus the immobile eleventh sphere—the empyrean heaven, inhabited by God, his angels, and those souls taken directly there after their earthly existence.

Clavius explains that the influence of a sphere's motion extends only downward. "Thus it is determined and certain that by its very slow eastward motion the ninth sphere draws with it the eight lower spheres, but it does not affect the highest sphere at all. In fact, according to the opinion of astronomers, any given orb carries by its motion the lower orb [that is] contiguous to and concentric with it, but the higher orb is not [affected by the given orb]."[24] Thus the sphere of Saturn and all the lower spheres exhibit the compounded motions of the tenth, ninth, and eighth spheres.

Yet this does not account for the motions of all the lower spheres, because the proper motion of a given planet has its own peculiar period that does not propagate downward to the lower planets. How can one understand this *motus* that affects some orbs by putting them in a westward diurnal motion while intervening ones follow eastward motions of various periods and about other axes? Clavius does not offer a complete explanation, but the materialized constructions do allow him to offer a plausible partial answer rooted in the distinction between the concentric planetary sphere, the *sphaera tota,* and the eccentric orb, or deferent, that bears the planet.

He continues, "But no lower planet is moved according to the proper motion of a higher planet, because [the lower planet] is borne around a different center by its own locomotive powers."[25] The key is that the total sphere of any planet, judging by its inner and outer surfaces, is concentric with the cosmos and thus shares the motions of the higher concentric spheres. So the planets, like the stars, rise and set each day and participate in precession and trepidation. But the deferent—the carrying orb—of the planet itself is eccentric (as is, for that matter, any given epicycle). For Clavius this distinction has a dynamical significance, as he later elaborates.

> We see the spheres of all the planets and the ninth heaven, at the same time as the firmament, carried from east to west in the space of twenty-four hours according to the motion of the primum mobile. On the other hand, we know that those same spheres of the planets, together with the firmament, are drawn from west to east according to the motion of the ninth sphere . . . and it has been observed that all the heavens of the planets are moved slightly according to the

motion of trepidation, or the advance and retreat of the eighth
sphere. . . . When the individual motion in [each of] the planets is
measured, [it is found] that the proper motion of no other [planet]
is communicated to the lower planet, so that not the smallest part
of that motion [of the higher planet] can be noticed. . . . (Jupiter
has absolutely nothing from the thirty-year motion of Saturn; and
likewise to Mars nothing is conveyed from the twelve-year motion
of Jupiter, and so it is with the rest, as all agree.) It is clearly seen
that the carrying orbs [deferents] of the planets are not concentric,
otherwise the motion of any given higher planet would be conveyed
to all the lower planets.[26]

If the planetary deferents were not eccentric, the proper motions of the
planets would compound in them the same way that the motions of the
tenth through eighth spheres build up. The basic principle is that concentric
spheres are coupled in some unexplained way and can transmit their motion to all the lower concentric spheres. Thus the primum mobile gives
its diurnal motion to the ninth, eighth, and all lower concentric spheres.
The ninth gives its precessional motion to all of its inferiors. The eighth
likewise gives its trepidational motion to all the lower spheres. The seventh
sphere (Saturn's) and lower do not contribute any peculiar motions but
only manifest the motions of the higher concentric spheres. The eccentric
orbs, on the other hand, are not coupled either to each other or to the
concentric spheres and thus exhibit peculiar motions that do not propagate.
This is as close as late Ptolemaic cosmology comes to a mathematical—or
better, geometrical—celestial dynamics, and it is a major reason why it
was crucial to maintain the concentric nature of the planetary spheres.
Peurbach's materializations permitted eccentric and epicycle constructions
to meet the requirement of concentricity.

The Order of the Planetary Spheres

The third element of late Ptolemaic cosmology is the acceptance of the
Ptolemaic order of the planets or, rather, the planetary spheres.[27] This was
not an issue of great concern to medieval *Sphere* commentators. Sacrobosco himself listed the planets in their Ptolemaic order but, as is typical
of him, did not attempt any justification of it, and the question was rarely
raised in commentaries on his text.[28] In the *Almagest* (probably Clavius's
single most important source), Ptolemy himself did not treat the subject
extensively, and the discussion is not quantitative.

Ptolemy noted first a general agreement among astronomers that the
sphere of the moon is closer to Earth than the spheres of the sun and

planets, which are themselves closer than the sphere of the stars. He cites also the consensus that Saturn is the most distant planet, then Jupiter, and next Mars but acknowledges disagreement over the relative ordering of Sun, Venus, and Mercury. Earlier astronomers, according to Ptolemy, thought that after Mars (counting inward toward Earth) came the sphere of the sun, followed by Venus, and then Mercury. But more recently others had proposed that Venus and Mercury must lie beyond the sun, because no one has ever seen either of those planets transit the solar disk. Ptolemy declares his preference for the older order, observing that the two planets need not always be along our line of sight when they pass below the sun. Finally he notes that, barring actual parallax measurements that would allow us to know the distances to the planets, the older order placing the sphere of the sun between the spheres of Venus and Mars appears more plausible, because then the sun itself separates the planets that can appear anywhere in the sky with respect to the sun (Mars, Jupiter, Saturn) from those that never stray far from the sun (Venus and Mercury).[29]

Clavius, in contrast, presents an extensive review of opinions followed by several arguments in support of the Ptolemaic order. He argues first that the Ptolemaic order of the planets is supported by measurements of diurnal parallax (though he uses the older term *diversitas aspectus*) meaning the angular difference between the "apparent place" of an object measured by the observer and the "true place," which is where it would be seen from the center of the earth.[30] Clavius asserts that the successively smaller parallaxes of Moon, Mercury, Venus, and Sun establish that segment of the planetary order. The rest of the planets, he tells us, are too far away to exhibit a measurable parallax, so their order cannot be determined by that method.[31] In fact, however, only the lunar parallax is measurably large, as even Ptolemy knew well and stated in the *Almagest*. Ptolemy had calculated the solar parallax (but not observed it) by means of his eclipse diagram.[32] Apparently Clavius, who seems to have been following Maurolico in this matter, accepted Ptolemy's solar parallax as if it were observational data and presumed that the parallaxes of Venus and Mars must be between the lunar and solar values.[33] This reasoning, which is circular in retrospect, might be attributed to the fact that Clavius and Maurolico were largely concerned to gather authorities rather than data, and to be persuasive rather than rigorous. The parallaxes are more anecdotal than analytical, which suited their rhetorical purposes and may help explain why they did not see how circular the reasoning is.

The order of the superior planets, which all agreed show no parallax, may be inferred from the speeds of their motions according to the principle that the more a sphere lags behind the primum mobile, the lower it is.

Thus because the moon's proper motion (from west to east, opposite to the primum mobile) is the fastest of all the planets—which is to say it lags most behind the stars—it must be farthest from the primum mobile and lowest (closest to the earth). Similarly, the planet with the slowest proper motion, Saturn, must be the highest. This argument then establishes the descending order of Saturn, Jupiter, and Mars, but leaves uncertain (as Ptolemy said) the order of Mercury, Venus, and the sun, since they all traverse the ecliptic in the same amount of time on the average. Clavius notes that there have been various opinions on the relative order of the sun and inferior planets, but he does not bother to reconcile them, perhaps because the earlier parallax argument has taken care of that issue.[34]

Clavius also draws on solar eclipses and lunar occultations to defend the Ptolemaic order of the planets. (In a solar eclipse, the moon moves in front of the sun. In a lunar occulation, the moon moves in front of a star or planet.) These phenomena show, and "there can be no doubt," that the moon is lower than the sun, planets, and stars. Nor does the fact that Mercury and Venus have never been observed to pass before the sun mean that they are above it, because, as Ptolemy and Regiomontanus agree, at lower conjunction with the sun, these planets may pass north or south of the sun without occulting it, just as there is not a solar eclipse at every new moon. Moreover, Albategnius (al-Battānī, d. 929), Thebit (Thābit ibn Qurra, d. 901) and others agree that the visual diameters of Venus and Mercury are so small that if they do pass before the sun, we would never notice the fact.[35] So, summing up, "Although no one of these [arguments] sufficiently defines this order, nevertheless all of them taken together confirm that the heavens are so arranged. For from the parallax the order of the moon, Mercury, Venus, and the sun are infallibly determined, and above those four the positions of Mars, Jupiter, Saturn, and the firmament are easily figured out from the speeds of their motions. . . . Finally the order of all the planets cannot be firmly determined from eclipses, but we can place the moon in the lowest position and all the planets below the firmament."[36]

Having wrung all he can from the phenomena, Clavius proceeds to argue on other grounds. He argues for the suitability of the sun's middle position between the superior and inferior planets[37] because the solar motion is, in a way, a common measure of the motions of both groups: the solar period (one year) is found in the epicycle motion of the superior planets and in the deferent motion of the inferior. However, all such details are explained in the *Theorica planetarum*. (This could be a reference to his own *Theorica planetarum*, but he might be referring to the genre as a whole since this is pretty standard fare.) It is also fitting that the sun should share its light evenly among all the planets and thus hold the middle position among them. At this point he cites Ovid's story of Phaethon as if it were evidence for the sun's position with respect to the planets and thus completely leaves philosophical argument behind. He further buttresses his position by suggesting that the order of the days of the week supports the sun's intermediate position and finally appeals to the wisdom of the Egyptians. Clavius tops off this section with a typical illustration showing "the number and order of the bodies [spheres] composing the universe,"[38] thus summing up nicely the cosmic order of late Ptolemaic cosmology (the eleven-sphere cosmos shown in fig. 7).

We must note here a seemingly significant change of wording that occurred between his first *Sphaera* (1570) and the *nunc iterum* edition (1581). After the discussion just described, aimed at explaining how astronomers determine the order of the planets, the 1570 edition states confidently, "Thus from all of these things we have said, it is apparent that the order set out by our author (i.e., Sacrobosco) is true and in greater accord with skillful astronomers, while other [orders] are not at all." But in the 1581 edition he altered *verum* to *veriorem,* that is, he changed "true" to "truer," and he dropped *alios autem minime* (while other [orders] are not at all).[39] So we might now translate the passage, "Thus from all of these things we have said, it is apparent that the order set out by our author is truer and in greater accord with skillful astronomers." William Wallace suggests, quite reasonably, that this expression of probability is characteristic of Jesuit philosophical teaching in the period and in contrast with the more absolutist Aristotelians of the sixteenth century.[40] Does the change from *verum* to *veriorem* signify a significant change in Clavius's thinking? Perhaps not. The sense of relative truth is already there in 1570 with the use of *magis,* "greater." It seems there is little change in the import or nuance of his statement beyond a possibly stronger sense of probability. Perhaps significant is the disappearance of *alios autem minime,* which might be seen as a reflection of Clavius's growing familiarity with the Copernican theory. He seems to have studied Copernicus's work extensively during his work on calendar reform, between roughly 1572 and 1582—in the years leading up to his *Sphaera* revision of 1581. As Copernicus was able to offer an argument for the order of the planets that is as compelling in its own way as those of Ptolemaic cosmology, perhaps Clavius decided to exercise more prudence in his claims about the coherence of alternative orders. That is not to suggest that he found Copernicus's argument acceptable, for as we shall see Clavius had plenty of reasons—astronomical, physical, logical, scriptural—to reject Copernicus's cosmological model. I suggest only that, astronomically speaking, he may have found Copernicus's order of the planets as much in "accord with skillful astronomers" (of whom Copernicus was one) as Ptolemy's.[41]

The Dimensions of the Cosmos

The first three elements (acceptance of Ptolemaic constructions, materialization of those constructions, and acceptance of the Ptolemaic planetary order) along with the principle of parsimony, make possible the fourth element, namely a consensus on the dimensions of the celestial spheres and the overall size of the cosmos. The principle of parsimony is implemented in medieval cosmology through the "nesting principle," which originated in Ptolemy's *Planetary Hypotheses* but entered Western cosmology in Arabic astronomical texts, mainly those of al-Farghānī (Alphraganus) Thābit ibn Qurra (Thebit), and al-Battānī (Albategnius).[42]

The Arabic astronomer al-Farghānī (ninth century, Baghdad), whose astronomy text was translated into Latin and widely known in Europe, stated the principle succinctly enough. "In this judgment [of the sizes of the planetary spheres] we presume that there is no empty space between the circles [i.e., spheres]."[43] The nesting principle thus prescribes that the successive celestial spheres are nested one within the other in such a way that the outer surface of one just touches the inner surface of the next with no intervening space or matter. Thus there are no nonfunctional, or "vain" components in the universe, hence the parsimony: nothing has been created in vain. Ptolemy himself introduced the principle into his scheme of celestial orbs saying, "It is not conceivable that there be in Nature a vacuum, or any meaningless and useless thing."[44] So the principle of parsimony, in essence, dictates that the universe composed of the planetary mechanisms will be the minimum size possible while still remaining functional. Clavius will, however, express a caveat on this line of argument.

The outer and inner surfaces of a given planetary sphere are defined by the maximum and minimum distances (apogee and perigee) of the planet from the center of the cosmos, that is, the center of the earth. Since the materialized Ptolemaic mechanisms determine the relative apogee and perigee for each planet, and since there is an accepted order of the planets, we have everything necessary for a universal scheme of relative cosmic dimensions. Given a measurement of some absolute distance—by the lunar parallax, for instance—the actual distances to all the celestial spheres and the overall size of the universe can, theoretically, be calculated.[45] Thus the Ptolemaic mechanisms and the principle of parsimony are the constraints that allow a calculation of the size of the universe. The materialized Ptolemaic mechanisms establish the minimum dimensions into which the whole ensemble will fit, and parsimony demands that the universe be no larger than those minimum dimensions.

Al-Farghānī calculated a set of cosmic dimensions according to this scheme, and his version became the most popular in the West, finding its way into many medieval and late Ptolemaic textbooks. The table of celestial dimensions printed by Clavius in his *Sphaera* was taken immediately (but with minor changes) from Maurolico's *Cosmographia*.[46] But the ultimate source for the table was al-Farghānī, whose numbers Maurolico had used—again with minor changes.[47] Al-Farghānī's scheme gives the distances, in units of the earth's radius, from the earth's center to the inner and outer surfaces of each celestial sphere. How big was the Ptolemaic cosmos? Clavius gives the distance to the firmament (the limit, in essence, of his physical world) as 22,612 earth radii. If we use the modern accepted value for the size of the earth (which is not so terribly far off from the value Clavius himself would have used), we can calculate that the distance from Earth to firmament would be 143 million kilometers—or a tad less than the modern distance between Sun and Earth. In other words, Clavius's entire observable universe, if centered in our solar system, would not extend beyond Earth's orbit. But the exact numbers are not important for our present purpose. What is significant is the wide acceptance of this scheme in Clavius's day. "[Clavius] was expressing a consensus that still reigned, a quantitative picture of the cosmos that was built into the minds of all students during their undergraduate training. In a cosmos still predicated on the concept of place, every place was quantitatively known with great precision by all university graduates who could remember the numbers. Those who could not nevertheless had a good idea of the orders of magnitude involved."[48]

Late Ptolemaic cosmology, as presented in the *Sphaera,* provided its audience with a full account of the large-scale structure of the universe. In addition to making possible mathematical predictions of celestial motions through the Ptolemaic models, the materialized Ptolemaic constructions provided mechanically plausible explanations of those motions. They are mechanically plausible because the positions of all the component bodies, the rigid sphere constructions, were well defined with respect to each other. Further, the nested sphere scheme defined the relationship between the total spheres of the various planets. Thus, and perhaps most important, Ptolemaic cosmology produced an accepted set of actual figures on celestial dimensions that demonstrated the scale of creation and the place of humankind within it.

Ptolemaic Astronomy and Aristotelian Physics

The fifth component of late Ptolemaic cosmology, and a very important one for Clavius, is the integration, or at least reconciliation, of the Ptolemaic planetary system with Aristotelian physics. That there is a potential conflict between the Ptolemaic planetary constructions and certain principles of Aristotelian physics was obvious to the earliest scholastic commentators, including Roger Bacon, Albertus Magnus, and Thomas Aquinas. This conflict was also a point strongly emphasized by Averroës and evident to Islamic scholars. Medieval Latin commentators, however, seem to have seen this question as more properly pertaining to natural philosophy than to the mathematical discipline of astronomy, and consequently such discussions occur most often in questions or commentaries on Aristotle's *De caelo* and *Metaphysics*.[49]

The root of the problem is the Aristotelian tenet that the natural motion of celestial bodies is circular and centered on the earth, in contrast to terrestrial matter, which moves naturally either toward or away from the earth's center in rectilinear motion. A planetary eccentric is a problem, then, because by definition the center of the motion is some point other than the earth. A further objection can be raised in that the eccentric causes the planet it bears to be sometimes closer to the earth and sometimes farther away. This motion, the critic may say, is in effect a rectilinear motion, because the planet alternately approaches and recedes from the earth. Aristotle teaches that this is not natural to the heavens. Clavius will raise and refute these objections in the course of his exposition.

Epicycles are even more vulnerable to the same objections. At least the Ptolemaic apologist can argue that the eccentric's motion produces a planetary trajectory that, if not strictly centered on the earth, at least encompasses the earth, and this may be amenable to Aristotle's doctrine, broadly construed. But the apologist is in deeper trouble with the epicycle, because the motion of the planet on the epicycle alone results in a planetary path that encloses no other body—and definitely not the earth. Thus some medieval philosophers, Jean Buridan, for example, would reject epicycles but allow eccentrics.[50]

Clavius argues at some length in his *Sphaera* that Ptolemaic constructions are compatible, if not with Aristotle himself, then with Aristotelian natural philosophy as the sixteenth century understood it. His treatment emphasizes that the traditional Ptolemaic cosmology is not less credible on account of criticisms by the philosophers. One of the primary themes in his *disputatio* on eccentrics and epicycles is that those mathematical constructions are compatible with Aristotelian natural philosophy, a position strongly denied by the Averroists.

Clavius responds to eight objections raised ''by the adversaries,'' and some of his responses are particularly interesting. The first objection, for instance, is the familiar one that eccentrics and epicycles imply motions of up and down in the heavens, where, according to Aristotle, there should be only circular motions. Clavius responds,

> The fact that by this motion [eccentrics and epicycles] now approach the earth and now recede from it is not absurd because this approach and recession do not take place in a straight line, which alone Aristotle denies to celestial bodies, since they are fitting only to the elements, which are heavy and light. If anyone should contend that Aristotle thought the contrary, it will be conceded to him, for Aristotle spoke only about these motions, which were known in his time, namely motion from and to the center by a straight line and about the center of the world. If the motions of the eccentrics and epicycles had been known in his time, I do not doubt that he would have spoken quite differently concerning motion about the center.[51]

As we see, Clavius first attempted to restrict to a rectilinear path those motions that Aristotle denies to celestial bodies. But at the same time he prepared a fallback position, which is to consider Aristotle as a historical author rather than an omniscient authority. So he pointed out that Aristotle knew only those things that had been discovered in his day—which is to say before the work of Hipparchus and Ptolemy that produced successful celestial models using the geometry of eccentrics and epicycles worked out by Apollonius. At that point Clavius is free to follow the time-honored procedure of putting words in Aristotle's mouth.

The second Averroist objection takes issue with the concept of the materialized eccentric, maintaining that the virtual eccentric (the eccentric *secundum quid*), because its thickness varies, is not a perfect sphere, which Aristotle says all celestial bodies must be. Clavius's response is interesting. ''We may solve the second objection if we say that all eccentric orbs, both virtual eccentrics and epicycles, are quite perfect spheres according to their proper centers. For the outermost surfaces of all these spheres are, in all their parts, equally far from their centers. Nor does it matter that the orbs of the virtual eccentric are thicker in one part than another, because no natural argument can prove that all celestial orbs must be uniform and of equal thickness. If, indeed, Aristotle teaches the contrary then we do not believe him in this matter.''[52] Here, as before, Clavius wants first to argue that the invoked doctrine of natural philosophy is not as far-reaching as the adversaries would have it—that no natural principle requires spheres of uniform thickness. But now the fallback is not to reinterpret Aristotle but to be prepared to disagree with him.

One last example deals with the question of centers. The adversaries argue that if eccentrics exist, they must be moved about their centers. But Aristotle says, they claim, that for every moved thing there must be something stationary, which for celestial motions is the earth. So however many eccentrics there are, there must be as many stationary Earths, which is absurd. Since there is only one Earth, all spheres must be concentric. Clavius's response this time reveals an interesting attitude toward the mobility of the earth. "Concerning the seventh objection, it is to be denied that the earth is necessarily stationary in any given center so that the celestial orbs can be moved around it. Although certainly God could have either removed this Earth completely or pushed it to some place other than the center of the world, nevertheless up till now the heavens have been carried by diurnal motion around the center of the world."[53] His main point is simply to deny that the earth is where it is so that the celestial spheres may move as they do. The circumstances of the earth are not as they are in order to suit the purposes of the celestial orbs. In fact, as he said, we must recognize the possibility that the cosmos could be constituted, if God willed it, such that Earth were not at the center. We can only find out how God made the world by examining the world, and we have good reason to believe that the earth is, in fact, at the center of the cosmos.

Clavius's concern is not necessarily to achieve complete agreement with Aristotle—though since it is better to reinterpret Aristotle than to contradict him Clavius generally tries to achieve some reconciliation. However, he will reject Aristotle if necessary. His goal is to bring Ptolemaic cosmology within the broader confines of a natural philosophy that is Aristotelian in the sixteenth-century sense. That is to say, in the sense of an eclectic philosophy capable of grafting onto the philosophical trunk "foreign" branches (such as eccentrics and epicycles) that seemed superior to their more conventional Aristotelian counterparts.

A Christian Vision

Sixth and last, late Ptolemaic cosmology is rounded out and completed by a Christian vision of the cosmos. This component, like the other five, is not original with Clavius but is common in varying degrees to many astronomical texts from Sacrobosco to Clavius.[54] It is, however, remarkably clear in Clavius. The Christian vision emerges not only as a matter of language—for instance, frequent references, usually in passing, to God (*Deus Optimus Maximus*). Nor is it just the occasional use of sacred writings as evidence in philosophical arguments. Though both of these contribute, the vision encompasses more and is best illustrated by examples.

An important example comes from Clavius's discussion of the number of celestial spheres, or heavens, in the universe—a major cosmological issue in many medieval and Renaissance astronomical texts. Clavius counts one major sphere for each planet, one for the fixed stars or firmament, and an additional two (three after 1593) above the firmament to account for the motion of the fixed stars. That makes ten (later eleven) spheres. But the Christian vision of the universe allows, or rather requires, the astronomical consideration of the empyrean heaven. "Beyond the eleven moving heavens, theologians such as Strabo and Bede and all the rest affirm that there is another heaven, which they call the empyrean. It is not a heaven with stars, but rather it is the happy seat and home of the angels and the blessed."[55]

Thus, the number of spheres in the cosmos is not ten, but eleven (later twelve). And the empyrean heaven has a place, an actual location, with respect to the other celestial spheres. It is, of course, the last and highest of all the heavens. Even though no celestial motions or luminous bodies (the clues to the existence of all the other spheres) attest to the empyrean's existence, it must be there to complete a Christian cosmology. Clavius did make a half-hearted (and nonastronomical) attempt to prove the existence of the motionless empyrean, but the uncertainty of its outcome had no effect on the general conclusion, "Thus do the astronomers of today determine that in the universe there are twelve heavens; eleven mobile and one, according to the opinion of the theologians, immobile."[56]

The degree to which the empyrean heaven was integrated into late Ptolemaic cosmology and the importance of its specific locus and possible natural functions might have constituted subtle reasons for resistance to the Copernican hypothesis. "The quantitative world picture was . . . an important part of the collective consciousness of educated Europe, and astronomical theories were judged in its light. Only the logic of his own system allowed, or rather forced, Copernicus to transcend it in the case of the fixed stars, and those huge distances made his system all the more objectionable to others."[57] Interpreted through the Christian vision of the Ptolemaic cosmos, the Copernican theory would have put the empyrean heaven, which was the desired destination of earthly spirits, at an indefinitely large distance beyond the realm of the fixed stars—perhaps an unsettling view for those who were accustomed to knowing exactly how far it was to heaven.

In Clavius's *Sphaera* we see a cosmology that would have been very satisfying to most of its students. It answered important questions about the station and situation of humanity by defining very specifically the places of the heavens and Earth. It demonstrated the ability of human reason to comprehend the natural world by showing that we can understand the construction and motion of the heavens so well as to know how they will behave in the future. Late Ptolemaic cosmology is not at all innovative, and that is one of its advantages. Novelty arouses controversy and is the antithesis of orthodoxy. Clavius wants to present a conventional view, to show that the established cosmology is backed by deep authoritative tradition, is consistent with contemporary ideas of mathematical sciences, is consistent with accepted physics and the methodology of natural philosophy, and is consistent with the teachings of the church. None of the alternative cosmologies that he will consider can measure up to all of these criteria.

We should not get the impression, however, that there were no loose ends in late Ptolemaic cosmology. One major piece of theoretical business left undone in the orthodox scheme was the determination of the exact number of spheres above the firmament. This problem was a common concern of sixteenth-century astronomical textbooks and gave rise to a minor genre of astronomical monographs, usually entitled something like *On the Motion of the Eighth Sphere.*[58]

As pointed out earlier, the number of spheres a theorist placed above the firmament depended on exactly what kind of movements he believed to have been observed in the eighth sphere (the sphere of the fixed stars) and what method he used to account for them. Some (for instance, Bellarmine) would reject the procedure of explaining complex and irregular motions by compounding hypothetical uniform circular motions and would maintain instead that the celestial bodies move in their own peculiar paths through a single fluid medium. Clavius will dismiss this idea as useless. Others (Agostino Ricci and Oronce Finé, for instance)[59] denied the necessity for any suprafirmamental spheres and thus remained content with eight. Clavius will dismiss this opinion for failing to take into account the painstaking observations of generations of astronomers but postpones a more detailed critique until his discussion of the nature of celestial motions. The Ptolemaic theory of a single suprafirmamental sphere also had to be rejected because later observations of the motion of the eighth sphere made it obsolete. For Clavius, in the earlier editions of the *Sphaera,* the Alphonsine solution of two suprafirmamental spheres is the only practical—that is, mathematically workable—theory that also takes into account the available body of observations. That makes ten moving spheres, or, adding in the immobile empyrean, an eleven-sphere cosmos. Clavius and others, including Giovanni Magini, would later add a third sphere above the fixed stars resulting in a twelve-sphere cosmos.

The phenomenon of variable precession, or trepidation, that gave rise to this problem was founded in observational errors passed on through the generations of astronomers between Ptolemy and Copernicus. But the difficulty of making sense of historical observations is great. Copernicus himself took them at face value and went to some trouble to construct an elaborate mechanism to account for the fictional phenomena that appeared in the observations.[60] Clavius and other Ptolemaic authors were doing the very same thing, and the fact that Clavius eventually adapted Copernicus's precession mechanisms in the later editions of the *Sphaera* (while managing to keep the total number of moving spheres at eleven) proves that this was still a lively scientific issue. Only the consummate observer Tycho Brahe had the experience and insight to realize that, aside from the precession of the equinoxes, those seemingly complex motions of the fixed stars were not real, but all stemmed from a lack of appreciation for observational error.

Another loose end of sorts appeared in the *Sphaera* of 1593 (*nunc quarto*) when Clavius introduced a substantial discussion of how astronomers determine the distances to the celestial spheres.[61] In the course of the discussion he takes the interesting step of pointing out that the principle of parsimony is only a working assumption.

> It should be carefully noted that the distances, thicknesses, and sizes of the orbs and stars discovered by the method we have described, although they may be enormous and seem, in a way, to be beyond human comprehension, are yet the minimum possible. Because astronomers lay out the eccentric orb [so that] the convex and concave [surfaces] of a given celestial sphere touch [the eccentric] in exactly one point. And in the same way so that the epicycle of whichever planet, and also the body of the sun, touch the convex and concave surface of the eccentric orb in a single point only.

Clavius makes the point here that the tangency to the eccentric, say, of the inner and outer bounding spheres is a deliberate choice on the part of astronomers, and he illustrates this construction with the standard cross-section of the virtual eccentrics seen in figure 15. He then goes on to make explicit that parsimony is not a necessary condition. "But it is conceivable that God could have made [the virtual eccentric] spheres thicker. . . . Given this, it is certain that the distances, thicknesses, and sizes of the heavens and stars could be greater than those found by astronomers. Thus we have demonstrated only by what means all these things may be inferred. For although God perhaps made those thicknesses and distances greater [than astronomers calculate], yet we can in no way know it by the motions [of the planets]."[62]

There is one final cosmological issue on which Clavius tentatively ventured to present speculative, though not novel, opinions. That issue is a natural one, given the conventional Aristotelian doctrine of a finite cosmos: what lies beyond the final, the empyrean, sphere? Is there space or not? Does it contain anything? Another world, perhaps? Medieval thinkers generally considered these questions beyond the scope of astronomy and treated them in philosophical works, such as commentaries on Aristotle's *De caelo,* or in theological treatises, such as commentaries on Peter Lombard's *Sentences.*[63] Clavius, however, recognizing the natural curiosity of his reader on this point, attempts to satisfy that curiosity, but (unfortunately for us) without defending his views. "Beyond this world, or rather beyond the empyrean heaven, no further body exists, but there is a kind of infinite space, so to speak, in the whole of which God exists in his essence—as the theologians affirm—and in which He could make an infinite number of other worlds, better even than this one, if He wanted."[64]

That those other worlds are hypothetical is emphasized by Clavius's marginal notation—new in the *Sphaera* of 1581 (*nunc iterum*)—accompanying this passage, which states, "Beyond the world there is nothing" (*Extra mundum nihil esse*). Since he has already admitted that beyond the empyrean sphere there is "a kind of infinite space," his annotation must mean "Beyond the world there is no other body," that is, there are no other worlds.[65]

Clavius does not say which theologians he had in mind as he cited their opinion about the possibility of the existence of other worlds. He was surely not thinking of Aquinas, whose teaching that the perfection of the world derives from its order and unity implies, almost by definition, that there can be only one world. Clavius's position on the question appears to be in the tradition of William of Ockham, and even more closely, Nicole Oresme, who also taught that there might be an empty extramundane space that is "infinite and indivisible and is the immensity of God and God himself, just as the duration of God called eternity is infinite, indivisible, and God himself." It is this space in which God *could* create other worlds.[66]

This is the cosmos that Clavius presented to his vast readership. A geocentric cosmos, finite in its observable extent but conceivably infinite in the realms astronomers could not probe. It was compatible with contemporary religious preferences and consistent with the prevailing Aristotelian physics. Finally, and perhaps most important for Clavius, mathematical astronomers could analyze its operations and reliably infer its detailed structure. But despite its successes and appealing features, other schools of thought sought to displace Ptolemaic astronomy, and Clavius saw it as part of his task to refute their challenges.

FOUR

The Rival Cosmologies

*May we not suppose that false opinion or thought is a sort of
heterodoxy; a person may make an exchange in his mind, and
say that one real object is another real object. For thus he always
thinks that which is, but he puts one thing in place of another;
and missing the aim of his thoughts, he may be truly said to have
false opinion.*

—Socrates, in Plato, *Theaetetus*

Standard histories of the scientific revolution depict the struggle between
alternative cosmologies as a contest in two stages with three competitors.
First, Copernicus challenged one of orthodoxy's founders, Ptolemy. Copernicus, through the clever agency of Galileo, dispensed with Ptolemy
only to be confronted then by Tycho. Tycho's theory, however, was a
clumsy compromise and unacceptable to most parties as a cosmology.
Thus, by the middle of the seventeenth century, the contest was concluded
in favor of Copernicus. We see this general outline in the presentations
of histories—which are standard and still widely used—such as *A Short
History of Astronomy,* by Arthur Berry, and *A History of Astronomy,* by
Anton Pannekoek. Neither mentions any medieval or Renaissance astronomical systems other than the Ptolemaic and Copernican.[1]

Such standard literature on the history of sixteenth-century astronomy
fails to recognize the diversity of cosmological options extant in that era.
Even the relatively reliable *History of Astronomy,* by J. L. E. Dreyer,
completely omits the concept of the fluid heavens as a cosmological alternative. The idea that the Copernican theory was the sole rival of the late
Ptolemaic cosmology during the cosmological debates of the later sixteenth century is a false dichotomy. It comes about easily enough, since
it was Galileo himself who set up that dichotomy in his *Dialogues on the
Two Great World Systems.* Galileo was not writing history but was en-
gaged in polemical combat, wielding dichotomy as one of his rhetorical
weapons. But dichotomy as a rhetorical technique is not suited to good
history. When we look into the background of Galileo's cosmological

debate (and Clavius is a major figure in the debate's early stages), we find that there was a broader spectrum of cosmological opinion than Galileo's dichotomy implies. We need only recall the importance of the Tychonic system, which Galileo makes little of because he cannot refute it, to realize what he was up to.[2]

Heterodox Cosmologies

Contrary to the impression we might get from Galileo, Clavius's *Sphaera* shows us that, in fact, there was considerable variety in the cosmological alternatives of the Renaissance. The great variety of cosmologies is usually unappreciated, but perhaps it should come as no surprise, because students of Renaissance Aristotelianism have found great variety in the philosophical realm, too. The same eclectic attitudes that allowed such Renaissance Aristotelians as Alessandro Achillini (1463–1512) and Agostino Nifo (1473?–1538) to blend occult sciences, Averroism, and Neoplatonism could also have encouraged the likes of Girolamo Fracastoro, G. B. Amico, and Bellarmine to revive and refurbish nonconventional cosmological approaches that they saw neglected since antiquity.[3]

The alternative cosmologies presented and refuted in the *Sphaera* range from the famous to the obscure. One, of course, is the Copernican system. But of even greater concern to Clavius is the system of homocentrics as constructed by Fracastoro. Another contender was the idea of a continuous fluid heaven. In this concept, the celestial realm is not divided into rigid, spherical shells. Instead, from the moon up to the firmament, the celestial region is an undifferentiated whole filled by some etherial substance with fluid properties. Devoid of rigid spheres, the whole fluid cosmos could rotate about the earth once a day, while through the cosmic fluid the planets independently pursue their courses "like birds in the air or fish in the sea." The fluid-heaven idea truly came into its own after Tycho claimed to have observational evidence that the solid celestial spheres could not exist. So this school of thought provided a congenial setting for Tycho's new ideas at the very time when the Ptolemaic cosmos became untenable and the Copernican was still suspect (and, for many, even prohibited). Clavius never included in his *Sphaera* a review of Tycho's famous alternative to the Copernican system, but some of his letters at least reveal his negative reaction to it.

Homocentrics

In contrast to his evolving reaction to the Copernican theory, Clavius's critique of homocentrics remained essentially constant from the first edition of the *Sphaera* to the last. Perhaps this reflects the moribund state of the homocentric theory—little or nothing had been published on the topic since 1538. Clavius, as champion of the Ptolemaic cosmology, singled out his primary opponent in Girolamo Fracastoro (1483–1553), the famed Veronese physician and would-be reformer of astronomy.[4] Clavius was perfectly aware of the very long and influential tradition, largely underappreciated today, in which Fracastoro worked. And it is Clavius's appreciation of that history that explains his great concern to address the issues raised by the homocentrists. He knew that homocentric planetary theories had their origin with Eudoxus and Callippus (fourth century B.C.), contemporaries of Plato and Aristotle, and that those theories antedated the models invented by Hipparchus (second century B.C.), which use eccentrics and epicycles.[5] Moreover, Aristotle himself had adopted the homocentric approach and presented it as the real structure of his cosmos. Yet from the beginning, the homocentric models failed to provide astronomers with a technically useful system.[6]

Homocentric theories were hard pressed to reproduce fully the details of planetary motions and completely incapable of explaining the brightness variations of the planets. Varying planetary magnitudes were most simply and naturally explained as a consequence of varying distances between planet and observer, but homocentrics by their very nature ruled out such varying distances. As a result of such problems, homocentrics were generally supplanted by the models of Hipparchus and Ptolemy. The Hipparchan and Ptolemaic models have the important virtue of making technically good predictions of astronomical phenomena, but at the cost of a philosophically suspect assumption—that a body can have uniform circular motion about points other than the center of the earth.

That philosophical compromise was unacceptable to the Aristotelian philosopher and commentator Averroës (Ibn Rushd, 1126–98) and led him to reject the Ptolemaic models and advocate a homocentric solution, though he did not offer one himself. Maimonides (1135–1204) also rejected Ptolemy's solution as physically unacceptable, though he seems to have stopped far short of accepting the physical reality of any alternative.[7] One of the few who went beyond mere criticism of Ptolemy's models to construct a homocentric alternative was al-Bitrūjī (Alpetragius, fl. 1190), who introduced a fundamentally different approach from that of Eudoxus and Callippus but still failed to provide a system that could compete quantitatively with Ptolemy's.[8] Nor were al-Bitrūjī's contributions unprecedented in the medieval Islamic world, for Thābit ibn Qurra and al-Zarqālī (ca. 1029–87) both constructed theories that, while not directly concerned with planetary motions, had a significant influence on later homocentric theorists.[9] Despite philosophical sympathy for homocentrics among the astronomers of the medieval Islamic world, the workaday basis for mathematical astronomy, for instance the Toledan Tables, was always Ptolemaic.[10]

In the Latin Middle Ages the Ptolemaic approach remained standard. Sacrobosco's *Sphere* accepted the Ptolemaic constructs without question. Philosophers, however, were well aware of the alternative approaches and often discussed their relative merits. Thomas Aquinas, for example, declined to choose between the homocentric and eccentric approaches for explaining the apparent irregularity of the planetary motions. Since both theories explained the phenomena on the basis of regular motions, Thomas, unconcerned with computational astronomy, was satisfied to show that "in fact, no motion of the heavens is irregular."[11] Roger Bacon, on the other hand, seems to have rejected eccentrics and epicycles (though he may not have taken a consistent stand) in favor of the homocentric approach of al-Bitrūjī on the grounds that Ptolemy's models imply mutually contrary motions in the heavens—a concept that was philosophically inadmissible for him.[12] A more enthusiastic backer of homocentrics was Henry of Langenstein, who devoted considerable effort to working out practical theories. But despite his displeasure with Ptolemaic mechanisms, he could not produce a workable alternative.[13] Bernard of Verdun, on the other hand, affirmed the reality of epicycles and eccentrics in the heavens while rejecting the homocentric model of al-Bitrūjī as incapable of explaining the phenomena. Jean Buridan took an intermediate path when he rejected epicycles as philosophically unpalatable but also rejected homocentrics. He retained only eccentrics, which were necessary for explaining the phenomena.[14]

There seems to have been less interest in the homocentric tradition in the fifteenth century, which is not terribly surprising given the lack of technical success of homocentric systems and the increasing sophistication of late Ptolemaic cosmology. The idea may have continued to hold its appeal for philosophers, but one searches in vain for fifteenth-century homocentric works comparable to Henry of Langenstein's fourteenth-century treatise, or the sixteenth-century books of G. B. Amico and Fracastoro. But there was some interest, because homocentrics seem to have been a stimulus to refinements in Ptolemaic theory. Peurbach, in the mid-fifteenth century, reportedly wrote his *Theoricae novae planetarum* in order to clear away certain objections to Ptolemaic theories—objections raised by homocentrists.[15] Finally, recent work leaves open the possibility of a positive role for homocentrics in the work of Peurbach's student and colleague Regiomontanus, whose supposed unfavorable view of homocentrics can no longer be maintained.[16]

Despite the dubious status of homocentric theories in the fifteenth century, the next century saw them regain popularity. In early sixteenth-century Padua there are signs of what Noel Swerdlow has called something of an "Aristotelian school of astronomy."[17] Whether or not it was truly a school, it is true that there was a significant succession of advocates of homocentric astronomy who had connections to Padua in the early sixteenth century. Fracastoro studied at Padua and taught there from 1501 to 1508 (during some of those years Copernicus was a student there). Perhaps as early as 1535 but at least by 1538, Fracastoro published his attempt to restore astronomy on purely Aristotelian principles in his *Homocentrica* (Venice, 1538). Fracastoro says in his prefatory letter to Pope Paul III (who had appointed him physician of the Council of Trent) that he was inspired by Giovanni Battista Della Torre, another would-be reformer of astronomy from Padua who never published his theory.

In 1536, Giovanni Battista Amico (1511?–36), who had also studied at Padua, published a book in which he also attempted to restore astronomy using homocentric principles.[18] Amico was probably too young to have studied with either Della Torre or Fracastoro in Padua, and his theory differs significantly from that of Fracastoro, so the general interest at Padua in homocentric astronomy would seem to have been more extensive than the published sources alone attest. The Paduan philosopher Achillini is loosely associated, perhaps mostly as inspiration, with the three theorists above. Achillini, in his *De orbibus*, drew on Averroës and Aristotle to attack Ptolemaic mechanisms but without offering models or explanations of alternatives.[19]

Neither Fracastoro nor Amico found satisfaction with trivial variations on the ancient homocentric schemes. Amico introduced a mechanism, also used by Copernicus, that produces linear reciprocating motion from two circular motions.[20] By this means Amico could produce oscillatory motions of the poles of the celestial spheres. Fracastoro, on the other hand, constructed homocentric models on the principle that the axis of each sphere should be perpendicular to the axes of the spheres immediately above and below it.[21]

Most, if not all, of the reformers ultimately took their inspiration and enthusiasm for the homocentric solution from Averroës, who recommended it in his commentary on Aristotle's *De caelo*.[22] The influential Paduan philosopher Nifo, like Achillini, drew on Averroës as he severely criticized the Ptolemaic system in his 1517 commentary on Aristotle's *De caelo*. Though he declined to offer a replacement system, he disparaged the Ptolemaic constructions as old wives' tales (*fabulas aniles*) and cited an extensive list of reasons (many of which Clavius would address) why Ptolemy's mechanisms could not be natural.[23] Nifo agreed with Averroës in condemning the physical reality of the Ptolemaic constructions but, like

Aquinas, opted not to commit himself to homocentrics. A later influential Aristotelian philosopher, Benedict Pereira, was similarly skeptical of Ptolemaic cosmology—and he was one of Clavius's colleagues in the Collegio Romano.

Clavius reacted strongly to the early sixteenth-century revival of homocentric astronomy. His *Sphaera* contains a long disputation, which first appeared in the 1581 (*nunc iterum*) edition, devoted to the examination of the question of how the planets are moved—that is, by what mechanisms and in what arrangement of spheres. He examined three major alternatives: the Ptolemaic, which, of course, he preferred; the Copernican, which he could not accept because of its implications for physics and contradictions of Scripture; and finally, the homocentric approach. Clavius unequivocally rejected the homocentric models on a variety of grounds and in the process lavished as much attention on them as he did on the Copernican alternative.[24]

Clavius was prepared to make one concession to the homocentric models: the approach works for appearances where nothing more than the speed of a planet's motion is concerned. It is, he says, only in this limited sense that the homocentric theories work at all. Clavius was overly generous in granting homocentrics even that much credit, because homocentric systems do fail even in simple planetary motion predictions. But by granting this to the homocentrics, he could move the critical discussion onward to their other shortcomings.

Clavius criticized the homocentric theories primarily for their inability to explain various phenomena and for the false consequences to which the explanations lead. The homocentrics, for example, cannot explain the variations in the apparent diameter of the moon and in the variations of the planets' brightness. The "adversaries," as he often referred to the homocentric Averroists, must resort to explanations having to do with seasonal variations in the thickness or clarity of the air. The Averroist argument generally held that some quality of the air—thickness or moistness—would spread out the rays of light from the celestial body and change the perceived size or brilliance of its disk. Clavius admitted that Fracastoro, at least, had come up with a new, though just as flawed, explanation. Fracastoro supposed an extra sphere just below the sphere of the moon. This sphere would have variable density and (like a nonuniform pane of glass) would produce optical distortions of the images of the celestial bodies, thus explaining their varying sizes and brightnesses. Clavius pointed out that both explanations, besides being ad hoc, are unreasonable and imply phenomena that are not confirmed in nature.

Clavius correctly traced the atmospheric explanation directly to Averroës and then pointed out the fact that the variations in size or brightness

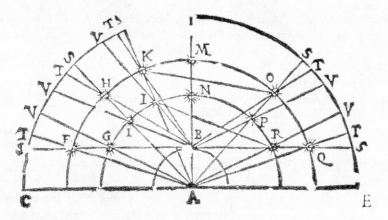

Figure 16. Explanation of variation in diurnal parallax. From Clavius, *Sphaera* (Venice, 1596), 66. Photograph courtesy of the University of Wisconsin–Madison Memorial Library.

of celestial bodies occurs at predictable times and at certain places in their paths. But atmospheric variations cannot explain this. We do not, Clavius objected, find that the air is so consistently clear or misty, or thin or thick, that the sizes of the sun or the moon should vary in a regular fashion. Moreover, he pointed out, there are extreme variations in the properties of the air from day to day, yet corresponding day-to-day variations in the size or brightness of the celestial bodies are not observed. Nor are things any better if we consider Fracastoro's explanation, for then not only would the density variations in the sublunar sphere affect the size and brightness of the sun, moon, and planets, but we should see the same variations in the images of the zodiacal stars. And this we do not find in nature. "This is an exceedingly subtle, but wholly futile fabrication. . . . If [the cause] were that [varying] density, the same fixed stars in the zodiac would appear to us at one time larger—when, of course, those denser parts are below them—than at another time, which conflicts with experience."[25]

Another example of an appearance that the homocentrics cannot explain is the variation in the moon's diurnal parallax or "diversity of aspects" (fig. 16 illustrates Clavius's own explanation of diurnal parallax). Symbols *O* and *P* (found in the right-hand half of the diagram) represent two different celestial objects, or the same object at different times. The parallax observed from point *B* on the earth is measured with respect to the center of the earth, *A*. The parallax of the closer object, *P,* is represented on the celestial sphere by the arc *SV*, which is larger than the arc *ST* representing the parallax of the more distant object, O. For a displacement of the observer equivalent to one earth-radius, the moon's parallax can be nearly one degree—a quantity so readily observable that it was measured as early as Hipparchus. If the moon were moved in an eccentric circle then it would have different distances from the earth at different times, for instance O and P, and would thus exhibit varying parallaxes, ST and SV. Clavius took the existence of such variations in lunar parallax for granted. Though lunar parallax can vary by as much as seven minutes of arc, a quantity measurable by a careful pretelescopic observer, it is very unlikely that Clavius measured the quantity himself. Clavius simply noted that the lunar parallax would not vary if the moon were always the same distance from the earth, as it would be if carried around in a sphere concentric with the earth, and thus the homocentric theories cannot explain any variation in lunar parallax at all.

In addition to showing how the advocates of homocentrics failed to save the phenomena of varying distances, Clavius, in good rhetorical style, also cited logical inconsistencies in their presentations. For instance, in order to isolate the motion of a lower planet from the gyrations of the higher planet, a procedure which simplified the theoretical job considerably, Fracastoro introduced a sphere between adjacent planets to "insulate" the lower from the higher. But he had also supposed that the diurnal motion of each successive planetary sphere could simply be attributed to the influence of the primum mobile, again simplifying the theorist's work. Clavius caught this inconsistency and objected that the insulating sphere should shield the lower planet not only from the higher planet, but from the influence of the primum mobile as well.

I know that authors of [systems of] concentric orbs construct between the spheres of the individual planets single orbs of restitution, which Fracastoro calls "circitores," whose duty is to restore to the lower planet as much contrary motion as is given to it by being drawn by the superior planet. Indeed this is seen to be a fiction. Besides the fact that this theory introduces great confusion into the motions, I do not see how the primum mobile can convey the diurnal motion to all the lower spheres when in the middle are those "circitores," which completely prevent the lower spheres from being moved along by the higher.[26]

Clavius also pointed out that Fracastoro supposed six orbs above the firmament, whereas the Peripatetics (Clavius does not say exactly whom he has in mind here), to whose principles Fracastoro was supposed to be adhering, maintain that the firmament is the highest sphere and that there can be no spheres above it. ''[The homocentrists] propose, as is made clear in Fracastoro's book, seventy-seven or seventy-nine orbs or mobile spheres—eight spheres containing stars [or planets] and the rest devoid of stars. Six of those [starless spheres] are located above the firmament. This opinion is not only opposed by a majority of astronomers, who heretofore have discovered only three celestial spheres above the firmament, but indeed it conflicts with all the Peripatetics, who, following Aristotle's opinion, do not want to admit even a single orb above the firmament.''[27]

Finally, Clavius indicted the homocentrists for the offense of being unable to agree even among themselves about the number, arrangement, and motions of the celestial spheres. Even if they convinced us to abandon eccentrics and epicycles, he says, we would not know which one of the homocentric theories to follow.

Clavius's extensive critique demonstrates that homocentrics had been a lively cosmological alternative in the early decades of the sixteenth century and would undoubtedly have been a familiar concept to an earlier generation of scholars—Clavius's teachers. Moreover, given the connections between Clavius's colleague Pereira and Paduan Averroists like Achillini, it is likely that Clavius actually had before him in Rome a sympathizer, and possibly even an advocate, of homocentric astronomy.

"Like Birds in the Air, or Fish in the Sea"

Homocentrics were far from the only rival cosmology with an advocate in the Collegio Romano, for the powerful Jesuit theologian Robert Bellarmine (1542–1621; professor of controversies, 1576–87; rector, 1592–94) was a strong advocate of what I have called the fluid-heaven theory. It is an old idea that the planets might not be rigidly fixed within celestial spheres, and indeed that the celestial spheres do not exist at all. In the conventional view, the planets were essentially bright spots in or on spheres that are otherwise perfectly transparent. The motions of the planets, therefore, would only be the visible consequences of the multiple movements of the spheres. The fluid-heaven view, however, would see the planetary spheres not as rigid shells but as concentric zones of a fluid medium through which the planet, an independent luminous body, moves on its own. Perhaps there is no real distinction between one planetary sphere and the next, which would then imply a single vast fluid heaven through which the planets glide. Possibly the fluid heaven is responsible for the common diurnal motion of all celestial bodies from east to west, while the proper motions of each, from west to east, are individual motions through the celestial fluid.

In the West, the basic idea of a fluid cosmos can be traced at least to the late thirteenth century. The early *Sphere* commentator Robertus Anglicus (fl. 1260–80) describes a concentric system of distinct fluid orbs, that is spherical shells, within which the planets move. The fluidity of the spherical shells allows the middle regions of the orb to move in one direction, presumably matching the motion of the planet, while the outer and inner surfaces of the orb can move in another direction, matching the motion of the surfaces of the neighboring orbs.[28] Another *Sphere* commentator, Andalo di Negro (ca. 1270–1340), also suggests a concentric series of distinct fluid orbs, each flowing at a different rate corresponding to the speed of that orb's planet, which he apparently likens to a fish's motion.[29] Andalo's account is the earliest reference to the piscine simile that I have found.[30] The philosopher Pietro d'Abano in his *Lucidator dubitabilium astronomiae* (ca. 1310) denied that the planets are moved on spheres and thus may have preferred the view that the planets move through a celestial fluid.[31]

It is not clear whether these early authors saw the fluid-heaven idea as an exclusive alternative to the Ptolemaic constructions. Robertus Anglicus, for example, after describing the fluid nature of the orbs, states that there is some disagreement between the naturalists (philosophers) and the mathematicians about those orbs: the philosophers consider all the orbs to be concentric, but the mathematicians do not.[32] Robertus does not opine one way or another, so his neutrality suggests that he thought the fluid orbs to be amenable to either interpretation.

In comparison to Aristotle's celestial spheres, the concept of the fluid heavens bears a resemblance to the cosmological ideas of the ancient Stoics. Peter Barker has recently concluded that the influence of ancient Stoicism, known through Cicero's *De natura deorum*, was growing in western Europe beginning with Petrarch in the late fourteenth century.[33] In the Stoic cosmology, the heavens were filled with the animate fluid substance called *pneuma*. The celestial bodies, which possessed intelligence, could direct their own motions through these fluid heavens.[34] Whether Stoic influence can be identified as early as Robertus Anglicus will require further studies, but Barker argues that the Stoic ideas achieved wide circulation by the late sixteenth century and that they play an important role in the formulation of Bellarmine's cosmological thought.[35]

By Clavius's time the concept of the fluid heavens, along with the idea that the planets move through them like fish in water or birds in air, had become clear rivals of the late Ptolemaic framework. Giovanni Pontano (1426–1503) had written in *De rebus coelestibus* that the planets move according to their own free wills like fish or birds, and thus that it is natural that they should move occasionally back and forth in their courses.[36] This *free will among the planets!*

is a significant step beyond a fluid-heaven concept like that in Robertus Anglicus, for example, because Robertus, while suggesting that celestial matter might be fluid, stopped short of discussing causes for planetary motions. But if irregularities in the motion of the planets are attributed to free will on the part of the planets as in Pontano, then the mathematical regularity of eccentrics and epicycles clearly has no explanatory role.

In the sixteenth century, the fluid-heaven idea continued to have strong appeal for some, at least in Italy. The Platonic philosopher and Clavius's close contemporary, Francesco Patrizi (1529–97), was an advocate of the idea that the planets moved under their own initiative through a fluid medium rather than as captives of rotating spheres. In fact, he rejected the very concept of the celestial spheres. Patrizi's primary cosmological exposition is *Nova de universis philosophia* (1591), also called *Pancosmia,* in which he states that the planets "fly within a liquid sky."[37] Patrizi is also remarkable for the notion that the diurnal motion results from a daily rotation of the earth from west to east, rather than a motion of the heavens from east to west. Patrizi's views were well known. They came, for example, to the attention of Kepler, who, in his note to the reader in book 4 of *Epitome astronomiae copernicanae,* refers to Patrizi in the same breath with Fracastoro as the founder of a cosmological system.[38] Kepler also criticized Patrizi's views severely in his *Defense of Tycho* and singled him out again in his *New Astronomy.*[39] In 1592 Pope Clement VIII brought Patrizi to Rome as a professor of Platonic philosophy.[40] The idea of the fluid heavens must have been in the air during Clavius's career, for Patrizi only published his views in 1591, but Clavius had been criticizing the concepts (but without mentioning the name of Patrizi or anyone else) since 1581. Thus the fortunes of the fluid heavens were rising in contrast to homocentrics, which apparently culminated decades earlier.

Though it is only a guess that the homocentric view was advocated by Pereira, it is certain that the Collegio Romano had an influential believer in a fluid-heaven cosmology, namely, Bellarmine.[41] Bellarmine's cosmological ideas, while perhaps informed by Stoic teachings, were guided most strongly by his reading of Scripture. His objective was to find a cosmology that explained most simply the bare phenomena while agreeing most closely with the statements of Scripture on the matter. The words of Scripture, for example, justify only three "heavens": the airy (the realm of birds and clouds), the sidereal (the realm of planets and stars), and the empyrean. "It is correct to allow at least three heavens, the airy one, the starry one, and the empiraeum. . . . If there are other heavens beyond the three mentioned, that is something to be examined by the philosophers and the astrologers more than by the theologian."[42] But Bellarmine is skeptical that the existence of the eight or more spheres of Aristotelian cosmology can be proven by the philosopher or "astrologer." He makes a particular example of the suprafirmamental spheres.

Such a heaven above the firmament is purely imaginary; the existence of a heaven above the eighth sphere has been deduced from the two or three motions which appear to belong to that sphere; but, leaving aside the fact that some of our contemporaries have proven that the eighth sphere has only one motion, even if we should allow a ninth sphere, how do these people come to know that this is the clear heaven, completely transparent and devoid of stars? As a matter of fact we neither see all the stars nor can anyone claim that all of the stars we do see are in the eighth sphere.[43]

Ugo Baldini and George Coyne interpret this passage (correctly, I think) as impugning the procedure of resolving observed motions into unseen uniform circular motions. Because the spheres that produce this circular motion have no basis in human experience, Bellarmine rejected them thus: "That the heavenly bodies are moved by the heavens does not appear to be valid at present, because many laughable and incredible consequences come from it." He gives several reasons, the most interesting of which are that this idea would imply "that one and the same star describes contrary movements, a notion difficult to clarify and even more so to uphold" and that "such complex and extraordinary structures as epicycles and eccentrics are dreamed up so that even the astrologers are reticent to speak about them."[44] The first of these is interesting because Clavius would level the same criticism at the concept of fluid heavens. The second is particularly significant because it shows us the degree of Bellarmine's skepticism regarding the constructions of late Ptolemaic cosmology.

For Bellarmine, the scriptural distinction of the airy and sidereal heavens is no more than one of location or proximity to the earth, so Bellarmine also rejects the Aristotelian material distinction between the sublunar and supralunar realms, as well as the doctrine of the immutability of the heavens. On the other hand, he held fast to the concept that the cosmos was geocentric and geostationary, but based that belief on the agreement of Scripture and human experience and not on any dogmatic Aristotelianism.[45]

Bellarmine's cosmos was, in outline, a geocentric structure bounded on the outside by a single solid sphere bearing the fixed stars. Between the sphere of the fixed stars and the earth a fluid substance, perhaps (though not necessarily) airlike, fills all of space. But because we know nothing of the nature of celestial motions (since they are remote from human experience and Scripture is silent), we also know nothing of their causes. Thus, "if we wish to hold that the heaven of the stars is one only . . . [which is] more in accord with Scriptures, we must of necessity say that the stars are not transported with the movements of the sky, but they move of themselves like the birds of the air and the fish of the water."[46] The use of the standard simile cannot be by chance. It places Bellarmine's thinking, though not his approach, squarely in the tradition of Pontano and Andalo di Negro. Moreover, both Clavius and Pereira use the simile disparagingly in criticizing the very thesis that Bellarmine was advocating. Bellarmine, a theologian, never published his cosmological ideas, and Clavius never refers to him as a proponent of the fluid-heaven theory. Nevertheless, Bellarmine did not hide his ideas (he was at least willing to reveal them in letters), and it seems very likely that his colleagues at the Collegio Romano, including Clavius, would have associated him personally with the fluid-heaven cosmology.[47]

The first reaction in Clavius's *Sphaera* to the fluid heavens appears in the section "On the Number of Celestial Orbs."[48] As is his pattern, he reviews a number of opinions, the first of which is "the opinion of those who suppose a single heaven," that is, "that the sun, moon, and all the rest of the stars exist in one and the same heaven." This opinion, held by some ancients and a few "recent imitators," rests on a single assertion, that "all our knowledge, according to the doctrines of the philosophers, arises from the senses."[49] Thus, whenever we lift our gaze toward the sky, we do not perceive by sight a multitude of heavens, but only one heaven containing all celestial bodies. This strictly empirical attitude corresponds exactly with Bellarmine's views.

But, Clavius tells us, this opinion can in no way be defended. No body (in this case, the unitary heaven) can simultaneously be moved with opposite and contrary motions; that is, no body can both rise and descend at the same time, nor move both toward and away from the same place (that is, e.g., move both eastward and westward) simultaneously. "Nevertheless, diverse and opposite motions are observed in the stars; and, as Aristotle holds and as we will demonstrate a bit later, stars do not move themselves like fish in water or birds in the air, but rather [they move] according to the motion of the orb in which they are fixed, like a knot in a board moves with the board or a nail in a wheel moves with the wheel. So more than one heaven ought to be conceded."[50]

What we see here is a fundamental difference in how an observation can be interpreted. Both Bellarmine and Clavius see the stars and planets rise and set, and both would agree that the observed periodicity of the stars' motion is different from that of the planets. In Bellarmine's view, all that rising and setting is simply a westward motion of the heavens—the planets are not moving eastward in any sense, they are simply moving westward more slowly than the stars. In this view there is no contrariety of motions because all the rotations are in the same sense, westward. But for Clavius the planetary motion is a compounding of two motions, each more fundamental than that observed, namely, a westward motion with the same period as the stars and a much slower eastward motion peculiar to each planet. Each of those motions is a result of a simple rotation of a celestial sphere. There is no contrariety in the motions because no sphere has a proper motion that is anything other than a simple uniform rotation. (To the modern reader, accustomed to the idea of considering motions relative to different frames of reference, this debate may seem vacuous. But we must again resist our modern prejudices and try to comprehend the dispute as the serious matter it was in the sixteenth century.)

Each party sees the other's theory as implying contrary motions in the heavens. Bellarmine sees Clavius saying that some spheres move westward while others move eastward at the same time that they are (in some communal sense) moved westward. He thus accuses Clavius (or rather Ptolemy, since he never mentions Clavius) of suggesting the unthinkable—that the same body can move in two directions at once. Clavius sees the advocates of the fluid heavens (he never mentions Bellarmine) saying that a given planet is moving only westward, yet he knows that the planet progresses eastward against the stars. Thus he accuses his adversaries of requiring that the same body move in two directions at once.

The view that Clavius is attacking—that the heavens are a single, unified body consisting of both planets and stars—is only one variation on the fluid heavens. Bellarmine allowed that the stars (but not the planets) are indeed fixed in their own (presumably rigid) sphere and that the fluid realm, containing the planets, extends from the stellar sphere down to the airy region. Clavius simplified the exposition and ignored any such distinctions, perhaps in order to contrast it with his point that the celestial region is composed of multiple parts that are distinguished by their motions.

The fluid-heaven question rises again in Clavius's discussion "On the Motions of the Celestial Orbs." After dismissing the opinion that the heavens do not move at all (all appearances being supposedly explained by a rotating Earth, perhaps a reference to Patrizi), he addresses "the opinion of those who say that the heaven is at rest and the stars are self-moved" and also "the opinion of those who say that the heaven is moved from east to west while the stars move themselves from west to east."[51] In the first case, "Some assert that while the heaven and the earth remain at rest, the stars move themselves, like birds in air or fish in the sea, from the east toward the west. But by this argument the planets cannot be moved by a dual motion, which conflicts with experience. For we not only see the planets moved from east to west but also from west to east."[52]

Clavius is refuting here the view that the planets move themselves through "the heaven," which in this case is a fluid medium, stationary with respect to the earth, that fills the cosmos and through which the celestial bodies, including the stars, circulate daily about the earth. His refutation of this opinion reveals to us how fundamental is his belief in the intrinsic plurality of planetary motion. In other words, he believes that each planet moves both westward (the diurnal motion) and eastward (the proper motion), and the phenomena (such as rising and setting times, but also stations and retrogradations) are a result of the compounding of these multiple motions as well as others. Clavius's argument treats his conviction about the plurality of planetary motion as if it were an observable fact. "We not only see the planets moved from east to west, but also from west to east." As far as Clavius is concerned, the simple westward motion posited by this first fluid-heaven concept is contradicted by the phenomena.

Next, he presents a more promising (but still wrong) alternative. "There are others who claim that the heaven is moved from east to west and the stars go around with it, while individual stars each have an individual motion from west to east. Thus it comes about, they say, that all the stars complete a daily rotation in the same time but are observed to move from west to east in unequal times."[53] In this case "the heaven" is still a space-filling fluid medium, but it is no longer stationary. It now circulates daily about the earth and carries with it the planets (and possibly also the stars), thus replacing the *motus raptus* of the Ptolemaic cosmology. This solution has potential, because it seems to provide separate explanations for both of the fundamental motions of celestial bodies: the diurnal, caused by the general rotation of the heaven, and the proper, caused by the peculiar motion of each body. These separate causes might seem to overcome the objection that he has made and will continue to make. Clavius only promises a rebuttal later. "We will demonstrate below that it is impossible that the stars move themselves . . . but rather it is necessary that they are borne around exactly according to the motion of the orb in which they stand."[54]

Nearly thirty pages later Clavius fulfills his promise to "demonstrate" the impossibility that the stars are moved by themselves.

Take any star or planet that someone says moves itself. This star is moved by motions that are, in a certain sense, opposite, as we said earlier. It is moved, that is, simply and continuously from east to west and, at the same time, continuously from west to east, just as we explained and demonstrated earlier. But no single body itself can be urged into disparate, opposite motions at one and the same instant. It implies a contradiction that one and the same body should, at the same time, progress from east to west and in the same moment from west to east in such a way that neither motion interrupts the other, but that both proceed uniformly without any cessation, unless the other motion moves the body as if conveyed in a vehicle. . . . This is more strongly confirmed in the case of the planets. They are moved by many motions, two of which are the westward and the eastward. And they are moved now faster toward the east, and now slower; sometimes they stand still, and sometimes retrogress toward the west, and so on, as is explained in the *Theorica planetarum*. If, therefore, the stars moved themselves, it would not be an adequate explanation of these variations. If, however, they are said to be moved according to the motion of the orb, then it is an easy matter for all the appearances to have a place, as will be explained in the *Theorica planetarum*.[55]

Clavius is invoking a corollary to the Aristotelian principle that a simple body, like a celestial orb, should have only a single motion, namely a uniform rotation. The corollary is that a celestial body cannot simultaneously have two contrary motions. Thus, if a planet appears to exhibit two contrary motions then at least one of the motions must be caused by another body, namely, the orb that acts as the vehicle for the orb bearing the planet. His argument is founded on a belief about the nature of motion of simple bodies like celestial spheres. He does not even attempt to defend this physical notion, which was almost axiomatic in those days.

That argument, however, is not Clavius's last shot at the theory of the fluid heavens, for it comes up once more in his *disputatio* on eccentrics and epicycles.

Indeed, if [planets] are so moved [like birds or fish] and not rather a result of the motions of the orbs in which they stand, then we can have no certain knowledge about those motions. So when, as stated in the observations above, the planets are sometimes more or less far from the earth, or sometimes moved faster or at other times slower, when they are now seen to stand still, and then to move forward in the zodiac from west to east, and then again to move backward, who is there who cannot see that the planets, if moved like fish or birds, ought sometimes to abandon their own circles, which go from west to east, so that they can recede further from the earth or approach closer to it, that sometimes they ought to neglect their courses and instead tend to recede to the opposite side, or finally sometimes all ought to halt in their paths in the heavens and be thoroughly unmoved? If they acted [like that], in what way, I ask, could their periods have been defined and by what reasons [could it be] known where in the heavens the planets will be diverted highest away from Earth and [where] turned back toward it again, etcetera?[56]

His first demonstration refuted the fluid-heaven theory by arguing that the kind of motion it implies is inadequate to account for the motions of the planets. But in the second argument Clavius has switched to criticizing it essentially for its ad hoc nature, that is, it can't predict anything, and it can't explain anything, because it reveals nothing of the causes of planetary motion. The fault is a problem not shared by the homocentric and Copernican theories, both of which agree with the Ptolemaic in assuming the mathematically tractable uniform circular motion (which agreement does not preclude, of course, disputes about whether the equant, for example, truly meets the definition of uniform circular motion).

The second fault—the failure to provide causal, predictive power—is an important theme in Clavius's evaluations of astronomical theories. When he criticized the homocentric theorists for not being able to agree on any particular arrangement of spheres, he was, in part, impugning the practicality of the idea for making predictions. But in principle, at least, the mathematical nature of homocentrics makes such predictions plausible. The fluid heavens, lacking any grounding in mathematical concepts like uniform circular motion, are incapable of yielding causal predictions even in principle. If the stars move like birds or fish (i.e., of their own power and will), then the only kind of prediction possible would be an extrapolation based on past behavior. But such a prediction carries little or no assurance because it is not based on knowledge of the causes of the motion. Predictions founded in causes are, of course, not a problem for the Copernican theory. Later, Clavius will criticize the Copernican theory not because of any failure to yield causal predictions but because, he will claim, the predictive power of the Copernican theory is merely derived from the Ptolemaic.

That the Stars Move in Channels

There is one more cosmology that Clavius takes the trouble to refute (aside from the Copernican, considered in chap. 5). He associates it loosely with the fluid heavens, though he identifies it as, in some ways, superior to the fluid-heaven theory. It is "the opinion of those who say that the stars [or planets] are moved in channels." Unfortunately, Clavius does not tell us just who espoused this theory, but it must have had some following in the sixteenth century for it to deserve a significant refutation. There is a rebuttal of this same theory as late as 1688 in *Universalis cosmographiae elementa* by the Jesuit Nicola Partenio Giannettasio. He described the celestial channels as tunnels through which the planets supposedly run *rabbits !* back and forth like rabbits.[57] Giannettasio, who favored the fluid heavens, had no serious interest in the celestial channels and dismissed them. He seems to know no more about the idea than Clavius and may have brought it up primarily because Clavius, whose views he follows closely in many other matters, had considered it.[58]

"They say," Clavius tells us, "that there is a single heaven that is moved by a single motion from east to west along with all of the stars. But the stars are borne by their own proper motions from west to east unhindered by the celestial orbs. Not, however, like fish in the sea or birds in the air, nor by penetrating or shearing the celestial bodies, but through channels of a sort."[59] Why he refers above to multiple orbs after saying that this theory posits a single heaven is not clear. It may just be a slip into the natural plurality of celestial orbs that he is accustomed to. As is often the case with astronomical language of this period, it is not always clear how Clavius means the word "star" (*stella*) to be interpreted. Sometimes context makes it clear, but not always.

Clavius explains that those who advocate this theory contrived (*confinxerunt*) the channels in such a way that each planet has its own channel suitable to the planet's proper motion, and that the size of the channel matches the size of the planet such that it just fills its channel (speaking in terms of the cross-section, presumably). Further, they place in this channel a kind of fluid substance like air that can yield to the star as it moves from west to east. It is not clear whether the proponents of this theory intended that only the planets should have channels while the fixed stars, perhaps, would be attached to a sphere. Clavius is little help on this question, either. He criticizes the theory for, among other reasons, failing to account for the threefold (daily, precession, and trepidation) motion of the fixed stars and adds that the celestial channels are incapable of explaining those motions. This, however, tells us precious little about the views of those who advocated the theory.

This idea of celestial channels leads to an interesting overall view of the celestial region, which Clavius describes thus: "So according to those authors, the whole heaven will be filled with channels in proportion to the multitude of stars, just like [the bodies of] animals, which are filled with various and many veins.''[60] If the multiplicity of stellar channels is comparable to the multiplicity of veins in animals, then the image suggests that there are many more channels than merely those of the seven planets. Perhaps, then, the stars are not fixed but follow parallel paths through the numerous "veins." This cosmology is particularly appealing, Clavius tells us, to those who do not like the idea, part of the late Ptolemaic view, that the primum mobile moves all the lower spheres through their diurnal rotation by its *motus raptus,* or transporting influence. Instead, the celestial arteries serve as a kind of vehicular matrix that is responsible for the common diurnal motion. In this regard the theory might be seen, from Clavius's view, as superior to the fluid heavens though he does not say so. "Nevertheless this opinion is both absurd and inadequate," Clavius continues,

> absurd because, without any necessity or even probable reason, it supposes that the body of the heavens is perforated by such channels and filled everywhere by that fluid substance that heretofore no philosopher seems to have admitted. Inadequate because it is impossible for this opinion to explain all the phenomena that diligent astronomers have observed in the celestial motions . . . [for] in no way can it explain more motions than those two that it allows them to have . . . [yet] the moon is observed to have at least six motions . . . [further] the planets, as is clear from the *Theorica planetarum,* are not always equally far from the center of the earth but sometimes appear closer, sometimes farther away, which can in no way be if the stars are self-moved in those channels, unless it be said that the channels are eccentric to the world. . . . [But if] the channels are fixed in the celestial bodies, than it must follow that a given planet would always reach its maximum distance from the earth in the same part of the sky, and that is quite false. . . . I omit the appearances of variation in latitude, of retrogradation, etc., which the aforesaid opinion can in no way explain. . . . It stands, therefore, that the stars are not moved by themselves, but according to the motion of the spheres in which they are embedded.[61]

To sum up, Clavius's textbook reveals a considerable richness of Renaissance cosmologies, five in all: the orthodox late Ptolemaic plus the four alternatives. Clavius's refutations of the four rival cosmologies, the homocentric, the fluid heaven, the peculiar celestial channels, and the Copernican theory, had all appeared in the *Sphaera* by the time of the 1581 (*nunc iterum*) edition. There were no further significant changes in his evaluations of these theories during the rest of the long publication history of the *Sphaera*. The amount of attention that Clavius gives to each of the rivals gives us some indication of how significant he felt each was as a threat to the cosmological orthodoxy. And what is most interesting is that, by that measure, he considers the homocentric alternative as much of a threat as the Copernican. Apparently less significant was the fluid-heaven cosmology, which nonetheless came in for considerable mention and constantly reiterated criticism. Finally, a distant fourth, follows the theory of the stellar channels.

Clavius's presentation of astronomical theory as he saw it was dominated by these debates over rival cosmologies. What are appropriate grounds for debate? By what criteria are theories to be evaluated? How important are saving and predicting the phenomena? How important are consistency with natural philosophy and proper methodology? In the following chapters I will consider these questions while examining Clavius's responses to Copernicus, Tycho, and the astronomical events of the late sixteenth century.

FIVE

Cosmological Debate and the Rebuttal of Copernicus

A thoughtless man who pays attention only to the numbers will think that the same result follows from different hypotheses and indeed that the truth can follow from falsehoods.
—Johannes Kepler, Apologia pro Tychone

Clavius's defense of late Ptolemaic cosmology was, in large measure, a reply to a group of alternative cosmologies that had been discussed for a rather long time. Fracastoro was only the most recent, and, as a famed physician and papal favorite, the most eminent advocate of the homocentric approach. But homocentrics in one form or another had been pondered seriously by many astronomers and philosophers—medieval and Renaissance, Latin and Arabic. As a result, the arguments and responses were fairly developed, the issues well defined. The theory of a fluid heaven and self-moved stars had similarly been a standard topic for some time, and the arguments and issues were well worked over.

On the other hand, the Copernican cosmology was new. Of course, Clavius and others recognized that Copernicus was following the general idea of a heliocentric system as originally proposed by Aristarchus. But there had been no serious consideration of heliocentric systems from classical antiquity until the time of Copernicus's work. Unlike the other major cosmological alternatives—the Ptolemaic, homocentric, and fluid—the Copernican cosmology had not accumulated a significant body of criticism and rebuttal beyond that which Copernicus himself had been able to anticipate in his book.

In the years between the publication of *De revolutionibus* in 1543 and Clavius's first *Sphaera* edition of 1570, the assimilation of Copernicus's work by astronomers had only begun.[1] Technical responses to Copernicus, whether critical or favorable, were rare, and the two mathematicians who might have influenced Clavius most strongly, Nuñez and Maurolico, offered no guidance. Nuñez appears to have mentioned Copernicus rarely at best, while Maurolico declined to refute the new theory with his now infamous comment that Copernicus was more deserving of a scourging than a rebuttal.[2] So in his presentation and critique of Copernicus, far from tilling well-worked soil, Clavius was plowing almost virgin ground.

In the first edition of the *Sphaera* Clavius refuted Copernicus briefly. But by his revised edition of 1581 the Jesuit had developed a fairly elaborate response to Copernicus, offering criticisms on physical, scriptural, and methodological grounds. His response, however, is complicated by the fact that it comes in two distinct parts: one cosmological and the other methodological. Clavius responds to Copernicus's cosmological assertions—the place and motion of the earth, the stability of the sun and stars—as part of a general refutation of all opinions that the earth moves or stands anywhere but at the center of the cosmos. In that discussion he does not even mention Copernicus explicitly, though it is clear that Clavius intended to apply this refutation to Copernicus because in an earlier passage he had promised to show that Copernicus's cosmological claims are false. The other prong of Clavius's attack on the Copernican system is a methodological assault on its status as a scientific theory. This tactic allows him to set the Copernican and Ptolemaic cosmologies in opposition despite their formal similarities in some areas (their common mathematical assumptions, for example).

Clavius's response to Copernicus does not include a detailed description of the Copernican system. Rather, there is a brief summary of the Copernican order of the planets and mention of the threefold terrestrial motion. This should in no way be considered a suppression of Copernicus. It is important once again to remember that the *Sphaera* was an introductory textbook. Therefore Clavius's goals in the *Sphaera* were generally not highly technical. This is especially true in regard to planetary theory, since planetary theory was normally handled in *Theorica planetarum* texts. Thus his nontechnical summary of Copernicus is suited to his ends and is no more elaborate than necessary. Nor is his cursory description of Copernicus's system atypical for sixteenth-century astronomical textbooks, which generally seemed to consider the Copernican theory too advanced a subject for introductory texts.[3] Clavius's lack of an extensive description of Copernicus's system may explain the incorrect statement by Houzeau and Lancaster on the *Sphaera* that "the author avoided speaking about the system of Copernicus"[4]—unless, of course, they simply did not read the book.

Cosmological Issues in the *Sphaera*

In the years between 1570 and the last *Sphaera* edition of 1611, many forces were at work altering the standards and issues of cosmological debate from the traditional ones we see in Clavius. There were several historical reasons for these changes. One was the increasing significance of Copernicus's work, apparent through its successful technical application in the Prutenic Tables, its utility as a source of data for calendar reform (specifically Copernicus's measurement of the length of the tropical year), and its value as a source of new ideas, such as its theory of the precession of the equinoxes. Another reason was the remarkable series of astronomical prodigies, new stars and extraordinary comets, that called into serious question some common assumptions of late Ptolemaic cosmology—primarily the impossibility of change in the celestial region and the solidity of the celestial spheres. Finally, of course, technology intruded in the form of the telescope and its brilliant applications by Galileo in 1609 and 1610. These events changed the nature and scope of cosmological controversy, but to see this we must begin by examining the earlier debates, and the first stages of the Copernican debate, as presented in Clavius's *Sphaera*.

The Nature of the Celestial Spheres and Their Motion

One of the pervasive issues lurking in Clavius's *Sphaera* is the number and order of the celestial spheres. It is, in fact, the issue of Copernicus's rearrangement of the celestial spheres that first brings him under Clavius's criticism. This is not surprising, for Sacrobosco addresses the issue; and questions on the number of celestial spheres and their order arose continually in other elementary astronomy texts of the second half of the sixteenth century.[5] Closely linked is the question of whether planetary motions, as assigned by Ptolemaic theory, implied anything other than the eternal perfection and circular motion attributed to them by Aristotelian physics.

Though the material nature of the celestial spheres was debated and far from certain in medieval discussions of celestial physics, there is little doubt in Clavius's *Sphaera* that the spheres are solid in the sense that they are dimensionally rigid and mutually impenetrable. Solid, that is, as opposed to fluid. Their solidity is one of the clearer doctrines of Clavius's cosmology. He explains the situation of a planet in its orb using the common metaphors of a knot in a board, or a nail in the rim of a wheel.[6] Clavius argues for solid (rigid) spheres by rejecting fluid heavens, which he sees as the only logical alternative. That is, the heavens are either fluid or solid. Since fluid heavens must be rejected for a number of reasons, we are left to conclude that the heavens are solid. Moreover, the idea of solid celestial orbs is perfectly consonant with Aristotle's teachings on the nature of the heavens, so the position is a generally congenial one.

A much greater threat to Ptolemaic cosmology than the fluid heavens was the argument that the motions of eccentrics and epicycles are not consistent with the teachings of Aristotle and thus have no place in any cosmological picture. Perhaps the severest critic of the Ptolemaic constructions was Averroës, who rejected them as incompatible with Aristotle's physics and thus inconceivable as part of the real world. The likes of Nifo, Achillini, Fracastoro, and G. B. Amico followed Averroës in their hostility to the Ptolemaic constructions. In addition there was the Jesuit philosopher Benedict Pereira, who taught at the Collegio Romano along with Clavius and who held views on many matters that were strongly opposed to Clavius's and similar to the skeptical Paduan Averroists. So when Clavius defended the legitimacy, from an Aristotelian standpoint, of Ptolemy's constructions, he was not only arguing against the remote figures of Averroës and Nifo (both of whom he names) but also against the views of at least one of his colleagues at the Collegio Romano. (Clavius never singles out Pereira by name or other direct reference, just as he never mentions Bellarmine when discussing fluid heavens.)

In his textbook on Aristotelian physics, Pereira presented views on the nature of celestial matter similar to those of some of the Paduan Averroists such as Jacopo Zabarella and Archangelus Mercenarius.[7] They held that celestial matter does consist of matter and form (as Aristotelian metaphysics teaches about ordinary things in our experience), but that celestial matter and celestial form are completely unlike their terrestrial counterparts.[8] This radical difference between celestial and terrestrial matter is a significant barrier to our knowledge of the former. Pereira makes the skeptical argument that, as a result of having no knowledge of the nature of celestial matter, astronomers cannot reason on the basis of causes and must therefore employ physical absurdities, like epicycles and eccentrics, to achieve their ends.

He goes on to take the radical position that not only is astronomy as a mathematical discipline incapable of dealing causally with celestial phenomena, it is, in fact, incapable of truly dealing with the phenomena at the observational level. Pereira nearly removes celestial physics from the phenomenal realm altogether by arguing that quantity (which is the proper object of the mathematical sciences) in celestial matter is not comparable with quantity in terrestrial matter because of fundamental differences in the two kinds of matter. Thus the astronomer's act of measuring the position of a celestial object is, in Pereira's view, a mathematical judgment about a visual phenomenon, but the result does not give the astronomer any information about the true place of the celestial body.[9] A true causal astronomy would have to be established a priori on knowledge of the nature of celestial matter. For Pereira that is a job for the physicist, not the mathematician.

Such a doctrine goes beyond mere criticism of the causes (eccentrics and epicycles) advocated by Ptolemaic astronomers. It would seem to undermine the foundations of any attempt at mathematical astronomy, even the homocentrics favored by the Peripatetics, because it denies that the astronomer can, by observation, determine the state of the heavens at any given time (beyond, that is, the assignment of angular positions on the sky). Even the Aristotelian homocentrist must have this information in order to construct tables for the purpose of calculating the positions of celestial objects—which is, after all, the purpose of the mathematical discipline of astronomy, strictly defined. If Pereira was thinking in this manner, then it would put him in the position of advocating an "astronomy without hypotheses" something like that desired by Ramus and, perhaps, Bellarmine.[10] All causal explanations of celestial phenomena would be left to physics, and, presumably, the physical causes would not be obliged to have any correspondence to the methods of conventional astronomical calculations.

This philosophical debate resonates strongly with the disciplinary squabbles among the philosophers and mathematicians of the Collegio Romano and helps to explain Clavius's elaborate apologetics on behalf of the general status and utility of mathematics and its component disciplines, such as astronomy. In the face of severe criticism from the likes of Pereira, he defended the Ptolemaic constructions as entities of the real world. They were necessary not only because no other theory provided adequate explanations of the phenomena, but because materialized eccentrics and epicycles function as a cosmological rationale for establishing the distances to the celestial bodies and the overall size and structure of the cosmos. To reject the Ptolemaic constructions, as the Averroists would do, would necessarily be to reject centuries of effort by astronomers and to destroy the quantitatively detailed cosmology and astronomy assembled by that effort.

In this context, Copernicus and Clavius have rather a lot in common (though Clavius rarely acknowledges this, and then only implicitly). Both proceed on the premise that one can reliably infer causes from effects—in this case that means inferring celestial mechanisms from the celestial phenomena. To put it another way, both believe that, on the basis of observations, astronomers can form valid conclusions about the real motion and arrangement of the celestial bodies. Both also concentrate on the mathematical foundations of astronomy (as opposed to the physical or philosophical). More specifically, and in contrast to, say, the homocentrists, both place themselves in the Ptolemaic astronomical tradition. Within that tradition, however, they took very different stances. Clavius chose to retain the conventional Ptolemaic cosmology, which had the advantage of remaining, on the whole, consistent with intuitive cosmological assumptions as well as the prevailing physics. He also accepted all the Ptolemaic constructions, except when observations required some revision, as in the case of the motion of the eighth sphere. Copernicus, on the other hand, while accepting most of Ptolemy's mathematical assumptions, attempted to reform Ptolemy by eliminating the equant and rearranging the planetary orbs so as to produce a more unified and mathematically harmonious system, and in doing so he departed from the ancient geocentric and geostatic cosmological assumptions. Nevertheless, in contrast to some of the skeptics, Clavius and Copernicus have considerable common ground on the matter of the number and order of the celestial spheres because, in their view, the mathematical astronomer has a great deal to say on the matter.

Although other *Sphere* commentaries had dealt with such questions as the nature of celestial matter and movers, Clavius generally avoided these issues. He did not defend Ptolemaic astronomy by entering into debate with Pereira or anyone else over the nature of celestial matter. Clavius is willing to enter into philosophical discussions about whether astronomical mechanisms like eccentrics and epicycles are consistent with doctrines of physics, but he believes that "the philosopher particularly strives to investigate the nature and substance of the heavens . . . but the astronomer deals with precise arguments concerning those same celestial bodies, which move around the center of the universe; namely, he determines the periods and variations of all of the motions."[11] And yet the astronomer's study of these motions does sometimes reveal certain things about the nature of celestial matter, its arrangement, and its movers, as we have seen in Clavius's rejections of homocentric planetary motions and planetary motions in a fluid heaven.

Clavius's belief that astronomers can infer certain things about the nature of celestial matter and its movers is illustrated by his response to yet another cosmological alternative. Al-Bitrūjī attempted to create a kind of homocentric system in response to Averroës's criticisms of Ptolemy. But Clavius does not treat al-Bitrūjī or his later advocate Alessandro Achillini in quite the same way as other homocentrists, for instance, Fracastoro and G. B. Amico. Instead, Clavius treats al-Bitrūjī as a proponent of a novel and mistaken view of the nature of celestial motion. Al-Bitrūjī, and Averroës before him, had criticized any astronomical theory that implied contrariety in celestial motions, since contrariety of any kind would be inconsistent with the uniformity, perfection, and eternity of the heavens.[12] In particular, the Ptolemaic models require that a given planet possess at least two contrary motions: its westward diurnal motion and its eastward proper motion. Now a simple body, in the Aristotelian view, can only have one natural motion, so only one of those celestial motions can be natural and the other must be violent, al-Bitrūjī argued. But celestial motions are perpetual, and perpetual motions cannot be violent, therefore there cannot be contrary motions in the heavens—a conclusion with which Clavius agrees.[13]

Al-Bitrūjī proposed that the planetary spheres were all moved from east to west only, and the apparent eastward motions of the planets with respect to the stars were a result of their merely slower westward motions. In his view the westward motive power has its source somewhere beyond the firmament, and this power propagates inward from the firmament to the earth, but with decreasing strength. The diminishing strength of the motive power causes Saturn to lag a little behind the stars, Jupiter still more so, and so on to the moon, which of all celestial bodies lags the most behind the fixed stars. The earth, at the center, moves very little, if at all.[14] Thus al-Bitrūjī's picture is, broadly, that all the celestial spheres perform "diurnal" rotations about the earth, but with progressively increasing periods that account for their apparent proper motions.

Clavius responds that the celestial motions cannot be of this nature, for if they were as al-Bitrūjī would have them, then why would the planets not move in the same plane as the stars, that is, the plane of the celestial equator? If the celestial bodies are all moved by the same force, then we should see the sun, moon, and other planets following paths that have as their poles the same poles as the firmament, namely the north and south celestial poles. Yet this is clearly not the case. Moreover, Clavius says, it makes no sense that the spheres of the planets should successively fall behind (*repedare*) each other, because there is no perceivable proportion in this retardation. "According to Ptolemy, the eighth sphere completes its circuit in thirty-six thousand years, Saturn in thirty years, Jupiter in twelve, Mars two, the sun one year, Mercury and Venus in about that same time, and the moon in twenty-seven days and eight hours; whence you see clearly that no definite proportion can be found."[15]

Finally, it is not true, Clavius says, that the celestial motions implied by the Ptolemaic motions are either contrary to each other or violent. They are not contrary, as he will explain again later, because they are not simultaneously proper to the same body, that is, each celestial sphere has its single, simple motion that is its proper motion. The other motions that it exhibits are a result, not of any complex proper motion, but of the compounding of its one proper motion and the imposed motions of the other spheres. Nor are the Ptolemaic motions violent. Even if we concede that one of them is somehow violent, it does not follow that it cannot be perpetual, because if the cause of a violent motion is inexhaustible, then the motion will likewise be perpetual.

Especially intriguing is Clavius's assertion that if the motions of the celestial spheres were a result of a single motive force spanning the cosmos, then some sort of mathematical proportion could be expected in the periods of the motions. In saying this, Clavius accepted the intuitive notion that the motive power diminishes with increasing distance from the mover, but he adds the interesting requirement that this decrease be shown to bear some mathematical proportion to the distance. Clavius seems to assume here that forces in nature will obey mathematical laws, though he did not formulate the thought explicitly as Galileo, Kepler, and others would later. In fact, this criticism of al-Bitrūjī leads one to wonder what kind of significance Clavius attached to the arguments in *De revolutionibus* that were based on the mathematical harmonies Copernicus had discovered. It is unfortunate that Clavius never commented in any detail on Copernicus's arguments. But in addition to Clavius's obvious respect for Copernicus's technical skill, the Jesuit's sympathy for seeking mathematical proportions in nature makes even more understandable the esteem in which he held his predecessor. Clavius's search for mathematical principles in the nature of celestial motion foreshadows the mathematical emphasis on physical and astronomical explanations that will become the scientific standard after the work of Galileo and Kepler.

The Capacity to Save the Phenomena

Another recurrent issue in the theoretical debates of the *Sphaera* is the capacity of a theory to save the astronomical phenomena. That is, the ability of the theory to provide convincing and consistent explanations of the observed behavior of the celestial bodies. One might think that the ability to explain the astronomical phenomena observed in nature would be a minimum requirement of any proposed astronomical theory. Yet, as was the case with Fracastoro's homocentrics, a theory's philosophical virtues sometimes were more important than its technical competence. Fracastoro was satisfied to warrant his theory on the basis of its philosophical compatibility with the Averroist version of Aristotelian physics, despite its inability to give accurate explanations of planetary motions or explain brightness variations without ad hoc hypotheses. Clavius placed great importance on the technical competence of astronomical theories and was quick to seize on the technical flaws of rival theories.

A related aspect in Clavius's evaluations of rival theories is the ability of a theory to go beyond explaining the already observed appearances and accurately predict phenomena of the future. Clavius cites this as a major problem, for example, with the fluid-heaven theory and its accompanying idea that the planets move through the fluid heaven "like fish in water or birds in air."

> Indeed, if [the planets] were so moved [i.e., like birds or fish] . . . then we could have no certain knowledge about those motions. . . . Is there anyone who cannot see that the planets, if moved like fish or birds, ought sometimes to abandon their own circles, which go from west to east, so that they can recede farther from the earth or approach closer to it, and that sometimes they ought to neglect their courses and instead recede to the opposite side or, finally, sometimes they all ought to halt in their paths in the heavens and be thoroughly unmoved? If they acted thus, in what way, I ask, could their periods de defined? And how could it be known, indeed, where in the heavens the planets would be diverted higher away from Earth or turned back toward it again?[16]

A further example is Clavius's emphasis, as he replied to the skeptics, on the ability to know the precise circumstances of celestial phenomena, such as eclipses:

> From eccentric orbs and epicycles, not only the appearances of past things already known are explained, but also future things are predicted of which no one now knows anything. So if I doubt whether, for example, at the full moon of September 1587, there will be a lunar eclipse, I can be sure from the motions of the eccentric orbs and epicycles that there will be an eclipse, so I will doubt no more. Indeed, from these motions I know at which hour the eclipse will begin and how much of the moon will be obscured. And in the same way all eclipses, both solar and lunar, can be predicted and also their times and magnitudes.[17]

There was, in fact, a lunar eclipse on 16 September 1587. (Though he published this well before 1587, Clavius never updated the example in editions later than 1587. His point, of course, is not the prediction of this particular eclipse, but that the details of any given eclipse—and by extension, planetary events such as conjunctions and oppositions—may be found by the Ptolemaic methods.)

Both homocentric and fluid-heaven theories are unable to predict future phenomena or even save the observed phenomena, but the Copernican theory can do both of these things with as much accuracy as the Ptolemaic. Yet Clavius will try to impugn its technical power by claiming that the Copernican theory only saves the phenomena, past and future, by virtue of the fact that the Ptolemaic theory has already solved all of the real predictive problems.

The Number and Order of the Celestial Spheres

The question of the number of celestial spheres and their relative order has already appeared several times and is an important and recurring concern for Clavius and other sixteenth-century astronomical writers. This is understandable particularly in the matter of the number of spheres. Astronomers, beginning with Hipparchus, had to take into account the motion of the eighth sphere (that is, the precession of the equinoxes) in order to make sense of observations significantly removed from their own time. This remained an important problem because the standard star catalog in use through the sixteenth century was the one in Ptolemy's *Almagest,* which was only superseded by Tycho's star catalog at the end of that century. Refinements in the theory of motion of the eighth sphere—most notably the introduction of trepidation by Thābit Ibn Qurra and its further refinement by the Alfonsine astronomers—generally took the form of an additional sphere or spheres above the firmament (the eighth sphere).[18] So the issue stems directly from the concerns of technical astronomers over the interpretation of historical data and the development of calculational tools. But once formulated as an astronomical issue, the question of the number and order of the celestial spheres has cosmological interest as well, and it is this aspect that we see most clearly in Clavius's *Sphaera.*

Clavius discusses the number of spheres at two levels of detail. In his commentary on Sacrobosco's first chapter, Clavius explained a variety of opinions on the number of celestial spheres and defended, naturally, the number arrived at in Ptolemaic cosmology (at first eleven, counting the empyrean, and later twelve in agreement with Magini). In that context, he was considering the complete sphere, the *sphaera tota.* Both the inner and outer surfaces of the complete sphere are concentric with the cosmos, and those surfaces between them contain all the component orbs required by the materialized Ptolemaic constructions, as presented by Peurbach. All the components, the various orbs, that belong to a given complete planetary model (the simple eccentric, virtual eccentrics, epicycle if required, etc.) go together to form a unit that counts as one complete sphere, no matter how internally complex.

In the course of the *disputatio* on eccentrics and epicycles (part of his chap. 4), Clavius counts spheres, in the sense of component orbs, when he compares Fracastoro's homocentric theory with the late Ptolemaic. "[Homocentrists] propose, as is made clear in Fracastoro's book, seventy-seven or seventy-nine orbs or mobile spheres—eight stellar [and planetary] spheres and all the rest devoid of stars, six of which are located above the firmament. This opinion is not only opposed by a majority of astronomers, who heretofore have discovered two such celestial spheres above the firmament, but indeed it conflicts with all the Peripatetics, who, following Aristotle's opinion, do not want to admit even a single orb above the firmament.''[19] Clavius goes on to use the large number of orbs in Fracastoro's theory as another reason to prefer the Ptolemaic system, then couples this issue with that of the relative capacity of the theories to save the phenomena, and finally reiterates the lack (as he sees it) of conflict between Aristotelian natural philosophy and the eccentrics and epicycles of mathematical astronomy. He continues,

> Those who propose eccentric orbs in the heavens escape such confusion, because they allow in the universe exactly thirty-three orbs, twenty-seven going around the earth, and six epicycles, all of which are beyond the earth.[20] Whence there will not be such a multitude of motions. . . . And also, according to the famous axiom of the philosophers, more would do in vain what fewer would do equally well; yet nearly one-third the number of eccentrics are proposed by us than concentrics are proposed by the adversaries. And all the phenomena are defended not only just as well, but much better by eccentrics than by concentrics, since the explanation for countless appearances cannot be given by means of concentrics, as is evident from what has been said [earlier]. Who will doubt that it is preferable that in the heavens eccentric orbs and epicycles be decided upon rather than concentrics? Especially since eccentrics do not conflict at all with natural philosophy as will be established from the solutions to the arguments of Averroës and his followers.[21]

In addition to the ever present issues of saving the appearances and the conflict of the Ptolemaic constructions with Aristotelian physics, Clavius attempts here to turn one of the Averroists' arguments back on them. One of their arguments, which Clavius will later cite and refute, was based on the idea of the economy of nature—really just an aspect of the principle of parsimony, which Clavius accepted implicitly. Nifo argued, so Clavius tells us, that the virtual eccentrics represent ''something unnecessary and idle''[22] in nature, because one or the other of the virtual eccentrics would be sufficient to define the simple eccentric deferent of the planet, thus the complementary virtual eccentric would be superfluous. This argument is very weak because it reveals a misunderstanding (on the part of Nifo) of the functions of the virtual eccentrics. Thus Clavius simply responds that the purpose of the virtual eccentrics is not merely to define the simple eccentric, but to render the entire planetary sphere concentric with the cosmos. Thus both are necessary and not at all in vain.[23]

In the quotation above, Clavius turned Nifo's argument around to use against the homocentrists and, by association, the Averroists like Nifo. If the economy of natural causation is a standard the philosophers wish to employ (and Clavius would undoubtedly agree that it is proper to do so), then how much better are thirty-odd eccentrics and epicycles of the Ptolemaic theory than the seventy-seven or more spheres of the homocentrists? Thus the issue of the number and arrangement of the celestial spheres appears in several contexts: not only in the general cosmological sense of the number and order of the complete, concentric sphere complexes but also in the rhetorical battles over alternative planetary mechanisms and, of course, on the subject of the Copernican theory.

The Status of the Earth

The issue of the number and order of the celestial spheres first brings Copernicus under Clavius's scrutiny:

> Among the ancients there were some, of whom the leader, four hundred years before Ptolemy, was Aristarchus of Samos (whom Nicholas Copernicus, among the moderns, follows in his work on the celestial revolutions), who concocted this order among the bodies making up the universe: Sun located in the center of the world and immobile, around which is the orb of Mercury, then that of Venus, and around that is the great orb containing the earth, the elements, and the moon, and around that is the orb of Mars, then the Jovian heaven, and after that the globe of Saturn. Finally follows the sphere of the fixed stars. But this opinion is shattered by many observations and the common opinion of philosophers. The earth ought to stand firm in the center of the whole world as we will demonstrate later by many observations and phenomena.[24]

This introduction, which stands unaltered in every edition of the *Sphaera* from 1570 to 1611, would lead us to expect that the refutation of Copernicus will turn on the issue of the stability and centrality of the earth. But that is not really the case. Clavius does indeed array a number of arguments in favor of the centrality and stability of the earth, but in the course of those arguments he never addresses any criticism specifically toward Copernicus.

It should be no surprise that Copernicus occupies a unique place in

Clavius's *Sphaera,* coming in for both praise and criticism. Were it not for Clavius's reaction to the cosmological doctrines of Copernicus's *De revolutionibus,* that volume would hold an unambiguous place of high esteem in the *Sphaera*—comparable to that of Maurolico, Regiomontanus, and Nuñez. The first (1570) edition reveals Clavius's double-edged treatment of Copernicus: Clavius pays his Polish predecessor the compliment of appropriating, with due credit, the stellar observations that appear in *De revolutionibus.* But when listing the various opinions on the order of the celestial spheres, he identifies Copernicus as a follower of Aristarchus and judges the heliocentric theory untenable without offering an extensive critique of it. In the light of this inconsequential critique, it is conceivable that Clavius was only superficially familiar with the Copernican system in 1570. He seems, rather, to have used it the same way that he used many other texts, that is, more as a collection of useful data than as an integral work with a unified argument. Copernicus was an important source of data (primarily for his determination of the length of the year) for the work of the Gregorian calendar reform commission, and it may be that it was the calendar reform work, which began around 1572 at the earliest, that first gave Clavius the opportunity or incentive to look seriously at Copernicus.

The 1581 *Sphaera* was considerably enlarged over the 1570 edition and contained more references to Copernicus, including an extensive and negative review of his heliocentric theory. By 1585 (the third revision), Copernicus was an important source for the *Sphaera*: Clavius cites Copernicus's determination of the length of the year, uses his observations to update Ptolemy's star catalog, prefers Copernicus's method of computing star positions over Ptolemy's method, praises Copernicus as an accurate observer while using his observations in calculating the motion of the eighth sphere, accepts his observations and conclusions on the value of the maximum solar declination, and cites his solution of a problem in spherical geometry.

The Centrality of the Earth

It is surprising that Copernicus does not emerge as a subject of criticism in the context of the centrality of the earth. Clavius recites all of the traditional arguments from Ptolemy, Sacrobosco, and others for the centrality of the earth with respect to the sphere of the fixed stars. He expounds at some length, for instance, on Sacrobosco's two arguments. First, because a given star appears equally bright at rising, setting, and when crossing the meridian (neglecting, he says, any obscurations in the atmosphere), we must be equally far from it at each point, and this could

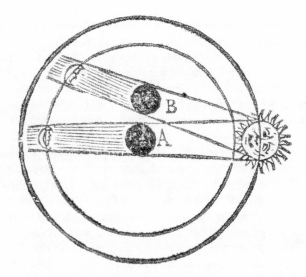

Figure 17. Demonstration of Earth's centrality by lunar eclipses. From Clavius, *Sphaera* (Venice, 1596), 142. Photograph courtesy of the University of Wisconsin–Madison Memorial Library.

only happen if the earth is at the center of the sphere of fixed stars. Second, if the earth were not at the center of the firmament, the horizon would not always divide the sky into two equal hemispheres. Yet the horizon does so divide the sky from every known place on Earth, so Earth must be at the center.[25]

Clavius then adds arguments from Ptolemy demonstrating that the earth can neither be displaced from the center along the axis of the world nor away from the axis in the plane of the celestial equator, for either shift would result in detectable consequences.[26] He then adds other arguments, for instance, that the shadows of gnomons (one of Clavius's favorite subjects) and the angular size of zodiacal signs would not always be symmetrical and equal, respectively, if the earth were not centered in the plane of the ecliptic.[27] These two preceding arguments are variations of those found in Ptolemy, as is the assertion (also employed by Clavius) that if the earth were not at the center of the cosmos, then lunar eclipses would not always occur with the sun and moon diametrically opposite in the sky (see fig. 17).

Only after nearly eight pages of observational arguments for the centrality of the earth does Clavius state, almost as an afterthought, physical reasons. For instance, according to Aristotle, because the earth is the heaviest body it tends toward the lowest possible place, the place farthest from the firmament, namely, the center of the cosmos. He closes his discussion of the earth's centrality on this note: "And indeed quite rightly has Nature placed the earth at the center of the cosmos, since such a worthless and crude body [as Earth] ought to be uniformly separated from all parts of the heavens, which is a body of highest excellence."[28] In this entire discussion Clavius has not mentioned or even alluded to Copernicus.

From that final expression of contempt for terrestrial matter, uttered while idealizing the celestial, it seems likely that Clavius's objections to Copernicanism—to placing the earth in the heavens—have deeper foundations than the merely mathematical or philosophical. Other historians have made the retrospective observation that Copernicus's stationing of the corruptible earth in the midst of the incorruptible heavens constituted an objection to his theory, but this disparaging prejudice against the earth itself was also evident to contemporary Copernican writers. John Wilkins, for example, while listing the reasons of anti-Copernicans for the earth's centrality, wrote that they argue "first, from the vileness of our earth, because it consists of a more sordid and base matter than any other part of the world; and therefore must be situated in the centre, which is the worst place, and at the greatest distance from those purer incorruptible bodies, the heavens."[29] Wilkins specifically notes Clavius as one of the principal spokesmen among the anti-Copernicans. Citing Wilkins, Arthur O. Lovejoy concluded, "It is sufficiently evident that . . . the geocentric cosmography served rather for man's humiliation than for his exaltation, and that Copernicanism was opposed partly on the ground that it assigned too dignified and lofty a position to his dwelling-place."[30] Indeed, Clavius considered it one of astronomy's greatest virtues that it raises men's minds to the contemplation of higher things, as he stated in the first sentence of the first edition of his *Sphaera*.[31] So we can understand that the resistance to placing the earth in the heavens ran much deeper than lack of imagination or rigid philosophical conviction—it tapped a deep prejudice concerning humanity's lowly status in the cosmos.

The Immobility of the Earth

In the question of the immobility of the earth, Clavius had to rely almost entirely on arguments from physics followed by arguments based on authority—in particular scriptural authority. Naturally he first discusses Sacrobosco's argument, which is that the earth cannot move because if it were moved away from the center of the cosmos that would necessarily be motion toward the heavens, that is, an ascent. But to ascend is contrary to the nature of the earthy element, therefore the earth cannot be moved from the center of the cosmos. Clavius observes that Sacrobosco's argument only rules out a natural, but not a violent, rectilinear motion of the earth. But a violent rectilinear motion is also impossible, because to move the earth by violence would require that it be displaced by some object of even greater heaviness, which would necessarily impell the earth with great speed. If this happened, "who cannot see that lighter bodies, such as the leaves of trees, chaff, and all other such things ought to be left behind in the air, for they could not follow [the earth's] very swift motion?"[32]

Clavius observed, further, that Sacrobosco's argument does not rule out a rotation of the earth, either about the north-south axis or about any other axis. The earth cannot rotate about the "axis of the world" (the north-south axis) in either sense (eastward or westward) because, "if the earth rotated about the axis of the world, it would come about that clouds, birds, and all things of the air would be seen to move in the opposite direction . . . thus it cannot follow that the earth could have such a swift motion as to be able to go around in twenty-four hours. Nor indeed could the air itself be carried around so swiftly with the earth, since it would flow about, hither and thither, just as if moved about this way and that by various winds, as everyday experience teaches us."[33] This line of argument, which cites the disturbances of the air that would result from the earth's rotation, is far from original. Clavius undoubtedly had it at hand in chapter 7, book 1 of the *Almagest*.[34] Clavius chose to portray the consequences of a twenty-four-hour rotation of the earth in much more dramatic terms than had Ptolemy—in order to highlight their absurdity, no doubt. In particular, no building could remain standing, and all would collapse in ruins.[35] Nor could he accept the argument that the very speed of the motion would prevent this from happening just as it prevents water from escaping from a vessel that is swung in a circle very fast, for this happens only if the impetus of the water is toward the bottom part of the vessel and not toward its opening. Indeed the impetus imposed on buildings would be away from the earth, and thus they could hardly remain standing, just as no water could remain in a vessel being swung in a circle if it opened outward from the circle.[36] Clavius's apocalyptic vision of collapsing buildings is not a recitation of anything found in Ptolemy or Sacrobosco. He may have been reacting, however, to Copernicus (whom he still does not mention here), who had denied in chapter 8, book 1, of *De revolutionibus* that the earth's diurnal rotation would violently disrupt all things on the surface.

"For the same reason," Clavius continues, "a stone or arrow projected straight upward with great force would not fall back to the same place, just as we see happen on a swiftly moving ship."[37] It is not clear whether by "for the same reason" (*pari ratione*) Clavius intends merely the hypothetical motion of the earth or more specifically the same impetus that would reduce buildings to ruins, but he probably intended the former, more general, sense. His clear, if mistaken, perception of the behavior of projectiles launched from a moving ship is notable since that very thought experiment would, nearly fifty years after he first printed it, play a prominent role in Galileo's debates over the motion of projectiles in moving reference frames. Galileo also employed the example of the vessel of water swung about in a circle.[38] Clavius, in fact, may have been the source of both arguments, either directly from his well-known *Sphaera* itself or additionally through the Aristotelian commentaries of the Jesuit philosopher Bartolomaeo Amico, whose source was probably also Clavius's *Sphaera*.[39]

Up to this point Clavius's case for the immobility of the earth has been based entirely on physical reasons, but he next introduces two arguments that result in astronomically observable consequences—just the kind of argument that Clavius, the astronomer, would prefer over arguments strictly from physics. If the earth should move, he says, then this motion must be either natural or preternatural (*praeter naturam*). It cannot be natural since a single, simple body is suited only to a single, simple motion, and we already know that the simple motion suited to the elements is rectilinear and directed toward or away from the center of the cosmos. "Nor can [the earth] be rotated preternaturally, namely, according to the motion of the heavens; for then we would always see the same part of the sky overhead and the stars would neither rise nor set—which is absurd."[40] By this preternatural rotation Clavius seems to intend a motion imposed and sustained by some motive power beyond the earth. The only possibility he sees for such a power is the same power that moves the concentric spheres of the planets in their diurnal rotations, the *motus raptus*. The line of argument must be (though Clavius does not state it explicitly) that since the surface of the earth is, like the complete spheres of the planets, concentric with the center of the cosmos, the *motus raptus* will move the earth (if it moves at all) in a diurnal rotation with the same period as the rotational period of the fixed stars. But the consequences of this would be manifest, as he notes, and so we see that the earth can move neither naturally nor otherwise.

Clavius has directed all of his arguments so far against the earth's rotation on the north-south axis defined by the celestial poles. But what of a rotation on another axis? Here, once again, easily observable astronomical consequences would follow. That is, in a given city the altitude of the north celestial pole above the horizon would vary as the off-axis rotation causes the city to approach and recede from the celestial pole. (In effect, the celestial poles would exhibit diurnal paths.) But this is false, he says, since we see at Rome, for example, that the north celestial pole has perpetually the same elevation above the horizon.[41]

Neither of these latter two arguments, which are based on astronomically observable consequences, is at all compelling. Even in their sixteenth-century context they only make sense if we already believe in the earth's immobility. In fact, compared to the many plausible astronomical arguments available for the centrality of the earth, the selection for the immobility of the earth seems rather impoverished. Given Clavius's extravagant claims for the power and utility of mathematical sciences, and astronomy in particular, the weakness of the astronomical arguments for this important tenet might be disquieting to the careful reader. Nevertheless Clavius had to conclude the scientific arguments for the immobility of the earth with the better part of his case resting on reasons from physics, not mathematics. "We conclude, therefore, along with the common opinion of astronomers and philosophers, that the earth is free of any motion, rectilinear or circular, but the heavens themselves are continually carried around the earth. [We conclude thus] chiefly because it is then much easier to defend all the phenomena, and no discrepancies follow from it."[42]

Though he may have exhausted his supply of physical and astronomical arguments, Clavius is far from done. On the question of the immobility of the earth there is a good supply of scriptural passages that can reinforce the less-than-robust scientific case for the earth's immobility.

> The meanings of the Scriptures affirm in many places that the earth is immobile and attest that the sun and the rest of the stars move. Thus we read in Psalm 103 [v. 5], *Which has founded the earth upon the stabilitie thereof: it shal not be inclined for ever and ever.* And again in Ecclesiastes 1 [vv. 4–6], *The earth standeth for ever. The sunne riseth, and goeth downe, and returneth to his place: and there rising againe, compasseth by the South, and bendeth to the North.* What could be clearer? And indeed quite clear is the testimony presented to us by Psalm 18 [vv. 6–7] where we read that the sun moves: *He put his tabernacle in the sunne: & himself as a bridgrome coming forth of his bridechamber. He hath rejoyced as a giant to runne the way, his comming forth from the toppe of heaven: And his recourse even to the toppe therof: neither is there that can hide him selfe from his heate.* And again, it is recounted among the miracles that God sometimes causes the sun to go backward and then go ahead as it is used to do.[43]

Those words bring to a close Clavius's case for the immobility of the earth. As in the arguments for the earth's centrality, he has nowhere mentioned the name of Copernicus, who is thus, at best, an indirect target by virtue of Clavius's statement elsewhere that Copernicus's views would be "shattered by many observations and the common opinion of philosophers" and the promise that he would "demonstrate later by many observations and phenomena" the centrality and stability of the earth.[44] It is especially interesting to note that Clavius does not here use scriptural testimony directly against Copernicus, as some had before and as still others would later during Galileo's troubles. In fact, one of the earliest responses to Copernicus, possibly as early as 1546, was written by Giovanni Maria Tolosani and rests almost entirely on scriptural objections. In that treatise one can find all the scriptural quotations used by Clavius, and many more, arrayed in a withering attack on the Polish astronomer.[45] It is significant that Clavius (whether intentionally or not) isolates Copernicus from scriptural arguments. Consequently there is no suggesting that Copernicus's astronomical propositions are improper or heterodox in any religious sense. They may contradict Scripture, but Clavius does not portray that, in itself, as harmful to the faith.

Having made the case for the immobility of the earth, Clavius approached the conclusion of this section by considering briefly the cause of that immobility. This discussion is a conspicuous excursion beyond the necessary scope of an astronomical textbook and into the disciplinary territory proper to physics. Clavius could have omitted the consideration of physical causes, but only at the cost of leaving incomplete his pedagogical program, namely, the establishment of the geocentric and geostatic foundations of Ptolemaic cosmology. He plunges ahead without apology.

Clavius recited first, as he was accustomed to do, several false causes proposed by philosophers—in this case all pre-Socratic Greeks. He dismissed the theory of Thales that the earth floats on water and the idea, which he attributed to Anaxagoras and Democritus, that the shape of the earth allows it to be held in place by air. He further rejected the idea of Anaximander that the earth is stationary because its tendency to move is equal in all directions, and thus it goes nowhere. This last proposed cause cannot be true, Clavius says, because a body cannot simultaneously tend to move in all directions. Besides, if this were the case, the earth's state would be maintained not naturally, but violently. In addition, we know that the natural tendency of parts of the earth, when unimpeded, is to move toward the center, not toward the heavens.[46]

Why, then, does the earth remain tranquil in the center of the cosmos? "There is no other reason," he says, "than that of its heaviness. It rests always in the lowest place, farthest from the heavens, namely, the center of the cosmos, and once situated there it cannot be displaced naturally, because any tendency to ascend is against its nature.''[47] As an example of the tendency of heavy bodies to come to rest at the center of the cosmos (which is the center of the earth), Clavius suggests a thought experiment: ''Suppose that a hole perforated the entire earth from one side to the other, and some heavy object were dropped into that hole. The object would reach its greatest impetus only at the center [of the earth], but [the impetus] would not then be toward the other side [away from the center], because then the object would begin to ascend, on account of the impetus of motion, as if to the beginning. It would then oscillate back and forth for a while until, having gradually lost its motive impetus, it would come to rest in the center.''[48]

This thought experiment is not at all original with Clavius. It comes from medieval treatises on the behavior of bodies in motion and is particularly associated with the fourteenth-century scholars of Merton College, Oxford.[49] In their hands, however, it was more of a tool for considering the limits to the natural motion of a body and the relationship of the parts of the moving body to the whole than for making any cosmological points. Clavius seems to have in mind a kind of medieval impetus that wears off with the passage of time, becoming ever less able to force the heavy body from its natural place. In this respect his treatment and conclusion is similar to that of Nicole Oresme, who, in considering the same situation, used his impetus theory to conclude that the body would oscillate about the center of the earth.[50]

It is curious that Clavius felt it important to include this hypothetical experiment but then left it without further comment (he goes immediately on to quote some verses from Manilius mentioning the immobility of the earth). Possibly he thought it a particularly striking example because of the intriguing concept of a hole passing completely through the earth. The idea does, however, also seem to echo his earlier statement that the earth is at the center of the cosmos ''and once situated there it cannot be displaced naturally.'' ''Once situated there'' might imply that, like the heavy object, the earth or its parts were previously somewhere else. Perhaps the idea that the earth and its parts, like the heavy object, move naturally to their proper place allows for an explanation of how God might have brought order out of the primordial chaos. However, Clavius nowhere shows a direct concern with cosmogony—that being a subject for theologians and not mathematicians.

Clavius has argued, then, that the centrality of the earth is well established by many astronomical arguments. The immobility of the earth is established by many arguments from physics and is the common opinion of astronomers, philosophers, and the Scriptures. But in none of his arguments has the heliocentric cosmology, Copernicus, or even Aristarchus been a target. The doctrines of the centrality and immobility of the earth are not, for Clavius, issues over which disputation is necessary in the same way that it is over questions like the philosophical legitimacy of eccentrics and epicycles or the number and order of celestial spheres. These latter questions have more than one side—there are theoretical assertions and alternatives that must be evaluated and criticized repeatedly. But the centrality and immobility of the earth called for a different, more positive approach that required no counterattack against rival theories.

The *Disputatio*: Grounds for Debate

Clavius's most insightful critique of Copernicus appears in the *Sphaera* in a completely different context from the general cosmological consider- ations that dominate chapter 1. It arises, instead, in chapter 4, which Sacrobosco had devoted to the theories of solar, lunar, and planetary motions and to eclipses of both kinds. Sacrobosco's treatment of these topics, however, was superficial, and its weaknesses gave rise to supple- mentary genres (the *Theorica planetarum* and *Canones* texts, for instance) almost as soon as Sacrobosco's ink was dry.[51] Clavius acknowledged Sacrobosco's shortcomings at the very beginning of his commentary on chapter 4:

> In the final chapter of this work [Sacrobosco] discusses the motion of other heavens [those of the planets], which go from west to east, and especially the motion of the sun and moon, so that he might show us the reasons for solar and lunar eclipses. But since he touches all these things very briefly, so we also will be very brief in this part, especially since this subject, if it is to be treated as it ought to be, demands longer discussion and is pertinent to the *Theorica planetarum,* which, God willing, we will bring to light soon.[52]

Having deferred the treatise on planetary theory, Clavius could instead devote his *disputatio* to the dominant theme in his *Sphaera*—the defense of Ptolemaic orthodoxy.

> I think it will be valuable (so that I may satisfy those who eagerly demand it of me) if here briefly I may draw together various observa- tions by which Ptolemy, Alphraganus, Thābit, and almost all other astronomers were strongly persuaded to believe that there are eccen- tric orbs and epicycles in the heavens. Then I will set out and try to destroy all of the strongest arguments of Averroës and his follow- ers with which they assail these kinds of orbs. Third and finally, I

will destroy their arguments and show them to be worthless, so that anyone may understand that the astronomers postulated these spheres in the sky not without reason, but with great diligence and incredible proficiency; and that the philosophers who follow Averroës, after attacking those orbs so much, have done so rashly.[53]

It would be most interesting to know who "eagerly demanded" Clavius's views on the status of eccentrics and epicycles. Clavius's reputation as an astronomical authority became widely established only in the wake of the 1582 Gregorian calendar reform. So since the *disputatio* appeared no later than the 1581 revision (*nunc iterum*), it is unlikely that such demands would have come from very far beyond Clavius's immediate circle.[54] On the other hand, this is just the sort of issue that would have arisen in the Collegio Romano, given the Averroist inclinations (in cosmological matters, at least) of Pereira and like-minded Jesuit philosophers. The contrasts between the views of Clavius and those of Pereira would have been obvious to the students, who were required to take courses from both. As the Society's first mathematical and astronomical authority, Clavius was the natural spokesman in those early days for those who were sympathetic to the search for mathematical causes of astronomical phenomena, and he would almost surely have been asked to publish his views. It is probable, therefore, that those who petitioned Clavius for his views on these matters were his students and associates at the Collegio Romano, who wanted his arguments written up and published. The students of Clavius who achieved any degree of notoriety as mathematicians (such as Maelcote, Grienberger, and Grassi) studied at the Collegio Romano in the 1590s or later.[55] But the anonymous petitioners probably came from among those present at the Collegio Romano in the 1570s. Thus Clavius's statement, coming in 1581 or earlier, is one of the earliest signs of the growing interest in mathematical science among the Roman Jesuits.

The Methodological Foundation of Astronomy

The *disputatio* in the fourth chapter of the *Sphaera* is the very treatise on eccentrics and epicycles that had been requested from Clavius. The arguments of the Averroists and Clavius's responses have already been discussed in some detail in chapter 3, above. They fit into the *disputatio* as a whole along with Clavius's responses to Fracastoro, Copernicus, and the other cosmological heterodoxies. Unique among the rivals is Copernicus, who enters the discussion primarily as an example of a mistaken theory arising from methodological error.

Clavius entered the methodological arena in the course of his defense of the Ptolemaic planetary constructions. The *disputatio* offers, in addition to refutations of the arguments of "the adversaries," three arguments in favor of eccentrics and epicycles. The first presents Clavius's dynamical distinction between the eccentric and concentric spheres and the corresponding distinction between the proper and diurnal motions of the planets. The second argues that the planets either move themselves (like fish or birds) or they are moved in concentric spheres or they are moved in eccentric spheres. Since he has soundly refuted the first two alternatives, the third must be true.

The methodological argument is his third and final defense of Ptolemaic mechanisms.[56] In beginning this lengthy discussion, Clavius asserted clearly that astronomy and natural philosophy proceed in the same way from observations to knowledge of causes.

> Now finally the proposition may be concluded. Just as in natural philosophy we may arrive at knowledge of causes by their effects, so it is indeed in astronomy, in which what happens to the celestial bodies goes on far from us. It is necessary that we come to knowledge of them, of their disposition, and of their composition through their effects, that is, from the motions of the stars perceived through our senses. Just as natural philosophers, along with Aristotle, inferred from the mutual generation and corruption of natural things [the existence of] prime matter as well as two other principles of natural transmutation[57] and also many other things, so even have astronomers, through the various kinds of celestial motion from east to west and west to east, sought out the exact number of celestial spheres. Some say eight because they recognized eight such different kinds of motion, still others say ten from the ten different kinds of motion they found.[58]

It is at least inconsistent, perhaps even hypocritical, that Clavius admits some differences of opinion among astronomers on the number of celestial spheres while elsewhere he blasted the homocentrists for their inability to agree on the number and arrangement of the spheres in their theories. But his point is that astronomers infer the number of celestial spheres from the kinds of motion they observe. Thus, astronomical phenomena should guide the development of astronomical theory. This is in contrast to the homocentrists who abandon the phenomena in single-minded pursuit of philosophical purity. He compares the position of the astronomer, who infers the existence of invisible eccentrics and epicycles from the motions of the planets, to the position of the natural philosopher, who infers the existence of unobservable prime matter from observations of generation and corruption in the elemental world. Given that some of his chief adversaries were natural philosophers, this is a useful stance, and he will exploit it a bit later. He continues, "Again by the same reasoning, through other phenomena, [astronomers] have established the order of the celestial spheres, as was fully explained in chapter 1. It is appropriate and most reasonable that from the particular motions of the planets and various appearances astronomers investigate the number of partial orbs, by which the planets are led around in such various motions, and also [investigate] their structure and shapes on the condition that the causes of all the motions and appearances be assigned correctly and that nothing absurd, which conflicts with natural philosophy, can be inferred."[59] Here Clavius broadens the basis of his case somewhat by referring to his explanation of the Ptolemaic order of the celestial spheres, a doctrine that he considers very well founded. The same method that leads to the well-established order of the spheres allows the astronomer, he argues, to investigate the components, the "partial orbs," that make Ptolemaic cosmology both mathematically workable and physically plausible.

In concluding this methodological sketch, Clavius reiterates the two criteria that eccentrics and epicycles must meet in order to be philosophically and mathematically acceptable:

> Therefore when eccentric orbs and epicycles are such that by them astronomers can account for all phenomena easily (as is clear in part from what has been said and in part will be explained more clearly in the *Theorica*), and when nothing follows from them that is absurd or incorrect in natural philosophy, as will soon be established by the solutions of the arguments that the adversaries of these kinds of orbs are wont to bring up, then have astronomers rightly declared that planets are conveyed in eccentric orbs and epicycles and not in concentric orbs, since by [the latter] we cannot explain such a great variety of planetary motions.[60]

Clavius does not generalize beyond the Ptolemaic constructions, so we cannot, without some reservations, assume that he would accept these same criteria as sufficient tests of the suitability of other astronomical theories. It seems reasonable, however, that he would have considered them necessary criteria for judging the suitability of other theoretical explanations for astronomical phenomena.

He does not elaborate on the meaning of his first criterion, that a theory should defend—that is, explain—the phenomena without strain (*nullo labore*). It may just mean that the theory should yield its results naturally and without ad hoc reasoning, which was precisely one of the problems

he identified with the homocentric theories. On the other hand, his reference, once again, to his never-published *Theorica planetarum* text may
indicate a more practical concern, that is, that the theory should have
an explicit and complete technical underpinning. That kind of technical
foundation (which for Ptolemaic theory comprised the *Theorica planetarum,* astronomical tables, and ultimately the *Almagest* itself) was lacking in the case of both the homocentric and the fluid-heaven models. But
the Copernican theory, which Clavius ignores here, did not suffer this
imperfection.

The second criterion, that nothing absurd or incorrect in natural philosophy should follow from the theory, is one of Clavius's major themes in
the *disputatio.* If applied to the rival theories it would have no impact on
homocentrics because, if anything, thay are even more compatible with
natural philosophy than the eccentrics and epicycles. This standard would,
however, seem to reject fluid-heaven theories, since Clavius elsewhere
argued that planets moving like "fish in water or birds in air" would
imply certain absurdities such as a single body moving in two directions
at once. It would also, of course, eliminate the Copernican theory, since
the earth's motion involves many absurdities, as Clavius has pointed out.

Now Clavius is ready to present the methodological criticism that he
must counter in order to establish the epistemological legitimacy of the
theories of Ptolemaic astronomers.

> Yet the adversaries try to weaken this argument by conceding that,
> having postulated eccentric orbs and epicycles, all phenomena can
> be explained; but [they say] it does not follow from this that the
> stated orbs are found in nature, since they may all be fictitous. [They
> argue] that on the one hand perhaps all the appearances can be more
> easily explained in a way that may now be unknown to us or, on
> the other hand, it may be that while the stated orbs do indeed account
> for the appearances, they may still be very much fictitious and by
> no means the true cause of those appearances, just as from false
> [premises] one may reach true [conclusions], as Aristotle established
> in his Logic.[61]

Clavius is about to address these criticisms in some detail, but first he
voluntarily adds fuel to the rhetorical fire of his adversaries so that its
later extinction will be all the more impressive.

> To these claims we may add confirmation in this way. Nicholas
> Copernicus in the work *De revolutionibus orbium caelestium* ex-
> plains all the phenomena in another way, proposing, namely, that

the firmament is motionless and fixed and that the sun stands at the center of the universe, and he assigns to the earth, which is in the third heaven, a threefold motion, etc. By these means eccentrics and epicycles are not necessary to maintain the planetary phenomena. Ptolemy, on the other hand, using the epicycle, attributes the cause of all appearances to the sun, [the phenomena of] which he defends by an eccentric. Therefore it cannot be concluded from our third argument that the sun is moved in an eccentric, because perhaps it is carried in an epicycle.[62]

Though there are actually two loosely linked arguments in this passage, his main point is obvious: the adversaries could argue that there are many different ways to save a given set of phenomena, and here are two examples (Ptolemy and Copernicus). The adversaries would then say that since both theories save the phenomena equally well, one cannot decide between them. Therefore we cannot claim that one corresponds to reality any better than the other. In building the case for the adversaries, the first argument offers the example of Copernicus, for whom "eccentrics and epicycles are not necessary to maintain the planetary phenomena." In fact, of course, Copernicus did employ eccentrics and epicycles. But Clavius's statement here is more hyperbole than misrepresentation, because he means to emphasize the major planetary epicycles of the Ptolemaic models, which were indeed unnecessary in Copernicus's theory and were one of the most prominent points of criticism by the "adversaries." He gives us two very different theories: one assigns each planet a major epicycle and the other does not, yet both explain the phenomena equally well.

The second argument is tacked on the very end almost as an afterthought: "because perhaps [the sun] is carried in an epicycle." This refers to the well-known fact that Ptolemaic solar theory can employ an eccentric or an epicycle and work just as well either way—the same problem as in the Copernicus/Ptolemy question. It is an important point for skeptical critics, because those, like Clavius, who would claim that the Ptolemaic constructions are real must be able to explain which alternative is the real one—solar eccentric or solar epicycle. Here Clavius has acknowledged the problem but kept its profile low. This is wise because the solution he will shortly give is rather weak.

Clavius's remark that "Ptolemy . . . using the epicycle, attributes the cause of all appearances to the sun" calls for a brief comment. He is referring to the feature of the Ptolemaic planetary models that each has some kind of linkage to the sun. In the superior planets, the period of revolution of each major planetary epicycle is exactly equal to the period of revolution of the sun around the earth. In the inferior planets, it is the period of revolution of the planetary eccentric deferent that is equal to the solar period, though Clavius does not mention that here. This would seem to be what he means by "attributes the cause . . . to the sun." However, Clavius's Latin for this statement is, "Ptolemaeus per epicyclum reddit omnium apparentiarum causam in Sole," and the verb *reddere* also can mean "to reflect" or "to restore to view." The implication would be that by the epicycles Ptolemy revealed that the cause of all the appearances is in the sun.

This reading has an interesting aspect, namely, that Clavius saw the solar factors in the Ptolemaic planetary models as indicating a cause (the sun) of the planetary motions—a cause that would be lost in the Copernican arrangement. In the search for a causal astronomy, relinquishing a well-established cause would be a definite step backward and, in Clavius's likely estimation, an important disadvantage of the Copernican theory. This interpretation finds support in the fact that when discussing the order of the planets, Clavius listed the common solar factor in the inferior and superior planets as a reason why the sun's position is between those two groups of planets. He was certainly impressed with the causal significance of this arrangement.

Clavius has now finished strengthening his adversaries' skeptical argument against the reality of the Ptolemaic constructions, and he is ready to reply.

> Nevertheless, our third argument remains sound, and the response of the adversaries proves nothing. First, if they have a more suitable way, they should show it to us, and we will be content and much in their debt. Indeed, astronomers support no other [theory] as either explaining all celestial phenomena or explaining them in a more satisfactory manner; whether this is done by eccentric orbs and epicycles or in another way. And since no more suitable way has yet been discovered than that which defends everything by eccentrics and epicycles, it is quite believable that the celestial spheres are made of orbs of this kind. Thus, if they cannot show us a better way, and if they do not want to destroy utterly so much natural philosophy, which is passed on in the schools, but would rather contribute to all the other arts that inquire into causes by means of effects, surely they ought to acquiesce to this way, [which is] assembled from such a variety of phenomena.[63]

Clavius here makes the understated but interesting historical argument that eccentrics and epicycles are the more plausible "since no more suitable way has yet been discovered.'' This argument seems to assume that, given enough time, some astronomer will find (or will have found) the real causes of celestial phenomena. Clavius was well aware of the long and reasonably continuous history of cosmological speculations in Western thought, and he believed, it would appear, that there had been enough time for all the useful alternatives to have been invented and evaluated. His view of the history of astronomy as one of exhaustion of alternatives gives added significance to his characterization of Copernicus as a follower of Aristarchus. For Clavius, Copernicus's theories were not really a novelty—they were a revival of a suggestion that had already been invented and found wanting in antiquity.

Clavius concluded the passage above with the assertion that rejecting the methods that affirm the astronomy of eccentrics and epicycles will destroy, no less, all scholastic natural philosophy. This is because, like mathematical astronomy, these other arts achieve knowledge of causes by effects. He goes on to explain,

> Whenever someone adduces some cause from evident effects, I will state absolutely that doubtless another cause unknown to us could perhaps produce those effects. . . . If, on that account, it is incorrect to adduce eccentrics and epicycles from appearances, because from falsehood one can attain truth, then it will ruin universal natural philosophy. For by this means, when someone concludes from an observed effect that this or that is the cause of it, I can say indeed it is not, since from the false one can derive the true—and thus all natural principles discovered by philosophers are destroyed. But that would be absurd, and the force and strength of our argument does not seem to be weakened by the adversaries.[64]

With good reason, Clavius claims that the same methods that lead natural philosophers to knowledge of the principles of physics also lead astronomers to knowledge of eccentrics and epicycles—that is, natural philosophers, like astronomers, reason from observable effects to unobservable causes and principles. Since his most important adversaries are natural philosophers, this defense causes them to appear to be undermining their own position. This stance also puts astronomy and physics on a par as disciplines and thus reaffirms another plank of Clavius's program—the establishment of the mathematical sciences as peers of the philosophical ones.

"Natural principles [that is, causes] discovered by philosophers" must be established by inferring their existence and nature from their effects, because we can only see the effects, not the causes. So given two equivalent causes, each of which explains the phenomena equally well (solar eccentric or solar epicycle, e.g.), how do we decide which one is the real cause?

This particular methodological issue was a major point of contention in sixteenth-century astronomy, and a variety of positions were taken on it.[65] Clavius's adversaries—Averröes, Nifo, Achillini, and the like—were skeptical realists. They did not deny the possibility of learning the true causes of celestial phenomena, but they did deny that the Ptolemaic constructions embodied the true causes. For this group, the motions implied by eccentrics and epicycles departed too much from strict Aristotelian physics to be judged real—the real causes had just not been discovered yet.[66] A small number of those holding this opinion—Fracastoro and G. B. Amico are the primary examples—actually attempted to construct "real" alternatives to the fictitious Ptolemaic constructions.

Some were much more skeptical, doubting even the possibility of ever knowing the real causes of celestial phenomena. Jardine has identified a small but varied group that entertained opinions of this order. It included Giovanni Pontano, Peter Ramus, Nicodemus Frischlin, and Nicolai Reymers Ursus (Nicolai Baer). Pereira, though closely associated with the Averroist viewpoint, occasionally makes statements that seem to cast doubt on the very possibility of a causal astronomy, which would admit him to the more skeptical group.[67]

The realists, like Clavius, Ptolemy, Copernicus, and later Kepler, were convinced that the phenomena can make a compelling case for the knowability of celestial causes. But to make that case, Clavius had to explain what was wrong with the criticism brought by his adversaries.

Indeed it should be said that the rule of the logicians, that *the true follows from the false,* is not to the point because it is in one way that truth is inferred from the false and in another way that phenomena are defended using eccentrics and epicycles. For in the former it is from the power of the syllogistic form that truth is derived from falsehood. Whence knowing the truth of some proposition, false premises can be arranged in such a form that by necessity, from the power of the syllogism, that true proposition is concluded. Thus, because I know that the animal is sensitive, I can concoct such a syllogism: All plants are sensitive, all animals are plants, therefore all animals are sensitive. But if I were to doubt anything concerning the conclusion, I could never acquire certainty of it from false premises even if it should be rightly concluded from the power of the syllogism, because I could easily prove anything in this way. Thus

if I doubt that all stars are round, I can justly infer it from the power
of this syllogism: All stones are round, all stars are stones, therefore
all stars are round. Yet never would I be made certain with respect
to my doubts about the aforesaid conclusion.[68]

So false premises can indeed lead to a true conclusion. But, Clavius says,
such a syllogism provides no certainty, because absolutely any conclusion
can be proved in that fashion. And this is just what the adversaries would
argue is the problem with eccentrics and epicycles. Clavius responds that
there is an important difference between the two cases—false premises
may indeed lead to a true conclusion,

> but from eccentric orbs and epicycles, not only the appearances of
> past things already known are defended, but also future things are
> predicted, the time of which is completely unknown. So if I were
> to doubt whether, for example, at the full moon of September 1587
> there will be a lunar eclipse, I can be sure from the motions of the
> eccentric orbs and epicycles that there will be an eclipse, so I would
> doubt no more. Indeed, from these motions I know at which hour
> the eclipse will begin and how much of the moon will be obscured.
> And in the same way, all eclipses, both solar and lunar, can be
> predicted and also their times and magnitudes; even though they
> have no particular order among themselves, such that a determinate
> time interval lies between two successive eclipses, but sometimes
> in a year two occur, sometimes one, sometimes none.[69]

Thus, the key is predictive power—the ability of the theory to yield something previously unknown. A false syllogism depends on prior knowledge
for its construction, like the prior knowledge that animals are sensitive or
stars round. But such a false syllogism reveals no more after its formulation than was known beforehand. Clavius treats eccentrics and epicycles
as premises of mathematical arguments that result in explanations and
predictions of astronomical phenomena. As such, eccentrics and epicycles
are not at all like the premises of a false syllogism. They not only explain
what we already knew—that is, they save the phenomena—but they can
predict things (in particular, the times and circumstances of future events)
that we can know in no other way.

To drive his point home, Clavius indulges in some rhetorical exaggera-
tion. In fact, he says, the astronomer can predict so well how the heavens
will behave in the future that if one did not know that he employed
eccentrics and epicycles, the uninitiated might think that the astronomer
were in some sense controlling the heavens himself. "It is not believable,

however, that we oblige the heavens to obey our fictions and move as we will, or as conforms to our principles. However we do seem to compel them if eccentrics and epicycles are fictions, as our adversaries maintain.''[70] Far from compelling the heavens to move as they wish, astronomers do the very opposite: they seek by their calculations to understand better the celestial motions they observe and record. But this work is arduous and long.

> This truly pertains to Nicholas Copernicus. He does not reject eccentrics and epicycles as fictitious and repugnant to philosophy. Indeed he supposes the earth itself to be like an epicycle, and in the moon he puts an epicycle on an epicycle. But [he does] this only with the intention of correcting the periods of the motions of the planets, which he has now discovered to be defective. It is very difficult to define the periods of motion in this way (so that they will not deviate from the truth over the course of many years), since no mortal has ever been able to determine the period of a single planet so that a few minutes are not either lacking or in excess—discrepancies which, in an interval of many years, lead to noticeable error.[71]

Clavius's statement that Copernicus worked ''only with the intention to correct the periods of the motions of the planets'' seems to indicate that Clavius read Copernicus in the spirit of Andreas Osiander's preface to *De revolutionibus,* in the understanding, that is, that his astronomical theories were created only with the intention of saving and predicting the phenomena but cannot be expected to reveal the real workings of the heavens. Did Clavius accept Osiander's preface as an expression of Copernicus's opinion? Baldini has pointed out that if he did, it would seem to conflict with Clavius's continual concern to reject the Copernican (or indeed any) motions of the earth as physically absurd.[72] Why, Baldini asks, would Clavius bother to reject a mere mathematical hypothesis as physically absurd? Moreover, Baldini argues, Clavius's colleague Bellarmine and student Grienberger both clearly interpreted the Copernican hypotheses as a physical theory, thus implicitly recognizing the contradiction between Osiander's preface and the rest of Copernicus's work. Baldini judges that ''one can conclude that if Bellarmine and his associates in the Collegio Romano did not, perhaps, take note of the *historical* question of the paternity of Osiander's preface, this does not mean that the *epistemological* question of its conformity with the logical structure of the work evaded them.''[73]

This view may well be applicable to Clavius, too, but another is possible. We must remember that Clavius's *Sphaera* is partly a polemical work of cosmological defense. Regardless of Copernicus's actual intentions, his theory, if interpreted cosmologically, would be a threat to Ptolemaic cosmology. Therefore, in the light of Clavius's statement above, it is possible that he could have read Osiander's preface as if it expressed Copernicus's real intent, yet replied to the theory as if it were a cosmological alternative—which it was. Clavius's partitioning, so to speak, of Copernican interpretations would also be consistent with his generous praise for Copernicus as the restorer of astronomy despite his rejection of the Copernican cosmology. In any case, Clavius treated the Copernican theories as if they had far broader consequences than a mere set of corrections to astronomical parameters.

The Methodological Destruction of Copernicus

At this juncture, Clavius has completed his defense of eccentrics and epicycles against the methodological criticism—the skeptical argument, in Jardine's terms—and is ready to turn that criticism on Copernicus. Here, Clavius will argue, is a case in which a false syllogism, depending on things already known and revealing nothing new, has been masquerading as real science.

> That Copernicus preserved the phenomena in another way [than Ptolemy] is not surprising, because by the motions of the eccentrics and epicycles he knew the time, quantity, and quality of the appearances, both future and past. And, as he was very ingenious, he could think up a new way by which those appearances could be more easily (so he thought) explained, and the periods of motion, which he had now found to be defective, corrected by some amount. This seems to have been his principal endeavor, as we have stated, just as in many syllogisms we can prove some already known conclusions even from false premises.[74]

Whether Clavius would have brought the same criticism against homocentric theories like those of Fracastoro and G. B. Amico is not clear, though it seems that they are subject to the same objection. Like Copernicus, they also had Ptolemaic theories at their disposal to set the standard for the rival theory's performance, past and future. In the sense that they only duplicated the Ptolemaic theory (and poorly at that), the alternatives could reveal nothing new and could offer no certainty—the failings of the false syllogism. Clavius's sense of historical development may explain why he did not apply the methodological criticism to the homocentrists. He knew that homocentrics, as proposed by Eudoxus and Callippus, were historical predecessors of Ptolemy's constructions. As such, they had been created (at least in their earliest form) without prior knowledge of the superior system of eccentrics and epicycles and were thus not susceptible to the charge that their explanatory power (meager though it was) was purely derivative.

Once he revealed the methodological flaw in the Copernican theory, but before he discussed its physical implications, Clavius endeavored to reiterate that eccentrics and epicycles do not suffer even if one accepts the Copernican theory. In this passage we see how far he will go to maintain the basic suitability of the Ptolemaic approach.

> So far off the mark is it that eccentrics and epicycles are destroyed by the doctrine of Copernicus that, on the contrary, they should [for that reason] be supposed. On this account did astronomers devise these orbs: because most certainly they perceived, by various phenomena, that the planets are not always borne at the same distance from Earth. This does Copernicus freely profess since, according to his doctrine, the planets always have unequal distances from the earth, as is clear from his stationing the earth out away from the center of the world in the third heaven. Only this is gathered from his position: that since many phenomena can be defended in another way, it is not completely certain that the structure of eccentrics and epicycles is such as Ptolemy makes it. Nor indeed do we try in this question to persuade the reader of anything else than that the planets are not borne always at equal distances from Earth, indeed either there are eccentric orbs and epicycles in the heavens in the arrangement set up by Ptolemy, or surely some cause of these effects must be posited that is equivalent to eccentrics and epicycles.[75]

"It is not completely certain" (*non esse certum omnino*): here is the core of Clavius's position. Clavius may have been a convinced Ptolemaic, but he was not rigidly dogmatic about it. He says the constructions of Ptolemy's *Almagest* are, in some sense, not certain. What seems to be certain, judging by Clavius's explanations, is that celestial bodies are moved in circular paths and combinations thereof. What is less than certain is the exact arrangement of circles, though some arrangements can be ruled out—homocentrics being the most important case in point. "Most certainly [astronomers] perceived, by various phenomena," that the planets do not revolve about the earth in concentric circles. A major thrust of his argument remains antihomocentric because homocentrics simply cannot save the phenomena. If Ptolemy's arrangement is incorrect, then there must be some alternative arrangement. Clavius is not clear whether "equivalent to eccentrics and epicycles" means some non-Ptolemaic arrangement of eccentrics and epicycles or some equivalent that does not employ eccentrics and epicycles—whatever that might be.

How far can an equivalent go? Is the Copernican theory, despite its methodological transgressions, not an equivalent formulation that success-fully preserves and predicts the phenomena? Clavius admits that it is, though it remains unsatisfactory for other reasons.

> Thus if the position of Copernicus involved no falsities or absurdities there would be great doubt as to which of the two opinions—whether the Ptolemaic or the Copernican—should better be followed as ap-propriate for defending this kind of phenomena. But in fact many absurdities and errors are contained in the Copernican position—as that the earth is not in the center of the firmament and is moved by a threefold motion (which I can hardly understand, because ac-cording to philosophers one simple body ought to have one motion), and moreover that the sun stands at the center of the world and lacks any motion. All of which [assertions] conflict with the common teaching of philosophers and astronomers and also seem to contradict what the Scriptures teach, as we treated more fully in chapter 1.[76]

Clavius's language here is moderate in regard to the testimony of Scrip-ture. This is the only place where he directly apposes the teachings of Copernicus and the statements of Scripture, and even here he speaks only of the appearance of conflict and stops short of any explicit interpretation, much less any theological judgment. Pierre Duhem reads this passage quite differently, however. After quoting the passage above, Duhem claims, "Thus, two conditions come to be imposed on any astronomical hypothesis that would make its entry into science: It may not be *falsa in philosophia*. It may not be *erronea in fide*, nor, a fortiori, *formaliter haeretica*."[77] The first conclusion is accurate; Clavius agreed that an astro-nomical theory could not be *falsa in philosophia*. The second conclusion is quite misleading and wrong.

The terms *erronea in fide* (erroneous in faith) and *formaliter haeretica* (formally heretical) are terms used by canon lawyers and theologians. Not coincidentally, they are exactly the judgments made by the theological consultors of the Holy Office concerning key cosmological tenets of the Copernican cosmology. In 1616, four years after Clavius's death, the Holy Office judged formally heretical the proposition that the sun is the center of the world and completely immovable by local motion. At the same time they judged erroneous in faith the proposition that the earth is neither the center of the cosmos nor immovable but moves as a whole and with a diurnal motion. "Formally heretical" means that an opinion has been found to be directly contrary to a doctrine of faith, and "erroneous in faith" means that an opinion is not in itself contrary to a doctrine of faith, but that it is opposed to a doctrine that pertains to the faith.[78]

The point is that both are theological judgments made by panels of specialists appointed specifically for the purpose as part of the operations of the Holy Office. The terms do not appear in Clavius's *Sphaera* nor do the judgments they represent. Clavius's conclusion is simply that Copernicus's assertions seem to contradict Scripture. For Clavius, whether they constitute a threat to Catholicism would be a matter of interpretation and thus a matter for theologians, not astronomers.

Clavius maintains his distance from theology despite his theological credentials, for we have seen in several cases that although Clavius will venture an opinion on matters of theology or philosophy, he generally concedes to the specialists the prerogative of authoritative judgments (at least as long as those judgments suit his purposes). He was, for instance, careful to cite the opinion of the theologians on God's ability to create extramundane worlds—even though this was hardly a controversial statement and well within his own theological competence. Thus Duhem's use, in this context, of the Inquisition's terminology is quite misleading. Duhem's implications that Clavius spoke as an ecclesiastical authority and applied theological tests in his evaluation of Copernicus is unfounded, and the insinuation of Clavius's complicity in the Holy Office's condemnation of Copernicus—which took place well after Clavius's death—is absurd.

We should also recall that in the first chapter of Clavius's *Sphaera* (to which he refers in the quotation translated at n. 76) he used scriptural testimony in a positive way—that is, to establish and support the stability of the earth and mobility of the sun—but without using it to attack Copernicus directly. He did, as he has just done above, point out that the Copernican theory is in conflict with the literal reading of certain scriptural passages, yet he never asserted that for that reason the Copernican theory constituted any kind of threat to the faith, which was, on the other hand, the very point of the Inquisition's ruling.

Clavius concludes his discussion of Copernicus by returning to the greater scheme of things yet now speaking in terms of probability rather than certainty. "On that account the opinion of Ptolemy is to be preferred to these inventions of Copernicus. From all of which it is clear that it is just as likely that the existence of eccentrics and epicycles ought to be granted as it is probable that eight or ten moving heavens be granted; since like the number of heavens, the [eccentric and epicyclic] orbs are discovered by astronomers from the phenomena and the motions."[79] In concluding, Clavius has returned to one of those major issues in the

Sphaera and other sixteenth-century astronomy textbooks, namely, the number of celestial spheres in the cosmos, while reiterating the connection between the theory and the phenomena.

Clavius's acceptance of probability in place of certainty is very interesting. Probable reasoning was a common characteristic in Jesuit philosophy. And in fact, Clavius's claims to absolute certainty in astronomical matters are relatively rare and limited. His previous assertions of certainty have focused primarily on certainty of predictions. For instance, he asserted that the Ptolemaic theory made it possible to be certain whether, at a given time, there would be a lunar eclipse. However, he recognizes that this does not translate into certainty about the celestial mechanisms themselves, so we can only choose the more probable. Yet this is not a great problem, for probability is measurable by particular tests that we see him apply in evaluating theories that save the phenomena equally well. One test is proper to the realm of mathematics—and he generally calls it "convenience." This standard requires both economy of explanation (homocentrics fail this test in addition to failing to save the phenomena) and practical utility. In these respects the Copernican and Ptolemaic theories are equally probable. But the Copernican fails the trial by extramathematical criteria, for it contradicts the deep geocentric and geostatic prejudices of the times, the dicta of Aristotelian dynamics, and the statements of Scripture. These failings render the Copernican theory much less probable, and therefore much less acceptable, than the Ptolemaic.

The methodological discussion has so far left one stone unturned, and beneath it is the ambiguity of the solar eccentric/epicycle in Ptolemaic theory. Clavius himself had brought up the issue in the process of fortifying the skeptical argument and had presented it in tandem with the Copernican example as a skeptical objection to the possibility of knowledge of celestial realities. It is only proper that Clavius should resolve the ambiguity, which he himself brought up, before taking leave of the methodological discussion.

We understood from Ptolemy that it is uncertain whether the sun is borne in an eccentric or an epicycle, for he explains solar phenomena by either epicycle or eccentric alone. But whatever is said, it is clear that the sun has varying distances from Earth and is not at all borne in a concentric orb, which is enough for us, as we have stated. However, Ptolemy chooses rather the eccentric orb for the sun, because the eccentric goes around and encircles the center of the earth. But already we have stated and refuted the argument of Averroës and his followers and thus it appears that eccentrics and epicycles are neither monsters nor portents and are in no way repugnant to natural philosophy, as the adversaries falsely think.[80]

One could hardly call this a satisfactory solution to the skeptical argument. In fact, he evades the real issue. Clavius offers no criteria for deciding between the alternative theories—solar eccentric or solar epicycle— offering instead only the vague judgment that a circle that includes the earth (the solar eccentric) is preferable to one that does not (the solar epicycle). Rather than dealing with this nearly insoluble problem (in fact, the two alternatives are geometrically interchangeable), Clavius instead uses the opportunity to return once more to his point that concentrics cannot be accepted in a realist mathematical astronomy.

Clavius's choice of words is interesting. He asserts that "eccentrics and epicycles are neither monsters nor portents" (*eccentricos et epicyclos non esse monstra, aut portenta*). Both *monstrum* and *portentum* imply something that is out of the ordinary course of nature.[81] This is consistent with what Clavius has argued at various points—that eccentrics and epicycles are consistent with the common course of nature. They move in uniform circular motion, which is natural in the heavens. They obey the principle that a simple body should have a single, simple proper motion. Nor are the allegations of the adversaries valid: epicycles and eccentrics do not move rectilinearly (that is, toward and away from the earth, rather than in perfect circles), they need not have uniform thickness nor be exactly centered on the earth, and they are no more complicated than is necessary to achieve their ends.

In summary, Clavius's response to Copernicus is twofold. On the one hand, the Jesuit responded to the cosmological implications of heliocentrism by emphasizing how a moving earth conflicts with the accepted physical intuitions of his time and, more technically, how that concept has unacceptable consequences in the context of Aristotelian dynamics. He gave astronomical arguments that favor the earth's centrality and he may have thought the Copernican arrangement less capable of providing mathematical causes since it obviates the solar component in the planetary deferents and epicycles. Clavius showed no appreciation for the mathematical harmonies in Copernicus's theory that would appeal to later generations of astronomers, though he did expect some sorts of general mathematical relationships to follow from cosmological theories (as, in al-Bitrūjī's theory, for example). He completed the cosmological arguments with reference to the authoritative teaching of Scripture, which rejects the mobility of the earth and confirms that of the sun.

On the other hand, Clavius also criticized the methodological integrity of Copernicus's theory. The two astronomers may have been of like mind in their confidence that mathematical methods can correctly infer causes from their effects—but that assumes, of course, that one follows a proper method. The Copernican system is, for Clavius, purely a derivative construction that yields neither new information nor greater certainty. He sees it as a product of flawed method, an example of drawing true conclusions from false premises. For that reason, the Copernican theory served as a perfect foil for the defense against the skeptical argument. Clavius argued that the astute mathematical astronomer could distinguish between observationally equivalent theories by employing sound methodology and by drawing on other branches of learning, such as natural philosophy and the testimony of Scripture. Clavius did admit that the Copernican system was just as good as the Ptolemaic at explaining and predicting the phenomena, but he noted at the same time that the physical absurdities it implies quash any lingering doubts about its fictional nature.

In several respects Clavius's reaction to Copernicus is similar to the "Wittenberg interpretation" of certain Protestant astronomers such as Reinhold and Caspar Peucer.[82] Clavius agrees, like the Wittenberg astronomers, on the great value of Copernicus's work as a source for observations, tables, and calculations. He is also interested in the possibility of adapting certain Copernican devices to a geostatic universe, although, unlike some of the Protestant astronomers, Clavius did not share Copernicus's hostility toward the Ptolemaic equant.[83] The Jesuit's efforts at adaptation were wholly concerned with transplanting the Copernican precession theory into the Ptolemaic cosmology. Like his Protestant counterparts, Clavius showed little or no interest in the Copernican system's ability to determine the order and relative spacing of the planets. As we have seen, he remained completely committed to the traditional (Ptolemaic) ordering of the planets and to the accepted method, based on the nesting principle, that yielded the absolute (or at least minimum) sizes of all the spheres.

Clavius's greatest contrast with the Wittenberg astronomers seems to be his methodological critique of Copernicus. Clavius's view is far from that presented by Osiander, namely, that astronomical theories were at best hypothetical. Clavius, on the contrary, insisted that proper method could reveal the real motions of the heavenly bodies. Thus his attempt to reveal the flaw in Copernicus's method can be seen as a way to show, a priori, that Copernicus's theories can have no bearing on the reality of the movements of the celestial bodies. By implication, and in contrast, the Ptolemaic cosmos, which is methodologically sound, has a claim on the proper representation of reality, in Clavius's view.

Clavius's response to Copernicus is important because the popularity of the *Sphaera* and the extent of his reputation ensured a wide audience for his views. But we must keep in mind the context of his response. In the *Sphaera,* the Copernican system was only one of the rival cosmologies that Clavius troubled himself to refute. If we judge by the amount of text he devoted to criticizing the rivals of Ptolemaic cosmology, then we would be obliged to conclude that Clavius saw Fracastoro's homocentrics as the greater threat to a realist astronomy. It is tempting to explain this emphasis on Fracastoro as an atavism—a textual relic reflecting an astronomical controversy that was no longer current in the last third of the sixteenth century. But Clavius's argument and the example of homocentrics remained relevant in that period, especially given the vitality of the skeptical camp. This is because, as he repeats so often, an astronomical theory must save the phenomena as thoroughly as possible. Acceptance of the Copernican theory, despite its ability to save the phenomena, was not only unnecessary, but inconceivable for a number of reasons. To retreat to the philosophically seductive homocentric astronomy, which is fundamentally incapable of saving the phenomena, is to abandon the search for physical explanations of the real celestial mechanisms and thus to abandon also the hope of true human comprehension of the created world. The triumph of skepticism, even in astronomy, could only be a defeat for an orthodoxy that claimed to know how the heavens go.

SIX

Strains on Ptolemaic Cosmology, Inside and Out

This year at Messina an unusual sign, more marvelous than comets, appeared, namely, a star of notable and exceptional brilliance in a place where not any star was noted. . . . Therefore I deem that [Electra] now comes to light again, hoping that the Empire of the Romans descended from the Trojans is to be restored, and that Constantinople is to be regained.

—Maurolico on the nova of 1572

Ptolemaic cosmology did not fare well in the late sixteenth and early seventeenth centuries. Rival cosmologies and skeptical critics sought to supplant the Ptolemaic, and in those decades events conspired to hasten orthodoxy's collapse and seal its niche in the edifice of time. Adding insult to injury, the heavens themselves brought forth phenomena—novas and comets—that challenged the defenders of traditional views. Moreover, there were difficulties internal to the traditional cosmology itself. By comparison with Copernicus's theory, for example, the Ptolemaic mechanisms needed improvements. In particular, Copernicus's theory of the precession of the equinoxes was superior to the theory of trepidation that was generally considered part of Ptolemaic cosmology. Clavius's *Sphaera* is a revealing source on these challenges to the traditional cosmology, because it shows us how one of the most respected astronomers of his time perceived and responded to these great events. The long publication history of the *Sphaera* proves to be a useful record of the development of Clavius's thought.

Insults from the Heavens

In the preceding two chapters, I have shown that Clavius perceived two kinds of threats to the orthodoxy of late Ptolemaic cosmology. One threat was from the alternative cosmologies—especially the fluid-heaven, homocentric, and Copernican theories (joined later by the Tychonic system). Clavius's critiques of the first two hinged largely on operational issues such as their ability to save and predict the phenomena as well as the sufficiency of their mathematical foundations. The Copernican theory, on the other hand, raised questions of consistency with doctrines of natural philosophy and sacred opinion but also entailed more esoteric problems of methodological soundness.

The other kind of threat came not from rival theories but from those critics who asserted that the methods of mathematical astronomers were either incapable of arriving at true knowledge of the heavens or, at the very least, had as yet been unsuccessful throughout the long history of astronomical endeavor. In the opinion of the Averroist critics of Ptolemaic mathematical astronomy, this lack of success was manifested in the failure to produce an astronomical theory that was unquestionably consistent with Aristotelian physics. The great flaw in the enterprise of mathematical astronomy was the methodological error of incorrectly inferring causes from effects. The inference was incorrect, said the critics, because the causes failed to meet Aristotle's requirement that causes in scientific explanations should be necessary to the effect being explained. And as the critics never tired of pointing out, neither epicycles nor eccentrics were logically necessary since, at least in some cases, they were interchangeable.

The grounds on which Clavius had to secure his defense of the Ptolemaic system in the face of skeptical criticism were the same ones from which he had to launch his critique of Copernicus—namely, the grounds of consistency with natural philosophy and soundness of methodology. Setting aside for the moment the crucial issue of sacred doctrine, it would seem that Clavius and Copernicus had considerable room for a temporary alliance in confrontation with those who would dismantle the astronomical structures built out of eccentrics and epicycles. Clavius seems not to have seen this common ground, though a partisan of the Copernican persuasion, Kepler, did later perceive and advance their common interests in his *Apologia pro Tychone*.[1]

During the more than forty years that Clavius revised and republished his *Sphaera* (1570–1611), the issues mentioned above were, at least for him, the arenas for cosmological debate. But during that interval a remarkable series of celestial phenomena and discoveries occurred—events that gradually undermined parts of Ptolemaic cosmology. These events were the nova of 1572, the comet of 1577, the novas of 1600 and 1604, and Galileo's telescopic discoveries of 1609 and 1610. In the case of the novas and comet it was not so much that the events themselves were celestial novelties; rather, it was their lack of detectable parallax that was significant. And Clavius was one of the most authoritative voices in establishing those results.

Although in retrospect we see these moments as significant steps in the process of discarding the traditional cosmology and the development of a new one, Clavius did not see them that way. Clavius did not perceive their disruptive potential for several reasons. For one thing, the capacity to encompass new phenomena had never been used as an argument in favor of a cosmological system. The systems of Fracastoro and Copernicus, for instance, had not been presented by their authors as ways of explaining previously unexplained phenomena but, rather, as superior ways of explaining the already well known motions of the stars and planets.

Comets, while common enough, had—following Aristotle—usually been considered meteorological phenomena and thus not a burden to the theories of astronomers. Though there were authors who had considered comets to be celestial, Jane Jervis's brief survey of early sixteenth-century opinions on the matter mentions only two: Girolamo Cardano and Jean Pena. On the other hand, she found that such authoritative astronomical authors as Peter Apian and Nicholas Copernicus, as well as lesser lights Johannes Vögelin and Girolamo Fracastoro, had agreed with Aristotle that comets are atmospheric.[2]

If one considers comets to be within the terrestrial atmosphere, then the only new phenomenon to present itself in the history of astronomy before 1572 was the apparent change in the rate of precession of the equinoxes. But the efforts first of Thābit ibn Qurra and then of the Alfonsine astronomers had succeeded in accommodating variable precession, or trepidation, with Ptolemaic cosmology through the introduction of invisible superfirmamental spheres to produce motions of the eighth sphere as required. Thus the nova of 1572—the first major astronomical anomaly to come along since Copernicus had widened the vistas of cosmological debate in 1543—constituted a unique challenge for Clavius and the astronomers of his day.

The Nova of 1572

Clavius published a *digressio* on the *stella nova* (new star) of 1572 no later than 1585.[3] By the time of the 1581 edition (*nunc iterum*), Clavius had probably been in possession of Maurolico's observations of the nova since his return from Messina in 1574—which would seem to be sufficient time to prepare his comments. After its appearance in the 1585 *Sphaera* (at the latest), the *digressio* was printed unaltered in all subsequent editions. The only exception is a short additional paragraph, discussed below,

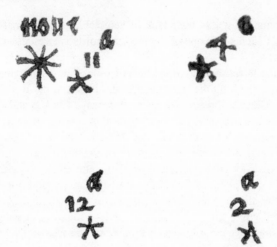

Figure 18. Location of the nova of 1572 in relation to the stars of the constellation Cassiopeia. The nova is in the upper left. The stars are numbered according to their order in Ptolemy's star catalogue (*2a* = *secunda,* *4a* = *quarta,* etc.). The modern designations of stars 2, 4, 11, and 12 are α, γ, κ, and β Cassiopeiae, respectively. From Clavius, *Sphaera* (Venice, 1596), 192. Photograph courtesy of the University of Wisconsin–Madison Memorial Library.

which appeared in the 1607 *Sphaera* (*nunc quinto*), commenting on the 1600 and 1604 novas.[4] The entire *digressio* with the added paragraph then appeared unchanged in the 1611 *Sphaera,* which was printed in his *Opera mathematica.*[5] Clavius incorporated into his own account a substantial set of excerpts from Maurolico's treatise on the nova and a shorter excerpt from the account of Paulinus Pridianus, "a physician and astronomer of Antwerp who observed the nova."[6]

Clavius's *digressio* on the nova of 1572 is largely concerned with refuting three suggested explanations of the phenomenon, each of which would have placed its cause in the upper reaches of the air surrounding the earth. Clavius does not identify the authors of these theories for us. The first theory, believed by "some but few,"[7] is that the nova was actually one of the thirteen already known stars that constitute the constellation of Cassiopeia, within whose boundaries the nova appeared.[8] The great increase of brightness observed in 1572, according to this first theory, occurred when some vapors or exhalations in the highest region of the air came between us and that previously dim star, thus bringing it to our attention. (Dreyer noted one German and two Italians who held that the nova was identical with κ Cassiopeiae.[9] The nova did, in fact, appear fairly close to the star κ Cassiopeiae, as shown in fig. 18.) The second theory also supposes that the nova was, in fact, a previously unremarkable star, momentarily brightened by interposed exhalations in the upper regions of the air. This second theory, however, supposes that the new star was not one of the thirteen stars known from Ptolemy's catalog but, rather, one dimmer than the sixth magnitude and thus invisible until magnified by the airy exhalations. The third and final theory suggests that the nova had nothing to do with the stars and was instead a comet in the highest region of the air.

The first theory must be false, Clavius says, because many astronomers in many different countries, including Maurolico in Sicily and Clavius himself in Rome, observed the nova often and found it always in the company of all the other thirteen stars of Cassiopeia. Clavius reports a particular occasion in December 1573—the nova by then had decreased in brightness to third magnitude—when at Rome not only he but several others noted all thirteen of Cassiopeia's stars in addition to the nova. And about half a page later he refers again to personal observations of the nova made in the company of others.[10] Unfortunately we know nothing else of the circumstances or results of these observations. They were perhaps informal, and indeed many of Clavius's observational experiences seem almost anecdotal. In any case, no records of these or any other early observations seem to have survived among Clavius's papers. Yet these stories are interesting, because if Clavius's company was composed of his students or colleagues at the Collegio Romano, as seems likely, then this is another bit of evidence for very early interest in science, and specifically astronomy, among the Roman Jesuits. It is a pity that there is no clue to the identities of any of the observers other than Clavius.

Finally, in response to the idea that exhalations in the atmosphere magnified the image of one of Cassiopeia's thirteen stars, Clavius notes that the nova was observed in Germany, Spain, France, Italy, and Sicily. If the exhalation covered an area of the sky large enough to interpose itself between the star and so many widely dispersed observers, then it should also have affected the brightness of the eleventh star (κ) of Cassiopeia, which is the closest to the nova itself. Yet that star became no brigher than it had ever been before and, he says, it remains of the same brightness today.

This same argument, Clavius continues, can be used against the second theory of the nova's origin, namely that exhalations magnified the image of a star previously unknown because it is intrinsically dimmer than the sixth magnitude. As before, if the exhalations magnified the very dim star, then they should have magnified all the more the eleventh star of Cassiopeia. Yet the latter has always remained a fourth-magnitude star.

Clavius refutes the third possible explanation of the nova by invoking a parallax argument or, rather, an argument based on the lack of observable parallax. The nova could not have been a comet, as the third theory suggests, because "skilled astronomers everywhere have noted that the [new] star kept the same place among the fixed stars . . . and furthermore they observed in it almost no parallax from various locations. In that case, who could doubt that [the star] is not in the highest region of the air, where the rest of the comets are formed, but indeed has its place beyond the moon?"[11]

If Clavius rejects all these alternatives, we may reasonably ask for his opinions on the matter. Where in the heavens is the nova? How did it get there? He wastes no time in telling us. "I think that the nova, whatever it was, existed in the firmament, where the fixed stars are. For it appeared in the celestial [etherial] region and not in the elemental . . . [as we know] because it showed no parallax. . . . I believe that the new star stood in the firmament and not in any other celestial orb . . . because neither I nor any other astronomer to my knowledge observed any other motion in [the nova] aside from those that we observe in the fixed stars."[12] That is, if the nova had been in one of the planetary spheres, then it would have exhibited the motion proper to that sphere. But it showed only the motions of the fixed stars, therefore it must have been in the same sphere as the fixed stars—namely, the firmament. This is the same reasoning—lack of parallax and any motion other than that of the fixed stars—by which Tycho Brahe and Michael Maestlin independently concluded that the nova was in the firmament.[13]

It is possible, though unlikely, that Clavius knew of Tycho's conclusions on the celestial nature of the nova. The few printed copies of Tycho's treatise on the nova were distributed to a select group, and only about twenty copies now survive, making it entirely possible that Clavius had no access to it.[14] Had he known of them, Clavius would likely have cited Tycho's results as confirming his own. Moreover, Tycho's parallax measurements were accomplished by a method different from Clavius's. Tycho used the conventional but difficult method of diurnal parallax (allowing the daily motion of the heavens to displace the object and create the possibility of parallax). Clavius compared the reports of various observers scattered around Europe. Thus he measured, in effect, a geographical parallax. Given the wide use and reprinting of the *Sphaera,* it is probable that Clavius's announcement, no later than 1585, of the celestial nature of the nova reached far more readers than Tycho's account, which was not republished until the edition of his *Progymnasmata* in 1602.

How did Clavius explain the appearance of the nova? He had two suggestions. "I am convinced that either the nova was created by God in the eighth sphere to presage some great thing (though what this thing might be is still unknown), or at least that the eighth sphere can produce comets as [they are produced] in the air—though [the celestial comets] happen less often.''[15] It is not clear which of these explanations Clavius favored, nor is it even clear whether he considered them mutually exclusive. The Jesuit was not alone in suspecting divine involvement in the appearance of the nova, for Tycho, too, finding no satisfactory explanation for the star's appearance, concluded that God had created it for his own purposes.[16]

Clavius's explanation of the nova as a miraculous event attracted severe criticism from the English Copernican John Wilkins. Responding in 1638 to Clavius's suggestion, Wilkins said ''To fly unto a miracle for such things were a great injury to nature, and to derogate from her skill; an indignity much misbecoming a man who professes himself to be a philosopher. *Miraculum* (saith one) *est ignorantiae asylum.*'' This is fair indeed. But Wilkins then went on to insinuate that Clavius was a dogmatic Aristotelian whose only concern was to preserve Aristotle's authoritative opinion. ''But here is the misery of it, we first tie ourselves unto Aristotle's principles, and then conclude that nothing could contradict them but a miracle; whereas it would be much better for the commonwealth of learning if we would ground our principles rather upon the frequent experiences of our own, than the bare authority of others.''[17]

In that last statement, insofar as he meant to implicate Clavius, Wilkins was being quite unfair, because Clavius's miraculous explanation was closely associated, as we have seen, with the naturalistic hypothesis that comets might be generated in the heavens, and this contradicts the Aristotelian doctrine that comets are creatures of the upper atmosphere. And if that were not enough to dissociate himself from dogmatic Aristotelians, Clavius goes on to make a very interesting statement at the same time that he explicitly contrasts his views with those of the Peripatetic philosophers. Referring to his suggestion, which he admits is not original, that comets might be generated in the heavens, Clavius continues, ''If this is true, then the Peripatetics ought to consider how they can defend Aristotle's opinion concerning the matter of the heavens. For perhaps it should be said that the heavens are not some fifth essence but, rather, mutable bodies—albeit less corruptible than bodies here below.'' After noting that this opinion has precedents among some philosophers and patristic writers (citing specifically Plato and Saint Ambrose, Saint Basil, and Saint Gregory of Nyssa) he goes on, ''Whichever it finally is (and I do not insert my opinion into such matters), it is enough for me at present that, as demonstrated a little earlier, the star of which we speak was located in the firmament.''[18]

So by no later than 1585, Clavius was almost ready to relinquish celestial incorruptibility as a part of his cosmological teachings, but he stopped short of making a definitive statement on the issue. His position, we should note, was a result of his astronomical determination (through parallax) of the location of the nova of 1572 and not based on related concerns of physics or metaphysics. But religious concerns may have guided his thinking. Clavius's mention of the patristic writers on the question of celestial corruptibility is consistent with Edward Grant's suggestion that the growing importance of the writings of the Greek fathers in western Europe in the late fifteenth and early sixteenth centuries aided the movement away from the strict Aristotelian position of celestial incorruptibility.[19]

Clavius indicated some self-restraint about expressing his own opinion on "such matters" because the nature of celestial substance is a question for physics—for the natural philosophers, that is. His recognition, if not strict observance, of disciplinary boundaries is also apparent in the other place in which the question of the incorruptibility of the heavens arose— namely in the explanation of the properties of the celestial or ethereal region that he had presented earlier in the *Sphaera*, where he said, "According to the philosophers, the ethereal, or celestial, region cannot be altered, can be neither increased nor diminished, nor generated nor corrupted."[20] In fact, in the whole passage that enumerates the five properties of the ethereal region (that it encircles the elemental region, that it is luminous, that it is incorruptible, that its motion is perpetually circular, and that celestial substance is completely unlike terrestrial elements), Clavius was careful to note several times that these are the views of the philosophers. It is hard to say, merely from his recitation of the opinions, whether he believed them or not. In fact, the situation is mixed, because he held unswervingly his insistence on the circular nature of motion in the heavens—also a philosophical principle. On the other hand, he did at least question the doctrine of incorruptibility as a result of the 1572 nova. Further, he backed away from the fifth doctrine (that celestial substance is completely unlike terrestrial elements) when suggesting that rather than radical dissimilarity, celestial matter might be like terrestrial matter at least in being corruptible—only less corruptible. The difference was no longer of kind but of degree.

It is an interesting question whether Clavius's caution was a result of concern for disciplinary boundaries or, on the other hand, caused by his reservations about the truth of Peripatetic doctrines. In the case of celestial incorruptibility, a comparison between earlier editions of the *Sphaera* and those in which the 1572 nova is first discussed suggests that Clavius was backing away from the Peripatetic doctrines themselves. In the 1570 *Sphaera*, Clavius stated the third celestial property, incorruptibility, thus:

"The ethereal, or celestial, region cannot be altered, can be neither increased nor diminished, nor generated nor corrupted."[21] In later editions we read, "*According to the philosophers,* the ethereal, or celestial, region cannot be altered, can be neither increased nor diminished, nor generated nor corrupted."[22] This new attribution of the incorruptibility of the heavens to the philosophers puts some distance between himself and the doctrine. And it may indicate a real change in his views as opposed to a mere reiteration of disciplinary boundaries. Further, the appearance of this qualification in the same edition as his treatment of the nova of 1572 probably indicates that the nova was a cause of this significant change in his cosmological thinking.

The restraint Clavius showed when it came to his own opinions on matters of natural philosophy did not keep him from suggesting that the natural philosophers should consider the proper conclusions of the mathematical astronomers in their deliberations on the causes of such phenomena as the nova. In this case, the astronomers had concluded that the nova occurred neither in the elementary realm nor in any of the planetary spheres and must, therefore, have been in the firmament. So the philosophers should take this finding into account when discussing the nature of celestial substance.

Clavius did not overtly break with the philosophical doctrine of the incorruptibility of the heavens. In essence he favored a middle position in which celestial substances are corruptible (in direct contradiction to strict Aristotelianism), yet not as corruptible as substances in the elemental sphere. Thus Clavius retained some semblance of the Aristotelian idea of a metaphysical distinction between terrestrial and celestial matter while giving up the idea of pure celestial incorruptibility. The difference between the celestial and terrestrial natures became more of a distinction than a dichotomy—the chasm between the natures of the regions is no longer quite so large. Rather than a radical split between eternity and corruptibility we have a difference between degrees of corruptibility. This celestial-terrestrial distinction is still consistent with the idea, held by Clavius and a long line of other geocentrists, that the earth is at the center of the universe because of its vile and corruptible nature. Clavius shows us that it was possible to admit the possibility, even the likelihood, of celestial change without necessarily capitulating to the idea (implicit in Copernicus and explicit in Galileo) that celestial and terrestrial physics were one and the same. Clavius's views illustrate an aspect of that diversity and independence from traditional Aristotelianism that Charles Schmitt argued was characteristic of Renaissance Aristotelianism.[23] Clavius's views also indicate the general direction of thinking on the part of some Jesuit astrono-

mers, particularly Riccioli, who, in his *Almagestum novum* (forty years
after Clavius's death) fully accepted the concept of the corruptibility of
the heavens.[24]

Maurolico was a fundamental source in Clavius's treatment of the nova
of 1572. Clavius presented only minimal observational data concerning
the nova, though he says that he observed it more than once. He made
only the vaguest attempt to summarize its dates of appearance and disappearance, saying that it appeared in 1572 and disappeared in 1573 (or
1574–he is inconsistent, perhaps as a result of relying on reports of late
sightings). This vagueness is particularly curious, given that Maurolico
supplied (though Clavius did not print) the fact that the nova appeared on
6 November 1572 or earlier.[25] The Jesuit gives a verbal description of its
position, accompanied by a rough illustration showing the nova in relation
to the other stars of Cassiopeia (fig. 18). And he says nothing of its color
or other visible qualities, or of any changes in those qualities, except to
comment that, at its brightest (whenever that was), the nova surpassed
Venus in brilliance and to add in passing that by December 1573 the star
had faded to third magnitude.[26] Clavius does tell us that the nova's position
was sixty-six and one-half degrees from the celestial equator, yet this is
one of the several quantitative nuggets found in Maurolico's account.
Clavius supplied, in short, no quantitative information that he could not
have gotten from Maurolico. Perhaps he considered Maurolico's measurements superior to his own if, in fact, he had any of his own.

In addition, Maurolico anticipated Clavius in concluding that the nova
was among the fixed stars and not part of the earth's atmosphere.[27] Maurolico, like Clavius later, also interpreted the nova as some kind of omen
of great events to come. But the general tone of Clavius's account of the
nova is different from Maurolico's. Maurolico couched the whole treatise
in dramatic terms, opening with a quotation from Genesis and going on
at some length about the events that the awe-inspiring nova might presage.
With the great naval battle of Lepanto (which had taken place just the
year before) no doubt weighing on his mind, Maurolico offered a prayer
for the protection of Sicily and closed his accounts of the nova with a
quotation from Dante. Clavius printed the astronomical core of Maurolico's account, stripped of historical, literary, and allegorical trappings.
By comparison with Maurolico's treatise, Clavius's account is very busi-
nesslike and didactic. The only remaining trace of Maurolico's astrological
speculations is Clavius's simple remark that the nova could have been a
portent.

Clavius's major contribution—aside from printing in his *Sphaera* parts
of Maurolico's otherwise unpublished account of the 1572 nova—is his

synthesis of the reports of observers in many different countries. He recognized that they agreed closely enough on its position with respect to the fixed stars to enable him to put an upper limit on the nova's parallax. These observations also allowed him to rule out the possibility that atmospheric exhalations might explain the nova's presence. These conclusions made a convincing case for the celestial—as opposed to meterological—nature of the nova. Finally, there was his clear statement that the nova implied some kind of change in the celestial region and his challenge to the natural philosophers to reconcile that conclusion with the Aristotelian doctrine of celestial incorruptibility—a subject that Maurolico left untouched.

In the 1607 *Sphaera* (*nunc quinto*), Clavius added a paragraph commenting on the novas that had appeared in 1600 and 1604, and the addition appeared essentially unmodified in the 1611 *Sphaera*. Referring to his judgment on the 1572 nova, Clavius says,

> The same can be said concerning that new star, which (as was related to me from Germany) appeared in 1600 and has persisted until now in Cygnus next to that [star] that shines in the [Swan's] breast. And the same is to be said again concerning the nova that was first seen in October, 1604, between the seventeenth and eighteenth degrees of Sagittarius and having a north latitude of two degrees or thereabouts; although at the time of this writing it has diminished to the point where it is hardly visible. Both stars are found to be in the firmament by the same arguments, [which are that] they reveal no parallax at all, since observed from any location they are [always] the same distance from the other fixed stars. And also they reveal no other motions beyond that [motion] that we note in the fixed stars.[28]

So the reasoning and conclusion are the same for these novas as for the one of 1572. The lack of parallax assures us that they are beyond the orb of the moon, and the lack of any other motion aside from the diurnal motion of the stars is evidence that they must be part of the sphere of the stars, that is, the firmament.

Contrary to his personal account of the 1572 nova, Clavius gives no indication in the *Sphaera* of having observed the novas of 1600 or 1604 himself. But this may be deceptive, because it appears from other sources that Clavius was still observing in 1604, his sixty-sixth year. In a letter to the astronomer Magini, dated 18 November 1604, Clavius wrote, "The new star is visible here at Rome, and with instruments we have found [that it is] always the same distance from the fixed stars (such as Arcturus, Lyra, Cygnus, and others). So it seems that it must stand in the firmament."[29] Thus a month after its first appearance, the Roman Jesuit astronomers had concluded that the 1604 nova confirmed Clavius's previous findings concerning the novas of 1572 and 1600. That Clavius was still physically fit enough to be personally involved in these observations is made more likely by the testimony of another letter that Clavius was present five years later at observations of a lunar eclipse in 1609.[30]

Even if he were not observing personally, Clavius would have had plenty of other sources of observations at his disposal—enough, that is, to explain his statements on the latter novas. His former students and younger colleagues at the Collegio Romano—particularly Grienberger and Maelcote—surely observed these novas. In fact, Baldini identifies Maelcote as the author of the Collegio Romano treatise on the 1604 nova.[31] Moreover, Clavius received accounts of these events by correspondence from far beyond the Collegio. For instance, the Jesuit forwarded to Magini a copy of a remarkably detailed account of the 1604 nova's appearance first sent to Clavius by "a mathematical physician" (who has remained anonymous) of Cosenza in Calabria.[32] Through the agency of Mark Welser, Clavius had also come into possession of an account of the nova of 1600 written by an associate of Tycho.[33] Thus, even if the aged Clavius had been slowed by the years, his students, colleagues, and correspondence network kept him up to date on astronomical events. In fact, his wide-ranging contacts, made possible at least in part by his status as a Roman Jesuit, was one of his most important assets.

The Comet of 1577

Comets occupy a place of little importance in Clavius's *Sphaera,* and this is surprising given the popular interest in comets and their significance for astronomical theory in Clavius's time. Bright comets excited interest in antiquity and the Middle Ages, but historians generally agree that the comet of 1577 was truly a watershed in the history of astronomy.[34] Yet Clavius mentions comets only twice, and then incidentally, in the various editions of the *Sphaera.*

Comets make their first appearance in the *Sphaera* in one of those long expositions (so characteristic of later *Sphere* commentaries) of a passage in Sacrobosco. In a few lines of the original text Sacrobosco had explained that the four terrestrial elements, subject to generation and corruption, are disposed around the center of the cosmos. Closest to the center is the sphere of earth, which is surrounded by water (except in certain places so as to allow life on dry land). The sphere of water, in turn, is surrounded by the air, and the air, finally, is surrounded by the sphere of elemental fire.[35] This, of course, is not really a subject suitable to astronomy, but pertains more to meteorology or some other part of natural philosophy.

Nevertheless, since it is in the traditional primary text it is the prerogative of the commentator, Clavius, to expound on it. The result is a treatise, running some nine pages, on the elements, their properties, and their places.[36]

Since the elements were conventionally explained as combinations of Aristotle's fundamental and contrary material qualities (hot, cold, moist, and dry), and all other terrestrial substances as combinations of the elements, Clavius opted to devote about two of those pages (in the 1611 edition) to a mathematical digression on combinatorics—the same digression that Leibniz later cited as a significant contribution to the field.[37] This section of commentary concludes with a description, which appeared in every edition of the *Sphaera,* of the earth's atmosphere. "Philosophers divide the air into three parts, namely upper, middle, and lower," Clavius tells us. Then he mentions, almost in passing, that comets are a phenomenon of the upper atmosphere. "The highest air, which, as we see, sustains comets, is always hot because of its continuous motion (which it gets from the primum mobile), its vicinity to the elemental fire, and the continual passage through it of the sun's rays."[38] This simple and unequivocal location of comets in the upper air remained unchanged from the first through the last edition of the *Sphaera.*

In the *digressio* on the nova of 1572, Clavius revealed a somewhat more complex opinion about comets. "I am convinced that either the nova was created by God in the eighth sphere to presage some great thing (though what this thing might be is still unknown), or at least that the eighth sphere can produce comets as [they are produced] in the air— though it follows that [the celestial comets] are less common."[39] Clavius now maintains a twofold theory of comets. There are the common comets that are generated in air, as conventional Aristotelianism taught, but there might be comets generated in the eighth sphere.

This distinction raises a great many questions, but we do not have enough information from Clavius to formulate even tentative answers. Why did he offer as an explanation for the nova the idea that comets could be generated in the eighth sphere? Are novas a species of comet or vice versa? What characteristics would make a nova explainable as a comet? Clavius supplies no help, so we are on our own. A nova has no tail, does not move with respect to the fixed stars, and can be visible for years. Comets generally, though not always, have tails, always move with respect to the stars, and are typically visible for a matter of weeks or months. The only common feature is that both phenomena appear and disappear unexpectedly and thus seem to indicate some kind of change—generation and corruption of some sort, perhaps—in whatever region they inhabit.

If the nova were a form of comet, would there be some physical significance in their different kinds of motion? Both have the diurnal motion
of the primum mobile, but the atmospheric comet also has a proper motion—often very rapid. Is this related to their relative positions in the
cosmos? That is, does the nova move like the fixed stars because it shares
their sphere while the atmospheric comet has a proper motion in the same
sense that the sun and moon, for instance, have proper motions on their
own eccentric circles? Does that mean that the atmospheric comet has its
own sphere? Or is the proper motion of the atmospheric comet not comparable to celestial motion but, rather, a result of, say, its composition or
perhaps the violence of the hot upper air? Clavius offers us no help on
such matters.

Why did Clavius not mention the comet of 1577 when the subject of
comets came up in the 1581 or later editions of the *Sphaera*? Not only is
there no mention of the great 1577 comet in the *Sphaera,* but there is no
mention of it in any of his correspondence. If Clavius did observe the
comet, it may not have occurred to him to measure its parallax as Tycho
and Maestlin did. Perhaps his experience of the 1572 nova had not led
him to question the Aristotelian teaching that comets (at least those with
tails and proper motions) were atmospheric phenomena. Given his twofold
view of the nature of comets, the comet of 1577, with its bright tail and
great proper motion (in contrast to a nova), would have looked to him
like a typical comet of the upper atmosphere rather than a celestial phenomenon.

Yet by the time he published his views on the 1572 nova, Clavius
should have been aware, at least, of the observations of other astronomers,
who had measured the parallax of the comet of 1577, and of the cosmological implications of those measurements. Doris Hellman notes at least five
important astronomers who measured the parallax of the comet of 1577
and concluded that it must be supralunar: Tycho Brahe, William IV (Landgrave of Hesse-Cassel), Michael Maestlin, Helisaeus Roeslin, and Cornelius Gemma. The latter three of these had already published their conclusions by 1578, while the conclusions of Tycho and the Landgrave were
only published much later, 1588 and 1618, respectively.[40] Still others
measured the parallax of the comet and found it to be measurably large.
Hellman notes four who published their results by 1578: Thaddeus Hage-
cius, Bartholomaeus Scultetus, Andreas Nolthius, and Georgius Busch.[41]
Evidently the idea of measuring the parallax of the comet of 1577 was far
from rare, so it is curious that Clavius did not even comment on the results
of others. Furthermore, the general idea of measuring the parallax of a
comet was certainly current in the Collegio Romano soon after Clavius's

death, for one of his more famous students, Orazio Grassi, measured parallaxes in his work on the comets of 1618.[42]

It seems doubtful that Clavius ducked the comet issue because some of the parallax measurements led to results incompatible with Aristotelian philosophy. We have already seen that Clavius was willing to part company with the philosophers when he concluded that the nova of 1572 was probably in the firmament and that it called into serious question the Aristotelian doctrine of the absolute incorruptibility of the heavens. Might Clavius have ignored (some might say suppressed) the issue of the comet because it challenged, in some way, the Ptolemaic cosmology? This, too, seems unlikely. To begin with, comets in the heavens might imply additional, albeit previously unknown, celestial spheres—but adding spheres to account for newly discovered phenomena was standard procedure. There should have been no problem, in principle. Though it is true that the motions of the comet would later be seen as evidence for the nonexistence of the solid celestial spheres, this interpretation came later and should not, in any case, have affected Clavius's earlier revisions of 1581 and 1585. Moreover, Clavius always seemed willing to acknowledge results that might challenge traditional views as he did with the nova of 1572, and as he would do many years later with Galileo's telescopic revelations.

Clavius's situation in the period in question may shed some light. First we should observe that none of the astronomers mentioned by Hellman as important observers of the comet of 1577 was Italian, which might indicate a generally lower level of interest in the comet in Italy. Further, 1577 was fairly early in Clavius's career, and his contacts with other astronomers, who might have alerted him to the comet's presence and the questions it raised, were still relatively scarce. (Recall that he depended heavily on Maurolico for his treatment of the nova of 1572.) And we might observe that 1577 would have found Clavius in the midst of his calendar reform work and also having just resumed teaching after a break from 1571 to 1576, all of which may have precluded his spending much time on astronomical observations.

The question still remains, Why did Clavius never publish an evaluation of the astronomical significance of the comet of 1577 as he did for the novas of 1572, 1600, and 1604? We can speculate that the negligible parallax of comets, like that of 1577, would have presented Clavius with a difficult problem that he might never have resolved to his own satisfaction. The problem would have arisen from his twofold view of comets— that some are in the upper atmosphere, while others can be in the firmament (if the novas are a form of comet). Many of the well-publicized parallax measurements of the comet of 1577 indicated it to be supralunar and clearly not in the atmosphere, but its proper motion with respect to the stars could only mean that it was not located in the firmament where it would be obliged to share the motion of the fixed stars. Thus it had to occupy a position somewhere between the moon and the firmament. This conclusion would have demanded, in Clavius's view, that the comet either share the motion of a planetary sphere or else have a previously unknown sphere of its own. A difficult choice, and perhaps one he never brought to a decision.

Maestlin, led to this pass by the same logic, chose the former alternative, though he eventually formulated his response on the basis of the Copernican theory, as he tried to fit the path of the comet to the motion of the sphere of Venus.[43] Similar considerations on the comet of 1577 played an important role in Tycho's development of his own alternative to the Ptolemaic and Copernican cosmologies.[44] The conservative Clavius, unwilling to accept the Copernican system and either unaware or disapproving of Tycho's system, would have before him the difficult task of explaining the comet's motion in the Ptolemaic framework.

Finally, we should recall that Clavius never conceived of his own mission as fundamentally observational, in contrast to Bernhard Walther or Tycho, for example. Clavius's role was as a theorist, but even more, as an educator. The production and assimilation of new observations were not so important as evaluating the claims of rival theories, supporting the correct one, and resolving any remaining discrepancies. The comet of 1577 would have been important to Clavius the theorist primarily as a possible discrepancy, yet the variety of opinions on the location of the comet (the authority and accuracy of which could not have been nearly so clear to contemporaries as they are to us) left uncertain whether it was, in fact, a discrepancy. Clavius the educator may simply have decided that a debate on the comet was not something that would benefit his readers.

Accommodations to Copernicus

Comets and novas raised questions about the nature of the cosmos and which cosmological ideas described it best, although Galileo's telescopic discoveries were probably the most dramatic setbacks to the cause of those who would preserve the traditional geocentric cosmology. But there are earlier and less spectacular signs of strain on Ptolemaic cosmology that would have allowed a relative advance in the fortunes of the Copernican theory—or at least wider opportunities for it. One such early sign is the interest on the part of some Ptolemaic astronomers in adapting the successful Copernican planetary mechanisms to a geocentric cosmology. Another is the recognition, made explicit by Clavius in his *Sphaera,* that the traditional doctrine of trepidation was wrong and that Copernicus's mechanisms, appropriately adapted to a geocentric cosmology, are more successful in saving the phenomena.

We know little about Clavius's work in the more technical aspects of planetary theory. The *Sphaera* was, of course, an introductory text, so Clavius specifically avoided technical discussions of planetary theory and, instead, deferred such treatment to his *Theorica planetarum,* which, though often promised, never appeared in print. What little we can gather about his work in this area must come from some manuscript fragments, his few technical remarks in the *Sphaera,* and his correspondence with Giovanni Magini, who, unlike Clavius, actually published what amounts to a post-Copernican *Theorica planetarum.* If not exactly collaborators, Clavius and Magini seem to have shared some values and goals in matters of astronomical theory. Moreover, Clavius actually adopted Magini's Ptolemaic reworking of Copernican precession theory. Magini's work on adapting Copernican mathematical innovations to the Ptolemaic cosmology might serve as a rough guide to what Clavius had in mind but never published.

Magini's Adaptations

The Italian astronomer Giovanni Magini (1555–1617) spent most of his life teaching mathematics at Bologna and during that time published several theoretical astronomical works. Magini won the chair of mathematics at Bologna over rival claimants, one of whom was Galileo (who then later won his own post at Padua). To the end of his life, Magini remained skeptical toward Galileo and his opinions, and on occasion offered support to Galileo's adversaries.[45] Magini's extensive correspondence with major European astronomers, including Galileo, Tycho, Kepler, Clavius, and others was collected and edited by Antonio Favaro late in the nineteenth century.[46]

Magini's most interesting work, for my present purposes, was published in 1589 in his *Novae coelestium orbium theoricae congruentes cum observationibus N. Copernici* (New Theories of the Celestial Orbs Conforming to the Observations of N. Copernicus). The title, however, is somewhat deceptive, for Magini's goal was not strictly to update the Ptolemaic theories of the planets as required by Copernicus's observations but, in fact, to introduce major theoretical changes that would incorporate some Copernican innovations into a system that remained, in a sense, Ptolemaic.[47] (Though somewhat misleading about Magini's intentions, the title reflects the same attitude that is occasionally evident in Clavius, namely, that Copernicus's work was viewed as significant for its empirical contents.)

Magini's system was Ptolemaic in that it was geocentric, geostatic, used the basic Ptolemaic constructions (e.g., epicycle and eccentric as in the *Almagest* and *Theorica planetarum*), and preserved the original Ptolemaic models for the five planets. However, Magini did away with the old idea of trepidation and replaced the traditional Alfonsine trepidation mechanisms with mechanisms adapted from Copernicus's explanations for precession of the equinoxes and variation in the obliquity of the ecliptic—though with minor numerical changes. He arranged these mechanisms, however, so that the appearances were a result of the motion of the stars rather than motions of the earth, as in Copernicus's theory.

Magini also introduced significant changes in the Ptolemaic solar theory in order to explain two phenomena that Copernicus (like all of his contemporaries) had believed were real, namely, the variation in the length of the tropical year and the variation of the maximum equation of the sun. Copernicus had explained these phenomena through the motion of the earth, but Magini had to introduce extra orbs into the solar model, for a total of five. By comparison, the solar model in Ptolemy and Peurbach consists of a simple eccentric circle, or sphere, which yields a total of three orbs when the virtual eccentrics (the nonuniform spheres of materialized eccentrics) are taken into account.[48]

Magini also criticized Copernicus's lunar model, though not because it required any terrestrial motions. (Everyone agreed that the earth is the moon's center of motion, so lunar models should have been free of the debates over the earth's motion.) Copernicus's lunar model employs a double epicycle—one epicycle on the circumference of another. But Magini found this unacceptable because, he said, any epicycle should cause the moon to turn first one face toward the earth, then another, so that we would not always see the same face. This, of course, is not what we observe.[49] In place of Copernicus's theory, Magini proposed an arrangement of six orbs that employed only eccentrics. Not only does this ensure that we will always see the same face of the moon, but, he said, it is appropriate that epicycles be reserved to the planets alone, because they are the celestial bodies that exhibit retrogradations. It is interesting to see Magini attempt to provide a general principle that will make some sense of the Ptolemaic system, namely, that there are two categories of objects, those that exhibit retrogradations (the planets), and those that don't (the major luminaries, the sun, and the moon). The categories are distinguished by their kinematic properties. That is, the planets have epicycles, the luminaries do not. This kind of attempt to provide more general principles founded on the properties of the celestial mechanisms is reminiscent of Clavius's suggestion of the dynamical distinction between eccentric and concentric orbs. Both seem to be attempts to find mathematical causes in the Ptolemaic scheme, broadly construed.

Clavius was aware of Magini's work and very interested in it. In a letter in 1595, Clavius wrote to Magini that "I think that it would be an extraordinarily useful and noble thing if you would print the observations by which you composed your theories . . . I urgently entreat you to do it. And there is no need to await whatever Tycho the Dane will do, because it seems to me that he will never finish." Clavius returned to the subject, after an unrelated passage, in the closing of the letter: "And I remind you again of the observations regarding the theories. May Our Lord keep you in his holy grace."[50] Magini never complied with this request to publish his observations.[51] It is interesting to note the air of expectation that seems to have surrounded Tycho's promised work on the theories of the planets, which, as Clavius correctly predicted, never appeared.[52] Clavius may have been correct in his suggestion that Magini's reticence was owing to Tycho's pending theoretical work. Tycho and Magini did correspond with each other on the subject, and there exists at least one letter from Tycho to Magini explaining, among other things, that the Dane had still not completed his work on the theories of the planets.[53]

Magini's view of Copernicus was, like Clavius's, mixed. Both admired Copernicus's theoretical skills, appreciated his useful observations, and accepted his mathematical reforms in certain areas, especially in his replacement of Alfonsine trepidation theory. Both, however, saw Copernican cosmology and its assumptions—the rotation and revolution of the earth—as physically inadmissible. Their responses to Copernicus were also similar: both saw the need for what was essentially an up-to-date *Theorica planetarum* that could preserve the physical assumptions and general mathematical principles of Ptolemaic cosmology while also incorporating the mathematical reforms of Copernicus. Magini succeeded in publishing such a *Theorica planetarum emendata,* but Clavius, despite his many promises to do so, never finished his own.

Clavius's Rejection of Trepidation

Except when discussing Copernicus's ideas on the motion of the earth or the stability of the sun, Clavius is generally complimentary to his Polish predecessor. As already mentioned, Clavius praised Copernicus greatly as a restorer of astronomy. But praise is cheap, and it was a particularly good bargain in the sixteenth century. It is not at all an empty gesture, however, that the Jesuit accepted Copernicus's new value for the length of the tropical year and willingly incorporated the corrections Copernicus had made to the positions of some stars in Ptolemy's star catalog. Clavius also accepted Copernicus's innovation of measuring celestial longitude from the "first" star of Aries, γ Arietis (named Mesarthim), rather than, as was traditional, from the "first point" of Aries, which is the vernal equinox.[54] (This step had the result of freeing that stellar coordinate from the effects of precession.)

In 1575 Clavius designed and built (or had built) a celestial globe using star positions taken from Copernicus's star catalog. He corrected the stellar coordinates in the catalog, which were by then fifty years old, using Copernicus's average value for the motion of the eighth sphere.[55] But Clavius's real accommodation to Copernicus came in the 1593 *Sphaera* (*nunc quarto*) when he rejected the Alfonsine theory of trepidation, which he had somewhat equivocally defended in the previous four editions, and embraced a geocentric-geostatic version of the explanation that Copernicus had given for the putative motion of the fixed stars.

Until the time that he adapted Copernicus's precession theory, Clavius had presented the conventional Alfonsine version of the motion of the eighth sphere.[56] From the first *Sphaera* edition (1570) until the 1585 edition (*nunc tertio*), he explained how astronomers since the time of Hipparchus had observed that the stars slowly slip eastward with respect to the equinoxes. That is why Ptolemy concluded that the eighth sphere (the sphere of the fixed stars) could not be the highest sphere. Instead, Ptolemy invoked a ninth sphere to provide the diurnal east-to-west motion and taught that the eighth sphere turned from west to east very slowly about the poles of the ecliptic, completing a full revolution in 36,000 years.[57] However, Clavius relates, Albategnius (al-Battānī) found a different rate, namely, one revolution in 23,760 years and the Alfonsines still another, one revolution in 49,000 years.[58]

note 58, p. 254 gives an important explanation

The stage was now set for the development of trepidation theory, which probably seemed like a reasonable extension of the procedure used by Ptolemy to explain the precession of the equinoxes. In order to explain the apparently variable motion of the fixed stars, Thābit Ibn Qurra introduced some major theoretical changes, which, slightly modified by the Alfonsine astronomers, became standard in medieval and Renaissance cosmology. Thābit suggested that the highest sphere is actually a tenth sphere rotating daily about the "poles of the world," (the north and south celestial poles) represented by axis *FC* in figure 19. The ninth sphere, rotating very slowly about the poles of the ecliptic axis (*AD*), produces the constant precessional motion that we see in the eighth sphere, which is the sphere of the fixed stars, or firmament. That is to say, the ninth sphere carries

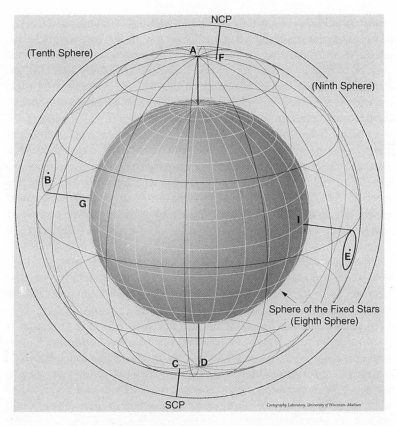

Figure 19. Alphonsine scheme explaining the variation in the precession of the equinoxes called trepidation. The north and south celestial poles are labeled *NCP* and *SCP*, respectively. Cartography Laboratory, University of Wisconsin–Madison.

the eighth through a revolution requiring some 36,000 years, according to Ptolemy's numbers. In order to introduce the necessary variations into this motion, Thābit suggested that the equinoxes of the eighth sphere (*G* and *I*) are moved in small circles (of radius nine degrees in the Alfonsine version) centered on the equinoxes of the ninth sphere (*B* and *E*). This oscillatory motion, or trepidation, had the effect of alternately augmenting and diminishing the precessional motion of the ninth sphere, which was just what seemed to be required by the historical observations.

By the time he issued the 1581 *Sphaera* (*nunc iterum*), Clavius was far more cautious in his presentation of the theory of trepidation than he had been in his first edition.[59] In 1581 he added this comment to the end of the paragraph in which he had first introduced the motion of trepidation, which he left otherwise unchanged: "But as we must truly admit, this kind of motion or something similar may be undoubtedly conceded on account of phenomena, or appearances, that we will present a little later. But it is quite uncertain whether it should be done as the Alfonsines teach. Indeed, many absurdities seem to follow from it, as we will explain elsewhere."[60] So Clavius has adopted a critical attitude toward the theoretical explanation of trepidation, though not toward the observational data on which the phenomenon was supposedly based. He is not yet ready to reject the Alfonsine mechanism of trepidation, but it is clear that his acceptance of it is no longer unconditional as it was in 1570. Moreover, he explicitly alludes to an alternative ("this kind of motion or something similar"), thus distancing himself from the doctrine of the Alfonsines and Thābit.

The same caution about trepidation appears once again in the 1581 *Sphaera*. When Clavius was about to explain what phenomena led astronomers to the theory of trepidation (as he had promised to do), he added a sentence (indicated by my italics) acknowledging that astronomical opinion was less than unanimous on this matter. "Now it is to be shown what phenomena, or appearances, motivate astronomers to attribute this motion to the heavens. *Some, in fact, think that the studies of astronomers should completely reject this motion as ridiculous.* First, therefore, we observe that the fixed stars advance nonuniformly from west to east."[61] As is so often the case, Clavius does not tell us to whom he refers in the italicized phrase. He is probably thinking, at the very least, of Copernicus, who had presented his own replacement for Alfonsine trepidation in his *De revolutionibus*. It may even be a reference to himself, for Clavius would eventually join Copernicus in rejecting Alfonsine trepidation and was clearly already leaning in that direction in 1581 or earlier.

Clavius drives home his doubts about the theoretical soundness of trepidation by adding, also in 1581, this concluding passage to the discussion of precession and trepidation: "Although it seems certain that either trepidation or something similar should be conceded in the eighth sphere on account of the stated appearances, the way in which astronomers explain it is quite uncertain. For truly many things that seem to conflict with the observations follow from the position that the heads of Aries and Libra in the eighth sphere describe circles around the heads of Aries and Libra in the ninth sphere, the radii of which circles are nine degrees. This we will fully explain in the theory of the eighth sphere."[62] The promised theory of the motion of the eighth sphere along with its explanation of the conflict between the observations and the theory of trepidation did not appear until the 1593 *Sphaera* (*nunc quarto*).

Why was Clavius questioning the accepted Alfonsine scheme for explaining the motions that he and his contemporaries, including Copernicus, thought had been observed in the eighth sphere? We get at least part of the answer from a comment that he made in the middle of the discussion of precession and trepidation. This comment, like the other remarks discussed above, appeared first in the 1581 *Sphaera*. He cannot, he tells us, pass in silence over two arguments proposed by "a certain very learned and noble man who flourished not many years ago." Unfortunately, if not unexpectedly, Clavius gives us no further clues about who this man was. The arguments purport to show that the west-to-east precessional motion of the stars does not exist, and Clavius digresses briefly from his presentation in order to refute this position. (These comments are not directed at Copernicus, who acknowledged the phenomena, though he explained them by the earth's motion.) The content of these obscure arguments is of no importance in this discussion, but what is important is Clavius's explanation that he came across these arguments in "certain texts . . . that I studied while on the commission that convened recently by order of the pope for the correction of the Roman calendar."[63]

The digression on those arguments is part of the same set of revisions as the statements in which Clavius began to back away from Alfonsine trepidation theory, namely, the revisions to the 1581 edition. His work on Pope Gregory's commission for calendar reform would undoubtedly have required extensive reading on the subjects of precession and trepidation theory—which are essential in calendrical work for the calculation of equinox dates, among other things. Thus it seems plausible that his reservations about Alfonsine trepidation theory began with his research for the calendar reform work. Since he would eventually prefer the geostatically adapted Copernican scheme for precession, perhaps it was Clavius's reading of *De revolutionibus* itself that stimulated his backing away from traditional trepidation theory. Clavius's use of Reinhold's *Prutenic Tables,* which the Jesuit employs in his *Romani calendarii explicatio,* would also have introduced him to Copernicus's alternative.[64]

In any case, by the time of the 1593 *Sphaera,* (*nunc quarto*), Clavius had abandoned Alfonsine trepidation completely and embraced a version of Copernicus's precession theory, which he then followed in all subsequent editions of the *Sphaera*. Clavius, it should be emphasized, did not see himself as a follower of Copernicus. Rather, he felt that he was presenting an explanation of precession that reconciled the acknowledged advantages of Copernicus's mathematics with the geocentric-geostatic constraints of the Ptolemaic cosmology, just as Magini had tried to do for the planetary theories. The nature of his dependence on Copernicus is expressed succinctly in the new clause of his 1593 *Sphaera's Ad lectorem* in which he advertises "a very useful disputation on the fourfold motion of the eighth sphere according to the periods discovered by Nicholas Copernicus, wherein the vanity of the motion of trepidation is confounded by most valid arguments and the eleventh sphere is shown to be the first movable sphere."[65] Thus Copernicus is important for the periods of the motions that he assigned to the eighth sphere, but not, of course, for his explanatory principles involving the motion of the earth.

The *Ad lectorem* advertises a confutation of trepidation, and Clavius makes this his first task in the *disputatio*. Clavius notes the difficulty of the undertaking, since to determine without doubt the real motions of the fixed stars would require observing a complete cycle of many centuries, but the world, he says, is only about one-fourth as old as the entire cycle. Lacking observations of a complete cycle, Clavius explains briefly how variations in the rate of precession led astronomers to formulate the theory of trepidation.

Yet, Clavius relates, this theory implies absurdities that conflict with other observations of astronomers. For one thing, the theory of trepidation implies that not all the stars would move in the same way, because near the equinoxes (where the small circles are) the paths of the stars would be nearly circular, whereas near the solstices the motion would be in straight lines east and west. But the actual phenomena refute this because all the fixed stars always have the same motion. Next, the sun's declination at the equinoxes should vary by as much as nine degrees north and south, but in fact the meridian altitude of the sun at the equinoxes is never seen to vary for a given location. Finally the theory of trepidation results in absurd and unheard of variations in the maximum solar declination and the length of the year. "It comes to this, that authorities ought not perpetuate the teaching of this [Alfonsine] trepidational motion from which [various quantities] can be calculated. For they used to think that calculations according to the motion of trepidation corresponded at least to the phenomena and observations, but it is argued in this matter that that motion does not exist in nature and is, rather, fictitious and concocted without any foundation."[66]

So much for trepidation. To replace it he will call on Copernicus, that "distinguished restorer of astronomy in our era," whose replacement for trepidation appeared in *De revolutionibus*.[67] Clavius explained that Copernicus devised a new scheme involving four motions (as opposed to the three motions in the Thābit-Alfonsine trepidation scheme) by which one can calculate the maximum solar declination, the nonuniform motion of the fixed stars, and the length of the year. Copernicus, Clavius explains, made two of the motions complete (*absolutos et perfectos*) in that they are full revolutions from east to west or west to east, and the other two incomplete insofar as they are librations over limited north-south or east-west ranges. "However," he adds almost immediately,

> just as we freely accept and welcome that fourfold motion of the eighth sphere according to the periods set out by Copernicus, so the way in which he uses it in his account we completely reject. For he assumes hypotheses that are, in a way, discordant and absurd, as with those latter two motions, or better, librations of the eighth sphere that he puts before our eyes. [His hypotheses] are foreign to common sense, indeed I say they are rash, as that the sun stands motionless in the center of the cosmos, but that the earth, endowed with multiple motions and along with the rest of the elements and the sphere of the moon, takes its place in the third heaven, between Venus and Mars.[68]

It is well known that Copernicus's theory called for the earth to rotate on its axis once a day and to revolve about the sun once in a year. In addition, he called for the earth to rotate slowly westward about the poles of the ecliptic so as to have the effect of keeping the earth's axis fixed in space. This was necessary because, in Copernicus's time as for millenia before, a planet was assumed to be fixed in the orb in which it is borne. (This is in contrast to the modern gyroscopic conception in which the axis of a rotating planet maintains a fixed orientation in space—in the absence of torques about its axis—as the planet completes its orbit.) If fixed in its deferent orb, a planet's oblique axis would complete a full revolution at the same time that the planet completes an orbit, so Copernicus had to give the earth a conical rotation (conical because it causes the earth's axis to generate a cone) in order to counteract this effect and render the earth's axis fixed in space. In fact, the earth's axis is not fixed but precesses slowly. So by defining the period of the earth's conical rotation to be a fraction less than a full year, Copernicus could explain the slow westward precession of the equinoxes.

Copernicus's two rotations of the earth, the daily and conical, are presumably the motions that Clavius refers to as "complete." The ones that he calls "incomplete" motions would be librations, or oscillations, of the earth's axis that Copernicus had introduced in order to explain phenomena that both he and Clavius (and essentially all other astronomers of the sixteenth century and before) thought to be real, namely, the variable rate

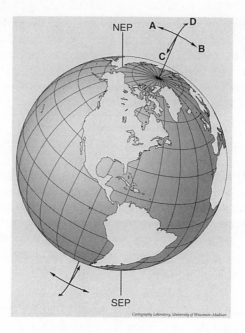

Figure 20. The librational motions of the earth used by Copernicus in his precession theory as a replacement for medieval trepidation theory. The north and south ecliptic poles are labeled *NEP* and *SEP,* respectively. Cartography Laboratory, University of Wisconsin–Madison.

of precession of the equinoxes, the variable length of the tropical year, the variable maximum solar declination, and the like that trepidation theory had been invoked to explain. In place of Alphonsine (so called for its codification in the Alphonsine Tables) trepidation theory, Copernicus suggested that the earth's axis undergoes two back-and-forth motions that he calls librations. A libration, for Copernicus, is an oscillatory motion of the earth's axis through a small angle. The Copernican librations are mutually perpendicular (see fig. 20). One oscillates north-south (that is, toward and away from the ecliptic poles along arc *AB*) through twenty-four minutes of arc with a period of 3,434 years so as to make the pole approach and recede from the solstices. The other oscillates in the east-west plane (perpendicular to the first libration along arc *CD*) through an arc of two degrees, twenty minutes with a period of 1,717 years.[69]

These librations of the earth's axis will cause each pole to trace out a figure 8 lying along the solstitial colure.[70] But for Clavius, to have the earth's axis move in such a way, or—as he would probably put it—by such a combination of motions, is almost inconceivable. "But who cannot see that this is self-contradictory? For if the pole inches its way up and down along the colure, who can understand how it can, at the same time, stray beyond the colure? Or if it does stray away here and there, how can it be moved at the same time up and down along the colure? In any event, I freely confess that I shall never be able to understand fully this contrariety.''[71]

How, then, is it possible to obtain the benefits of Copernicus's precession theory without the many physical and philosophical problems it raises? ''Concerning this issue the most learned Giovanni Antonio Magini of Padua, having rejected those [Copernican] hypotheses and retained the periods of the motions that Copernicus set up, endeavors to defend that fourfold motion of the eighth sphere through the customary hypotheses accepted by all astronomers and philosophers, namely, as common sense demands, he postulates this terrestrial globe, devoid of motion, in the center of the universe.''[72] Clavius explains that, just as the Alfonsine astronomers had posited two spheres above the firmament, in accord with their belief in three distinguishable motions of the fixed stars (precessional, trepidational, and daily motion), so Magini has concluded that there must be three spheres above the firmament, since Copernicus has shown the necessity of four motions of the fixed stars (precessional, daily, and the two librational motions). Clavius proposes to present Magini's theory noting, however, that he thinks his own explanation is somewhat more comprehensible than Magini's, as it has fewer lines and circles.

Thus, starting from the outside, we have the eleventh sphere, which rotates most swiftly (once in twenty-four hours) and is presumably the source of the *motus raptus* that carries all of the lower spheres along with that same motion (see fig. 21). Next is the tenth sphere, which rotates daily according to the motion of the eleventh but contributes the north-south librational motion along the solstitial colure. True to Copernicus's parameters, the tenth sphere oscillates through an angle of twenty-four minutes of arc with a period of 3,434 Egyptian years (along *AB*). The ninth sphere rotates with its superiors and moves according to the motions they confer, namely, the daily motion of the eleventh sphere and the north-south libration of the tenth. The ninth sphere adds the east-west libration, which spans, as in Copernicus, two degrees, twenty minutes of arc with a period of 1,717 Egyptian years (along *CD*). Finally follows the eighth sphere (not shown), the firmament itself, which exhibits the motions of all its superiors and further has the slow eastward motion about the poles of the ecliptic by which it completes a revolution in 25,816 Egyptian years. This, again, is Copernicus's number.[73] The arrangement and motions of these four spheres make it clear, Clavius tells us, why, at some

12th is empyrean see p. 179

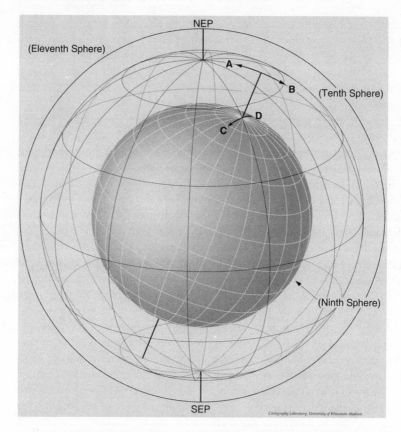

Figure 21. Clavius's scheme for transferring Copernicus's precession model into a geostatic framework. The north and south ecliptic poles are labeled *NEP* and *SEP,* respectively. Cartography Laboratory, University of Wisconsin– Madison.

times in the past, astronomers observed the fixed stars to move faster or slower than at other times. This theory also explains, he tells us, historical observations of variations in the length of the year and of the maximum solar declination.

It is unfortunate that Clavius did not see fit to explain some of the mechanical novelties of Magini's arrangement. For instance, the librations of the ninth and tenth spheres are, as Clavius himself points out, incomplete, presumably, in the sense that they are not complete revolutions. But those librational motions only bother Clavius as long as they are associated with the earth; once attributed to a celestial sphere, he seems to have no problem with them. Yet spheres with such linear motions have no exact counterparts in the Alfonsine scheme, and the novelty would seem to call for some justification. Despite the fact that Copernicus in *De revolutionibus* had felt compelled to justify the librations by showing that they could, at least, be derived from uniform circular motions, Clavius offers no such apology.[74]

Clavius's presentation of Magini's adapted Copernican scheme for the motions of the fixed stars can be seen in a historical light similar to Reinhold's use of Copernicus's planetary models in the *Prutenic Tables*. That is to say, by decoupling Copernicus's cosmological assumptions from his mathematical techniques, it became possible for those with geostatic and geocentric convictions to appreciate some of the virtues of Copernicus's work. The total acceptance of Copernicus's parameters by such eminent figures as Clavius and Magini could only help to build the Polish astronomer's reputation—for technical perspicacity, at least, if not philosophical sensitivity. Of course, the less technical level of Clavius's *Sphaera* made its audience very different from the one that used Reinhold's *Prutenic Tables*. Indeed, the introductory textbook nature of the *Sphaera* means that Clavius's considerable praise of and dependence on Copernicus would have been studied by, and perhaps even have influenced, a far greater number of readers than Reinhold's text.

Tycho Brahe generally gets credit for having dispensed with trepidation. Dreyer wrote, "The authority of Tycho Brahe was so great, that the mere fact of his having ignored the phenomenon of trepidation was sufficient to lay this spectre, which had haunted the precincts of Urania for a thousand years, and possibly much longer."[75] This is a curious kind of claim, and even if true, it is still an oversimplification. Clavius's 1593 renunciation of Alphonsine trepidation theory came almost a decade before Tycho's *Astronomiae instauratae progymnasmata,* published late in 1602. Clavius's rejection of trepidation while still accepting the reality of the phenomena represents an intermediate stage between the Alphonsine position and Tycho's, and it shows that the historical development was more than a simple matter of Tycho's negative opinion. Furthermore, Clavius's pronouncement probably helped prepare the way for Tycho's position, for Clavius's negative opinion on trepidation must have been much more widely known than Tycho's, at least before 1602 and perhaps for some years afterward.

Clavius's *Theorica planetarum*

In the medieval Latin tradition of astronomy education, the *Sphere* commentary was an introduction to the accepted cosmology and to certain basic concepts of spherical astronomy. Significant discussions of Ptolemaic planetary theory were routinely addressed in a different genre, the *Theorica planetarum*. The original *Theorica planetarum,* the author of which is unknown, appeared sometime between 1260 and 1280 and quickly became a standard member of the group of texts used for astronomy instruction in the medieval European universities.[76] In the middle of the fifteenth century, Peurbach composed his *Theoricae novae planetarum,* which corrected many errors and weaknesses in the older *Theorica* and introduced to European astronomy a complete set of planetary models based on materialized eccentrics and epicycles.[77]

It would be natural that Clavius should aspire to compose his own commentary on the *Theorica* as part of his program to produce a set of textbooks on mathematical subjects for use in the Jesuit schools. There can be no doubt that he intended to do so. In the very first edition of the *Sphaera* (1570), at the beginning of the fourth chapter, Clavius explained that Sacrobosco's treatment deals primarily with the motions of the sun and moon. But Clavius says that his commentary will consider these things only briefly, not only because Sacrobosco himself is brief but "especially because this subject demands a longer discussion and pertains to the *Theorica planetarum,* on which we will soon write a commentary, God willing."[78] Subsequent editions contain frequent references to a forthcoming *Theorica*: "as we will show in the *Theorica*" or "as will be explained in the *Theorica,*" and so on. In fact, Clavius's very last edition of the *Sphaera,* published in 1611 only a year before his death, still contains almost exactly the sentence quoted earlier from the first edition, except that instead of promising to write it he now refers to his *Theorica planetarum,* "which, God willing, we will soon publish."[79] So it certainly seems that Clavius had gone a long way toward completing his *Theorica*—far enough, at least, to speak of publication in the near future.

Further evidence of his intentions in this regard come from a curriculum document preserved in the Jesuit Archives in Rome. It is the middle member of a set of three, all written by Clavius. The first describes a curriculum for those who need only minimal training in mathematics, and the last details the subjects to be required of those who will specialize in mathematics.[80]

The middle document, written in Clavius's hand and entitled "Ordo secundus brevior," is a list of topics and texts for students needing a mathematical background of only intermediate depth. The right-hand column contains an item-by-item list of subjects to be taught, while the left column indicates what text or texts can be used for that particular subject. Nearly every subject in the right column has beside it in the left column the notation *noster* ("ours"), presumably indicating that Clavius either has or intends a textbook for that subject. Most of the items, in fact, correspond to some textbook published by Clavius at some time. Occasionally an alternative text appears on the list. Subject 16 is "Treatise on the motions of the planets and the eighth sphere along with the Alfonsine Tables, etc." The textbook column says "ours, or Regiomontanus's *Epitome of the Almagest*." This is one of only two items out of nineteen on the list that bear the notation "ours" but do not correspond to one of Clavius's published textbooks. The other subject is "Geographia." Clavius never published a geography textbook, although he did cite a *Cosmographia* as if he had written his own.[81] Perhaps, like his *Theorica,* his *Cosmographia* was planned but never made it into print.

The "Ordo secundus," probably drawn up before 1586, seems to indicate that Clavius had written a *Theorica planetarum* but was postponing its publication. This inference is reinforced by the corresponding item in the "Ordo tertio," (the last document of the set) where, next to the entry for "Tractatio de motibus planetarum," he has written, "this we are publishing." However, he follows with a remark that indicates some hesitation: "But perhaps it would be more authoritative if we would write a commentary on Regiomontanus's *Epitome.*"[82] Presumably Clavius worried that he ought to choose a text, such as Regiomontanus's, more modern than the medieval *Theorica planetarum.* In any case, as late as 1611, Clavius was still promising to publish his *Theorica.* If by then he conceived it more as a commentary on Regiomontanus than a traditional *Theorica planetarum,* he gave no such indication in the *Sphaera.*

Despite the strong indications that he had once written a *Theorica planetarum,* no trace of such a text by Clavius was known until very recently. Among the manuscripts preserved in the archives of the Pontifical Gregorian University in Rome, Ugo Baldini has identified two major sections of Clavius's *Theorica*—the entire section on solar theory and part of the lunar theory.[83] Baldini estimates that the manuscript version dates from about 1577 and judges that the text is not a draft intended for publication but, rather, a tract intended for the use of Clavius himself and his students. Neither the solar nor lunar theories expounded by Clavius diverge significantly from those of the *Almagest.* The theorical treatises seem less concerned with cosmology than with the establishment (by detailed and formal geometrical demonstrations) of the mathematical foundations of the Ptolemaic theories. In this way Clavius's unpublished *Theorica* resembles his book on the astrolabe, published in 1593, in which he placed great emphasis on the geometrical formalisms underlying the design and use of the instrument—a feature that distinguishes it from most other astrolabe books of its day. This occupation with the formal mathematical foundations of astronomy, whether Ptolemaic theory or astrolabe, is consistent with Clavius's lifelong goal of establishing the mathematical sciences as legitimate peers of the natural sciences.

One would not expect Clavius's *Theorica planetarum* to be a work of great innovation, and it is not. "He introduces no new phenomena, presents no new values for phenomena commonly accepted, presents no new observations, and does not modify in any substantial manner the traditional procedures of analysis and measurement." However he, "retained his primary goal of expounding the general patterns by which the existing astronomy was constructed and on which its cognitive status was based."[84] In other words, as with many of his other definitive textbooks, such as his *Euclid, Astrolabium,* and *Sphaera,* Clavius strove to present a grand summary of the traditional subject to his reader. And he did so in unflinching detail.

Clavius's reasons for withholding the long-promised text are unknown but probably not a great mystery. Though he apparently began the project, and completed substantial parts of it, in the middle 1570s, Clavius's work on the calendar reform and its defense (not to mention his steady teaching duties) would occupy him heavily for the next fifteen years or so. Moreover, as Baldini notes, Clavius awaited the forthcoming works of Magini and Brahe on planetary theory.

> Until about 1600, [Clavius] could think that a renovation of the observational base of astronomy would offer new solutions. . . . But after that date, the steady emergence of Brahe's work and the observations or new interpretations of astronomical phenomena (novas, distances of comets) would have had to eliminate, even in his view, this means of escape. One could offer a commentary on the *Almagest* as a historical, systematic encyclopedia of methods, problems, observations, and theories; but not as a conceptual synthesis that would resolve, in the traditional physical framework, the problems that the Ptolemaic model by now appeared constitutionally unsuitable to solve.[85]

Perhaps Clavius's goal was to produce a corrected and updated *Theorica planetarum,* even incorporating Copernican mathematical (but not cosmological) innovations along the lines of Magini's *Novae coelestium orbium theoricae* of 1589. Might he then have considered Magini's work sufficient in this regard to justify suppressing his own? This does not seem reasonable if for no other reason than that the three major revisions of the *Sphaera* that appeared after Magini's work (1593, 1607, and 1611) would have given Clavius ample opportunity to cite Magini instead of promising his own volume, and this he never did.

Could it be that in the course of preparing his *Theorica* Clavius found it impossible to resolve certain technical difficulties that, in turn, forced him to postpone publication? For an astronomer like Clavius who insisted that proper method could establish the reality of the mechanisms of mathematical astronomy, it was imperative that all the observable consequences of the mechanisms correspond to the observations. Yet Ptolemy's lunar model presents a notorious problem in this regard. It posits a lunar epicycle on a movable eccentric circle, which together produce such large variations in the moon's distance from the earth that the consequent variations in lunar parallax and apparent diameter should be dramatically larger than those actually observed. The discrepancies between the prediction of theory and the actual observations would be large enough to be visible to the most perfunctory of observers. Yet Ptolemy and most of his successors up to the time of Copernicus passed over the problem in silence.[86] The same can be said of Clavius, who never admits to such a severe problem with Ptolemy's lunar model despite his many assertions of the reality of Ptolemaic mechanisms. In fact, the lunar model presented in the surviving portion of Clavius's theory of the moon follows Ptolemy closely and does not address the problem of the lunar diameter.[87]

The vagaries of the moon's motion have always made it a stringent test of the accuracy of astronomical theories. Clavius would have faced a formidable task to create a lunar model that equaled Ptolemy's positional theory and also correctly predicted the moon's distance and thus its apparent diameter. And it is exactly this task that his own standards would have imposed on him. We have already seen that he was concerned to solve the difficulties with Alfonsine trepidation theory and found his solution in Magini's adaptation of Copernican precession and libration theory. In fact, a theoretical solution to the lunar problem was easily at hand in the same place, namely, Copernicus's *De revolutionibus*. Moreover, Copernican lunar theory, unlike Copernican precession, carried no objectionable baggage owing to the earth's heliocentric orbit. The theory solves the problem of variations in lunar diameter and parallax nicely while doing so with minimal violence to Ptolemy by using a double epicycle in place of a movable eccentric center.[88]

But it is a great oversimplification to suggest that Clavius might merely have adopted Copernican lunar theory had he truly wanted a better one. Copernicus's theory was available to Magini as well, yet he went to some trouble to construct a theory that is neither Ptolemaic nor Copernican. His reason was that he considered epicycles to be unsuitable for the moon's motion. This should remind us that theory choices often depend as much on the theorist's ideas of how things ought to be as on the predictive success of the theory itself. Perhaps in Clavius's view, it was one thing

see notes 86 and 88, p. 256

to reject Alfonsine trepidation theory, for which tradition allowed modification through the interaction of theory and observation, but quite another to challenge the planetary theories, which traditionally followed Ptolemy closely.

Clavius's letter to Magini, in which the Jesuit earnestly requested that his colleague publish certain observations, suggests another possible explanation for Clavius's failure to publish his *Theorica*. The letter specifically asked for the observations that Magini had used in constructing the planetary theories that he had published some years earlier. Perhaps Clavius simply wished to check Magini's theories against the data on which they were based. But it is also possible that Clavius needed observations in order to carry out his own theoretical researches. In this period, around 1595, although the Collegio Romano had moved into the grand palazzo that Pope Gregory XIII had provided for them, the Jesuit astronomers still had no access to a permanent observatory and thus were still using portable instruments on rooftops and terraces. These circumstances would not have been conducive to the accumulation of a respectable body of accurate observations, which would explain Clavius's interest in Magini's data. Magini's failure to share his observations would presumably have slowed Clavius's theoretical work.

Clavius's letter to Magini suggests yet another possibility for the Jesuit's delay in publishing his *Theorica*. In the letter, Clavius encouraged Magini to proceed without waiting to see if Tycho might yet publish something. If Magini, who had already committed his theories to print, was hesitant to divulge his data in anticipation of ''whatever Tycho the Dane will do,'' how much more reluctant might Clavius himself have been (despite his words to Magini) to publish a theoretical text that might be rendered superfluous or even obsolete by the forthcoming work of the formidable Tycho? On the other hand, the volumes of Tycho's *Progymnasmata* were published posthumously in 1602 and 1603, and if that had any effect on Clavius's inclination to publish his own *Theorica* there is no indication of it in the continuing promises to do so contained in the 1607 and 1611 editions of the *Sphaera*.

In summary, Clavius's *Sphaera* helps us see that Ptolemaic cosmology was under critical pressure in many ways. It was, of course, in direct competition with the Copernican and, later, the Tychonic systems in regard to its ability to explain and predict the familiar phenomena of astronomy. But Clavius's text reveals that, aside from Copernicus, there were still other cosmologies, especially the fluid heavens and homocentrics, that presented serious challenges to the priority of the Ptolemaic. Evidently, celestial phenomena, the novas in particular, demanded concessions from Clavius, the Ptolemaic defender, especially on the issues of corruptibility in the heavens.

Ptolemaic cosmology also had problems that were internal to itself—problems that prompted astronomers like Magini and Clavius to seek ways of adapting the Copernican mathematical devices to a geocentric-geostatic cosmology. The failure of Alfonsine trepidation theory to predict correctly the phenomena, which sixteenth-century astronomers believed to be real, constituted another pressure to which Clavius and Magini responded by adapting the Copernican precession and libration devices to Ptolemaic cosmology in the time-honored way. That is to say, they added another moving sphere, for a total of eleven (or twelve, counting the empyrean), to the structure of the cosmos and redefined the functions of some of the others.

Finally, Clavius's failure to publish his *Theorica planetarum* (and perhaps even to complete it) suggests to us the difficulty faced by an astronomer committed to Ptolemaic cosmology in reconciling the contradictory demands of cosmological tradition, Copernican mathematics, and the new celestial phenomena. For the historian of astronomy, of course, it would be very interesting to have Clavius's mature thoughts spelled out in a complete and published *Theorica planetarum*. But for Clavius himself, it might have been a good thing that he never brought it out, for in 1610 Galileo's telescopic discoveries would present the final and most severe challenge to Ptolemaic cosmology. Clavius would remark that Galileo's discoveries should give astronomers pause to consider how the new findings might be reconciled to the traditional views, and that reconsideration would naturally have applied to his *Theorica* had it already been made public. Clavius's involvement in Galileo's telescopic discoveries and the responses they provoked from him will be the subject of the final chapter.

SEVEN

Galileo, Tycho, and the Fate of the Celestial Spheres

Now it is our advantage, that by the help of Galileus's glass, we are advanced nearer unto them, and the heavens are made more present to us than they were before.
—John Wilkins, *Discovery of a World in the Moone* (1638)

In the last years of Clavius's life, another set of remarkable events—this time human rather than celestial—transpired to alter the discipline of astronomy. Galileo discovered that the telescope could reveal wonderful and unexpected things in the heavens. Tycho Brahe's rearrangement of the celestial spheres was gaining popularity and, along with it, the concept of the fluid heavens made a return.[1]

Clavius, Galileo, and the Telescope

During the year 1609, Galileo heard accounts of the instruments that we now call telescopes, learned how to make reasonably good ones, and learned how to use them for astronomical observing. By March 1610 he was able to publish his astronomical discoveries in his *avviso astronomico* (astronomical notice), as he occasionally referred to it, which he titled *Sidereus nuncius*. Confirmation of Galileo's observations—though not always his interpretations—accumulated steadily during the balance of 1610. In the forefront of this first generation of telescopic astronomers were the Roman Jesuit mathematicians: Clavius, his students, and his colleagues.[2]

During 1610 Clavius was in the process of revising his final edition of the *Sphaera* for inclusion in what would be the five-volume set of his collected works, entitled *Opera mathematica*. The third volume, which contains the last *Sphaera* he revised, was published in 1611.[3] The timing of the publication is fortunate. Had the *Sphaera* been issued in one of the volumes of the *Opera mathematica* already in print, it might have appeared too early for Clavius to include a notice of Galileo's discoveries. Had it been later, Clavius's health might have interfered with his writing and revisions. He fell ill for a time in mid-1611 and died in February 1612, just as the last sections of the final volume of his *Opera mathematica* were going to press.[4]

In that final version of the *Sphaera,* after enumerating Galileo's discoveries, Clavius made a statement that clearly recognized the significance of Galileo's observations for theoretical astronomy. However, there was (and is) some disagreement over exactly what he meant by his statement, "Since things are thus, astronomers ought to consider how the celestial orbs may be arranged in order to save these phenomena."[5] John Wilkins, the English Copernican, maintained that this statement showed that Clavius had abandoned Ptolemaic cosmology, while Alexander Ross, an English anti-Copernican, held that it merely called for some modification of Ptolemy. Galileo, also a Copernican of course, interpreted Clavius's statement as follows in his "Letter to the Grand Duchess": "I could name other mathematicians who, moved by my most recent discoveries, have admitted the necessity of altering the heretofore accepted system of the world, which in any case can endure no longer."[6] Paolo Antonio Foscarini, still another Copernican, interpreted Clavius as urging the search for a replacement cosmology.[7] The Jesuit Christoph Scheiner, who favored a geocentric cosmology of fluid heavens of a sort, also read Clavius as calling for fundamental change: "Christoph Clavius . . . advises astronomers that, on account of these new, and until now unseen, phenomena, they doubtless ought to seek another system of the heavens."[8]

Some modern historians of astronomy have interpreted Clavius's remark in a way similar to Scheiner's interpretation. In Robert Westman's account, Clavius said that Galileo's telescopic observations would have to be accommodated to a new system of the world, while William Donahue judges that Clavius's remarks suggested the adoption of the Tychonic system.[9] The most extreme modern evaluation has come from Pasquale D'Elia. D'Elia declared it beyond doubt that "before his death Clavius had come to see the necessity of considering the modification of the ancient positions in order to adopt the Copernican system."[10] Yet the scant and indirect evidence he offers for this startling statement is completely inadequate.

With all due respect to learned commentators, ancient and modern, it must be noted that Clavius's words themselves make no mention of any new system of the world. He speaks specifically of the arrangement of the celestial spheres and their consistency with the phenomena. In fact, were it not for the placement of this remark in association with Galileo's discoveries, the language is familiar enough that we might not be surprised to find such a statement in his accounts of homocentric spheres or trepidation theory. Jerome Langford was much more accurate when he stated that Clavius had admitted "that Ptolemy's system in *forma pura* was no longer tenable."[11] Ironically Galileo's evaluation, like Langford's, was pretty fair in interpreting Clavius's statement in the more moderate terms of alteration rather than revolution.

How did Clavius intend his own remark to be interpreted? Was he calling for only a minor adjustment in the Ptolemaic system, perhaps comparable in scope and significance to his own replacement of the traditional trepidation mechanisms with the adapted Copernican equivalents? Or was he actually stating that the Ptolemaic system was finished and agreeing to the necessity of a new approach, though without endorsing any particular one? No amount of scrutiny will yield an answer if we study only that single famous sentence. From early in 1610 until mid-1611, Clavius and the Roman Jesuits seriously evaluated and then publicly accepted Galileo's telescopic discoveries. In doing so, they influenced others to do the same. We must examine this sequence of events if we are to understand how it influenced Clavius and his colleagues. It is fitting that this process constituted the final chapter of Clavius's astronomical career, for by it, whether intentionally or not, he helped formulate the new cosmology that would replace the traditional one he had defended all his life.

Galileo was very familiar with Clavius and the Roman Jesuits. Adriano Carugo, Alistair Crombie, and William Wallace have investigated the strong influence of the Jesuit mathematicians and philosophers on Galileo. Wallace has detailed Galileo's heavy borrowing, in his early teaching years, from Clavius's *Sphaera*.[12] In fact, Galileo and Clavius were personally acquainted early in Galileo's career and well before the notoriety brought about by his telescopic discoveries. A letter of 8 January 1588 from Galileo to Clavius refers to a visit to Rome by Galileo, at which time they apparently met. This would be the time, probably autumn 1587, when Galileo left with Clavius some of his theorems on centers of gravity of bodies and asked the senior mathematician to give his opinion of them.[13] In another, much later, letter to Galileo, 18 December 1604, Clavius apologized for the fact that a copy of his book on the astrolabe, intended for Galileo, had not arrived. He promised to send another and, in the meantime, offered a copy of his *Geometria practica*, "although it is not worthy of you, but I do it in order to continue the friendship between us."[14] Clavius also mentioned the great excitement over the nova of that year and asked Galileo to tell him of any observations he has made of it. If their correspondence was infrequent, it was nonetheless amicable.

We do not know when Clavius first saw Galileo's accounts of his discoveries in the *Sidereus nuncius,* but the Jesuit almost surely saw it as soon as it became available in Rome—which was probably soon after its publication at Venice in March 1610. However, Galileo's book was printed in a single edition of 550 copies, which sold out almost immediately, so it is possible that it took some time for a copy to become available in Rome at the Collegio Romano.[15]

The first testimony of Clavius's knowledge of Galileo's telescopic discoveries comes in a letter from Mark Welser. Welser and Clavius had been correspondents since 1602, and they continued to exchange letters until the latter's death.[16] On 12 March 1610, the very day that Galileo dated his dedicatory letter in *Sidereus nuncius,* Welser wrote to Clavius that he had received reliable reports from Padua that Galileo had, with an instrument of his own invention, discovered four previously unknown planets, many more fixed stars than had been seen before, and wonderful things in the Milky Way. "I know very well that 'slowness in believing is the sinew of wisdom'; so I will not commit myself to anything, but rather I pray Your Reverence to tell me freely and in confidence of your opinion concerning this matter."[17] Clearly Welser's contacts had sent him fairly accurate reports of Galileo's activities in advance of the publication of *Sidereus nuncius,* so it is possible that similar reports had reached Rome and Clavius, and that this was not news to him.

Clavius seems not to have replied to Welser, or at least not to his inquiry on Galileo, for Welser wrote to Clavius again on 7 January 1611 and enclosed an extract from a letter that Welser had received from Galileo. Welser wanted Clavius to verify the accuracy of Galileo's account. "Seeing as I have been persistent in my reluctance to believe in the new planets, now I must waver because of the contents of a letter from Sig. Galileo of 17 December in this tone: 'Finally there have appeared a few observations of the Medicean Planets made by some Jesuit fathers, students of Father Clavius, and by Father Clavius himself . . .' I wish that Your Reverence would confirm this news, insofar as it concerns you and your students, so as to clear it completely of doubt."[18]

This time Clavius responded. Welser wrote to him again on 11 February 1611 acknowledging the reply he had received. "Your Reverence's letter leaves me convinced and assured in my enjoyment of the wonders found by S. Galilei around Jupiter, Saturn, and Venus. Because up to now, despite his many affirmations, I have always had some reservations in this matter, knowing how easy it is to deceive oneself."[19] Welser clearly placed great weight on Clavius's opinion on this matter, and Clavius confirmed Galileo's famous discoveries.

In the meantime, Galileo had offered his services to the grand duke of Tuscany, Cosimo II de' Medici, who had accepted, and by the end of May 1610, Galileo was preparing to move to Florence and take up his post as the grand duke's mathematician and philosopher.[20] Letters from the summer and fall of 1610 between Galileo, Clavius, and Grienberger reveal what kind of activity Galileo's discoveries had prompted in the Collegio Romano itself to make possible Clavius's verifications of Galileo's work.

It seems that Clavius's first reaction to Galileo's telescopic discoveries was, like Welser's, skeptical. In a letter of 1 October 1610, Lodovico Cardi da Cigoli (a well-known artist and one of Galileo's friends in Rome) wrote to Galileo of the gossip going around, "These followers of Clavius, all of them, believe nothing [of the discoveries]. Clavius, among others, and the head of them all, said to a friend of mine concerning the four [Medicean] stars that [Clavius] was laughing about them and that one would first have to build a spyglass that creates them, and [only] then would it show them. And [Clavius] said that Galileo should keep his own opinion and he [Clavius] would keep his."[21]

It should be noted, first of all, that Clavius's alleged remark falls into the category of hearsay. History, like the law, must interpret such statements cautiously, and the overtones of ridicule are especially to be mistrusted. That said, however, the gist of the remark could easily be reliable. It had certainly occurred to others, one of them Magini, that the telescopic images of the Jovian moons could be spurious reflections created, in fact, by the telescope itself.[22] Thus it would certainly be going too far to say that "Clavius had . . . been reported as saying that in order to see such things one would first have to put them inside the telescope."[23] Not only does Drake here trivialize what is a nontrivial optical problem, but his harsh interpretation suggests that Clavius imputed intent of fraud to Galileo, which is not justified by Cigoli's own words. On the other hand, there have also been apologetic interpretations, as when the historian of the Collegio Romano dismissed Clavius's remark as having been made playfully (*scherzosamente*), though there is no evidence in Cigoli's account of the tenor of the remark.[24] A broader view of events leading up to Clavius's remark will help produce a more balanced interpretation.

During the summer of 1610, Clavius and his colleagues had been attempting, unsuccessfully, to repeat Galileo's observations. They were probably motivated, at least in part, by Welser's request in March for a judgment on the matter. In a letter of 17 September 1610, Galileo had given Clavius some pointers on the effective use of a telescope for astronomical observation. Galileo says that he has heard from Antonio Santini (then a Venetian merchant) that the Roman Jesuits had not yet succeeded in seeing the Jovian satellites. He does not find this surprising and suggests that the difficulty may be owing to a telescope of insufficient quality or a poor mounting. Galileo notes that a firm mounting is quite necessary, for when held by hand, even though steadied on a wall or other solid object, the mere motion of the heartbeat or breathing can render the images unobservable. Galileo closes by expressing his hope to come to Rome and personally demonstrate the truth of his discoveries.[25] It would be six months before he could fulfill that hope.

Clearly, then, the Collegio Romano astronomers had been attempting to use a telescope to confirm Galileo's reports, and this must have occurred already during the summer of 1610 if the news had had time to reach Galileo in Florence by means of Santini. The telescope that they were at first using was probably made in the Collegio Romano itself. On 22 January 1611, Christoph Grienberger, Clavius's successor and former student, wrote to Galileo that when he had returned to Rome from Sicily in autumn 1610, he learned "from one of us [Jesuits], Giovan Paolo Lembo, that before hearing anything about [your instrument] he had made some spyglasses himself; not by imitation of others, but rather by the power of inference. He observed both the lunar irregularities and the multitudes of stars in the Pleiades, Orion, and other [constellations], but he did not see the new planets."[26] Grienberger's mention of Lembo's telescope is confirmed by a letter of February 1611, written by another Roman Jesuit, Paul Guldin.[27]

It is not particularly surprising that Lembo should have been experimenting with telescopic optics in the summer of 1610. Starting in the Netherlands, primitive telescopes began appearing all over Europe and had reached Venice and even Naples by the summer of 1609, when, of course, Galileo himself first became aware of the device. Moreover, anyone with modest skill could have constructed a rudimentary version. Albert Van Helden states that "the 'secret' of the composition of the new device was not difficult to figure out. The knowledge that it consisted of two lenses in a tube was enough to allow most investigators to duplicate it within a short period of time."[28] Whether Lembo made his first telescope "not by imitation of others, but rather by the power of inference" is much more doubtful but not much more unlikely than Galileo's (somewhat exaggerated) claim that he built his first telescope "on the basis of the science of refraction."[29]

After noting Lembo's first attempt at a telescope, which could not reveal the new planets, Grienberger's letter goes on to record Lembo's further efforts. "Afterward, with considerable work and diligence,

[Lembo] managed to make a spyglass of great perfection . . . by which at last we could make out the new planets, at least in a clear sky."[30] Grienberger continues telling that Clavius then received a much better telescope from Santini, which, at the time Grienberger wrote (22 January 1611), he had been using for two months.

Favaro was apparently skeptical that Lembo could have constructed a useful telescope without somehow depending on Galileo's success in doing so. In his index to the Galileo *Opere*, Favaro noted with reference to Grienberger's letter that Lembo "claims to have made celestial observations with the telescope before having received notice of Galileo's observations."[31] But as we have seen above, this is, in fact, a reasonable possibility, and it is entirely plausible that Lembo, like many other European optical experimenters in 1610, was building telescopes. Moreover, nothing could be more natural than that a student of Clavius, Grienberger, and Maelcote, should turn his new device on the heavens. So there is really no good reason to doubt Grienberger's account of Lembo's activity.

Moreover, in the letter to Galileo, Grienberger claims only the independence of Lembo's early work. The chronology of events in the letter is not clear, but it is possible Grienberger meant that only Lembo's first telescopes and observations were independent of Galileo's work. He specifically notes that the observation of the Jovian moons was a later stage of Lembo's work, and there is nothing in the letter to exclude the possibility that Lembo had learned of Galileo's work before he observed the Jovian moons. Grienberger does not dispute Galileo's claim of priority in the important discovery of the Jovian moons. In fact, the spirit of the letter is quite otherwise. Grienberger concedes the superiority of Galileo's telescopes and praises his work. And when he mentions Lembo's observations of the Jovian moons he is not contesting priority. On the contrary, it is as if their detection were the goal and true test of the Jesuit telescopes.

Now let us return to Clavius's remark that to see the Jovian satellites one would have to build an instrument to create them. There is nothing to indicate when Clavius's remark was made—if, indeed, he made it at all. But it seems that it might already have been old news by the time Cigoli related it to Galileo. The letter recording the remark is dated 1 October 1610. Yet Galileo's letter to Clavius of 17 September (recall that the news of Clavius's efforts had to come through Santini) suggests that Clavius and his colleagues had already been trying to reproduce Galileo's observations for some time—at least for some weeks—presumably using a telescope built by Lembo. And the telescope from Santini probably did not come into Clavius's hands until very late in 1610.[32] All this means that Galileo must have been well aware of Clavius's serious attempts to confirm the Jovian satellites by the time he received Cigoli's letter in October. If Galileo was disturbed at all by Clavius's reputed skepticism, perhaps it spurred him to expedite the delivery of a good telescope to the Collegio Romano. In any event, he does not seem to have taken offense.

Even if we accept Clavius's alleged remark at face value, his skepticism is understandable in the light of Lembo's work. We can only imagine the poor optics to which Lembo might have subjected his associates as he taught himself to make telescopes. Far from dismissing Galileo's work out of hand, Clavius might have been basing his remarks on the unsuccessful attempts to use Lembo's telescope. And in any case, Clavius's attitude was not at all dogmatic, since his skepticism dissipated as soon as he and his colleagues succeeded—whether using Lembo's telescope or the one from Santini—in seeing the Jovian moons.

Galileo at the Collegio Romano

Galileo fulfilled his promise to visit Rome and arrived on 19 March 1611, by which time the Collegio Romano astronomers had confirmed, in addition to the Jovian satellites, his observations of Saturn's "threefold" appearance and the phases of Venus.[33] It is interesting that Clavius, in the final year of his life, was still actively involved with these observations, looking through the telescope along with the rest of the group, as Grienberger wrote to Galileo in the letter of 27 January 1611.[34] Galileo remained in Rome until late May 1611. During that time he attended multiple fetes in his honor and conferred with prominent personalities of the city; and Galileo wasted little time in seeking out Clavius and the other Jesuit astronomers. As he described in a letter to the Medicean courtier and patronage broker Belisario Vinta, his constant correspondent, Galileo spent his first day in Rome meeting with Cardinal Francesco Maria del Monte, who had close connections to the court of the grand duke.[35] On his second day, 30 March, Galileo related that "I was with the Jesuit fathers and conferred for a long time with Father Clavius and two other very attentive Jesuit fathers [Grienberger and Maelcote] and their students. . . . I found that these fathers, having finally recognized the truth of the new Medicean planets, have been observing them here continuously for two months and continue to do so. And we have compared [their observations] with mine and found that they agree. They are still working to determine the periods of their revolutions."[36] About which, Galileo adds, they are in agreement with Kepler in judging it to be a very difficult and almost impossible task.

By remarkable good fortune, a set of observations of the Jovian moons from just that time was preserved in Galileo's collected papers.[37] Though occasionally attributed erroneously to Galileo, these observations (fig. 22) were, in fact, done by Jesuit astronomers of the Collegio Romano in the months leading up to and during Galileo's visit to Rome and presumably copied by Galileo while he was there. This rare surviving example of early Collegio Romano astronomical research contains observations from 28 November 1610 through 6 April 1611. When Galileo visited Clavius in late March 1611, the Jesuits had been carefully observing the Jovian moons for at least four months—twice the span for which he had given them credit. The observations were especially frequent in December and January, and one suspects that perhaps only winter weather prevented nightly entries. As it stands, they averaged slightly better than every other night during those two months. The identity of the observer or observers is unknown, but it seems likely to have included at least Maelcote, who stated that he had been observing the Jovian moons regularly since the beginning of 1611.

Two weeks after the meeting with Clavius and the others, on 14 April, Galileo met a group of eight interested observers in the vineyards of Monsignor Malvasia outside the Porta San Pancrazio in a high open spot for a demonstration of his telescope.[38] The only Jesuit and Collegio Romano figure named as being in attendance was Johann Schreck (also known as Terrentius) who at that time was probably an advanced student. He would eventually travel to China as a missionary and there publish a number of books, several astronomical, in Chinese.[39] Also present was "Piffari" of Siena, who surely was Francesco Pifferi, the cleric from Siena who, in 1604, had published an Italian paraphrase of Clavius's *Sphaera* and must have been well known at the Collegio Romano. Also named as present were Antonio Persio, a member of the Accademia dei Lincei, Giulio Cesare Lagalla, a philosopher at Rome, and Johannes Desmiani (or Demisiani) called "il Greco," who was Cardinal Gonzaga's mathematician. Persio was probably a supporter of Galileo's views.[40] Lagalla was an obstinate opponent of Galileo's discoveries. Gonzaga also was a critic, though a reserved one, of Galileo,[41] which may tell us something of where Desmiani stood—if, as is likely, his stance reflected that of his patron. Given this mixture of participants, receptive and hostile (and perhaps even some who were open-minded), the demonstration could not have been as impressive as Galileo had doubtless hoped it would be. The chronicle states that the group remained observing almost seven hours in the night but without ever agreeing among themselves concerning the observations. Galileo also took the trouble during his visit to show some sunspots to at least the Jesuits Maelcote and Guldin. On 11 December

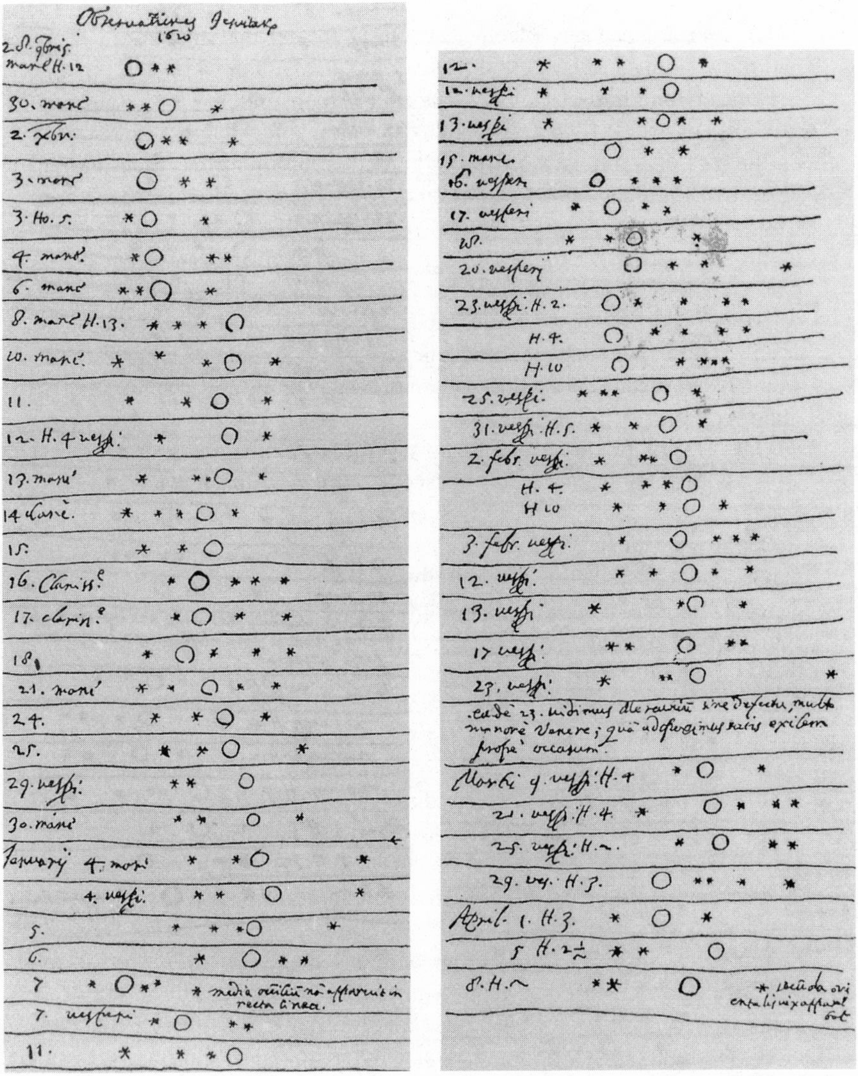

Figure 22. Galileo's copies of telescopic observations of the moons of Jupiter made by the Jesuits of the Collegio Romano between 28 November 1610 and 6 April 1611. From *Op. Gal.* 3, 2: 863–64. Photograph courtesy of the University of Wisconsin–Madison Memorial Library.

1612, over a year later, Maelcote wrote to Kepler that at Rome, Galileo had shown him sunspots with the telescope, and Guldin wrote to Christoph Scheiner telling him of Galileo's sunspot demonstrations.[42]

Galileo's visit to Rome succeeded in stimulating interest in his work on the part of those who had paid little attention up to that time. One of these was Cardinal Robert Bellarmine, long-time friend and colleague of Clavius. Bellarmine may have had his interest in Galileo's work stirred by the great fluttering in the Collegio Romano itself over telescopes, new planets, and other such marvels. In fact, he had been able to look at the celestial wonders through a telescope, though whether he used one of the Jesuits' telescopes is not certain. It is entirely possible that Galileo himself would have taken the trouble to demonstrate his discoveries to such an important man as Bellarmine.

On 19 April 1611, while Galileo was still in Rome and only five days after his marginally successful telescope demonstration outside the Porta San Pancrazio, Bellarmine sent a letter to "the Mathematicians of the Collegio Romano. . . . I know that Your Reverences are aware of the new celestial observations by a worthy mathematician using an instrument called a *cannone* or *ochiale*. By means of this instrument even I have seen some very marvelous things concerning the moon and Venus, but I wish that you would do me the pleasure of telling me sincerely your opinion concerning these things."[43] Bellarmine went on in his letter to ask specifically whether they agreed that (1) there are multitudes of fixed stars, invisible to the unaided eye and, in particular, that the Milky Way and nebulas are agglomerations of very dim stars; (2) that Saturn is not a single simple star but is three stars joined together; (3) that Venus changes shape, waxing and waning like the moon; (4) that the moon has a rough and uneven surface; and (5) finally, that around Jupiter four movable stars travel with various motions and speeds. Bellarmine concluded, "I want to know this because I hear various opinions spoken about these matters, and Your Reverences, versed as you are in the mathematical sciences, will easily be able to tell me if these new discoveries are well founded, or if they are rather appearances and not real." The cardinal was not asking for an elaborate or formal reply; he closed his letter saying, "If you want, you may reply on this same sheet of paper."[44]

The mathematicians replied promptly and, in the spirit of Bellarmine's request, succinctly. In only five days they drafted and sent a response that was signed by Clavius, Grienberger, Maelcote, and Lembo.[45] Their letter, dated 24 April 1611, confirmed that the *occhiale* reveals many stars in the "nebulas" of Cancer and the Pleiades, yet they declined to declare as certain that the Milky Way consists of minute stars, since it seemed that there were denser continuous parts. Yet, they added, it cannot be denied that there are in the Milky Way many minute stars, and given what one sees in the nebulas of Cancer and the Pleiades (that is, that they are composed of very small stars), it could be conjectured with some probability that the Milky Way is indeed a great multitude of stars, which cannot be distinguished because of their small size. In response to Bellarmine's concern to distinguish reality from appearance, Clavius and company carefully noted that the evidence did not show with certainty that the Milky Way is composed of stars, and thus that the proposition, however probable, was still a conjecture.

To Bellarmine's second query they replied they had observed that Saturn is not round, like Jupiter and Mars, but rather has an ovate and oblong shape that they illustrated crudely with a drawing of three circles side by side—the largest circle in the center and the small flanking circles of unequal sizes. Yet, they reported, the smaller "stars" on each side were not so distinct from the middle that they could say that there are three distinct stars. This is again a very circumspect evaluation. The Jesuit astronomers were very careful to go no further than their observations warranted.

To the question about Venus, they confirmed that it waxes and wanes like the moon. They stated that they had seen the planet nearly full in the evening sky and observed it as the illuminated portion, always facing the sun, diminished little by little to a crescent. Then, in the morning sky, after Venus's conjunction with the sun, they saw the planet again as a crescent with the illuminated portion facing the sun. At the time they wrote, it was growing ever brighter as its apparent diameter was shrinking. It is important to note that the Jesuit astronomers agreed completely with Galileo's observations. Their lack of any reservations about the accuracy of the observations is significant because the implication of Venus's behavior is inescapable: Venus must orbit the sun. Nevertheless, having been asked by Bellarmine only about the observations, they neither drew this conclusion nor explored any other implications of the appearances.

Clavius was the only one who felt called on to supplement the observations with a comment. Concerning the appearance of the moon, the report said, one cannot deny the great unevenness in it, but Father Clavius thinks that it is not the surface that is uneven. Rather, he considers it more likely that the body of the moon does not have uniform density and that the denser and rarer parts correspond to the well-known spots that we see on the moon with unaided vision. Others of the group think the lunar surface to be truly uneven, but so far, they lack sufficient certainty on the matter to be able to affirm it indubitably. So, although again unwilling to affirm any certainty in the matter, at least some of the astronomers agreed with Galileo. It seems likely that all three of the younger members concurred in this opinion, otherwise Clavius would not have been named as the sole advocate of his view. Galileo later drew the same conclusion in a letter of 16 July 1611 in which, after mentioning the objections of Clavius, he noted that "the other three fathers, who have carefully observed [the moon] on countless occasions, incline toward, or rather, completely follow, my opinion."[46] In a document written a short time later, another younger Jesuit at the Collegio Romano concurred with the Galilean opinion. Cristoforo Borro wrote in 1612 that it was certain that the moon was not perfectly round but has many uneven mountains and valleys. To him, the question had no need of reasoning or justification because Galileo's telescope had made this fact clear to the senses.[47]

Finally, the observers reported that moving around Jupiter there are four stars, which move quickly in a straight line—sometimes all toward the east, sometimes all toward the west, and sometimes some go one way and some the other. These stars cannot be fixed stars, because they have a very fast motion that is quite different from that of the fixed stars, and they are continuously changing their distances between Jupiter and themselves. This is, once again, an unrestricted endorsement of Galileo's observations. The astronomers close with a terse statement that they have presented what was needed to answer their inquisitor's questions.

The four Jesuit astronomers declined to endorse absolutely either the stellar nature of the Milky Way or the triple nature of Saturn—neither of which was a question of great cosmological import. Only Clavius had reservations about the unevenness of the lunar surface. The others, while not in agreement with their senior colleague, were unwilling to take a stand against him on the basis of their observations up to that date—at least not in the context of the letter to Bellarmine. The question of the nature of the lunar surface had considerable cosmological significance, since on it turned the issue of the similarity between the heavens and the earth—whether, that is, the celestial bodies were perfect spheres or were more like the earth in having variations in surface topography and composition.

On the last two points (Venus's phases and Jupiter's moons) the Collegio Romano astronomers expressed no reservations about Galileo's findings, though, of course, they also drew no conclusions from them. Nevertheless, the phases of Venus and the satellites of Jupiter had important and inescapable cosmological implications, for both observations clearly established centers of celestial revolution about bodies other than the earth—in direct contradiction of Aristotle, Ptolemy, and many generations of philosophers and astronomers. Although the Jesuit astronomers were aware of Galileo's observations of sunspots, it would appear that Bellarmine was not, since he did not inquire about them.

The promptness of the Jesuit astronomers' report to Bellarmine is easily understandable in the light of the events leading up to it. They had been experimenting with telescopes in the Collegio Romano for nearly a year and had, by April 1611, already observed all of the phenomena that Bellarmine inquired about. Furthermore, they had only recently had an opportunity to discuss all the issues with Galileo himself, who had been at the Collegio Romano at the end of March. They may also have had other opportunities to confer with Galileo, perhaps less formally, since he remained in Rome until the end of May. Thus, when Bellarmine's letter arrived, the Jesuit astronomers were well prepared, and perhaps even eager, for the opportunity to explain their recent researches. This was a chance to illustrate their utility to the Society of Jesus and to show that they were up to date on the latest questions of science and philosophy.

The climax of Galileo's visit to Rome in 1611, at least as far as Clavius and the Roman Jesuits were concerned, was surely the triumphant celebration in honor of the Pisan that convened at the Collegio Romano on 18 May. Villoslada recounts that the arrangements for the ceremony were entrusted to the students of mathematics and astronomy, who were commonly called "Clavius's academicians" (gli accademici del P. Clavio). Present at the festa were the students Johann Adam Schall, Paul Guldin, Nicholas Zucchi, and Gregory St. Vincent, among others. Many years later, St. Vincent related to Huygens that "Galileo entered the grand hall of the academy . . . and we, in his presence, expounded the new phenomena before the whole university of the Gregorian College. And we demonstrated with evidence, though to the scandal of the philosophers, that Venus circles about the sun."[48] The chronicle states that "with this public demonstration, Galileo will return to Florence much reassured and, in a manner of speaking, with an honorary degree conferred by the universal consensus of this university."[49]

A major part of that festa Galileana at the Collegio Romano was an address, written and delivered before Galileo himself by Maelcote, entitled Nuntius sidereus Collegii Romani.[50] It is, as its title suggests, essentially a review of Galileo's Sidereus nuncius of the previous year. Using the same order as Galileo's book, Maelcote's speech describes Galileo's discoveries in the moon, the fixed stars, and the planet Jupiter and then adds discussions of Galileo's later discoveries (not mentioned in his Sidereus nuncius), namely, the phases of Venus and the peculiar tripartite nature of Saturn. Maelcote reiterated many of the same points that he and his three colleagues had made in their letter to Bellarmine little more than a month before. However, in contrast to the conservative tone of the letter to Bellarmine, Maelcote is willing in this forum to express his own opinions about the significance of Galileo's discoveries.

Maelcote's discussion of the observations of the fixed stars and Jovian satellites reveals little more than Galileo's own accounts, except to show that Maelcote, at least, had been observing the Jovian satellites regularly from the beginning of 1611 until the time of the *festa*. He declined, however, to draw any lessons from the motions of Jupiter's satellites as Galileo had done. (Galileo had written that the Jovian satellites removed one of the objections of those—unnamed by him—who found the Copernican system unbelievable because it implied that the earth and the moon had to revolve about the sun together. Jupiter and its moons, Galileo said, showed that this was perfectly possible.)[51] Maelcote mentioned Saturn only by quoting from Galileo's description in his letter to Clavius of 30 December 1610.[52] He quoted Galileo's description of its peculiar shape and his observation that its shape was more apparent some months before when the planet was brighter.[53]

Maelcote's most striking thoughts are on the issue of the lunar irregularities and the phases of Venus—the former being the one issue on which the four Jesuit astronomers had not agreed and the latter being the matter of greatest cosmological import. Maelcote described in vivid terms and at some length what the telescopic observer sees as the lunar night recedes and the sun's light slowly fills in the details of the lunar surface.[54] Then he presented his conclusion, which was that "the lunar body is bounded by a figure that is in no way perfectly spherical, but is rather a rough and uneven surface." While leaving no doubt about his own opinion, apparently Maelcote thought that a concession to the opinion of Clavius and those of like mind was called for because he followed almost immediately with the statement that "if any of you should come to think that the cause of this appearance in the lunar body is variation of density and rarity, I will not obtrude my own opinion. It is enough for me to have told, just as the Nuncio [Galileo] did, what I saw and learned from the heavens concerning the lunar features. You be the judge on the outcome of the matter."[55]

On the subject of Venus, as with Saturn, Maelcote quoted Galileo's description of the sequence of phases observed as Venus approaches and recedes from the sun. Then he repeated Galileo's declaration, "Behold, now it is clear to you that Venus is moved around the sun (and the same can indubitably be said of Mercury) as though [the sun were] the center of the greatest revolutions of all the planets. And further it is undoubtable that the planets shine only by the borrowed light of the sun, which I do not judge to be true of the fixed stars.''[56]

There are two significant assertions here. The second—that the planets, unlike the stars, shine only by reflecting the light of the sun—need not disturb any fundamental cosmological principles and conflicts with none of the traditional doctrines that Clavius enumerated in his *Sphaera*. After all, the moon at least was generally assumed to shine only, or at least primarily, by the reflected light of the sun. But the first is a contradiction of Ptolemaic cosmology. Galileo's declaration of the sun as a center of planetary motion, endorsed by Maelcote without apology or reservation, is a clear break with both Aristotle and Ptolemy and opened the way for Tycho and Copernicus.

Maelcote soon resumed a more cautious tone. "It is not given to me (being neither a prophet nor a judge of such matters) nor to this occasion (which is nearly over) to determine or seek out whether [Venus] will be seen [again] as a circle like the full moon, or whether these variations will turn out to be from its circular motion around the sun or some other center.''[57] Maelcote seems to have been exercising the better part of valor. He was so discreet as to suggest that it remained uncertain whether Venus would again exhibit the "full" phase at its next superior conjunction. The fuller phases of Venus were the crucial ones, because we would never see Venus as other than a crescent of varying dimensions if it were always below the sun, as the Ptolemaic arrangement would have it.[58] He also left open the possibility that motion about some other center than the sun might be found to explain the observed variations. But after his triumphant explication of Galileo's work, such caveats seem merely formal, perhaps intended to offer a graceful escape for the many who could not follow the young astronomers in their abandonment of Ptolemaic cosmology.

Clavius's Reaction to Galileo's Discoveries

As he left Rome at the end of May, Galileo could only have been pleased by the consensus in his favor among the Jesuit astronomers. Only the resistance of Clavius to the reality of the lunar surface features would have been of significant concern to him. As might be expected, Clavius's opinion on this matter received varying reviews. Lodovico delle Colombe, a prominent Florentine philosopher and opponent of Galileo, wrote to Clavius toward the end of May expressing approval of Clavius's resistance to the idea of an uneven lunar surface and referring explicitly to the letter to Bellarmine. Colombe explained that his own opposition was based on his belief that the celestial bodies must be perfectly spherical in shape, which led him to conclude that the features revealed by the telescope must be submerged below the true lunar surface. That true surface must be perfectly clear and smooth (therefore invisible from Earth) and thus allows us to see what appear to be shadows and prominences of those "internal" lunar structures.[59]

We do not know whether Colombe's views are a good clue to what Clavius had in mind, for we have no document from Clavius or any of his associates that explains his views. The only source is the letter to Bellarmine, and it, of course, does not explain in any detail how density variations in a perfectly spherical moon could account for the telescopic observations. In July and August 1611, word had it in Rome that Clavius had written a defense of his opinion concerning the lunar surface. Letters to Galileo from Cigoli (as willing as ever to pass along a rumor) and Federico Cesi warned Galileo of this forthcoming contribution to the debate with Colombe. Cesi even reported that Clavius's treatise had already been printed.[60] Yet it would appear that it was never printed, if, indeed, Clavius ever wrote it. He may well have intended to write it, and that in itself may have given rise to the rumor. If written, it was not referred to by his colleagues or close associates and seems not to have survived among his manuscripts. Galileo continued to correspond with Grienberger concerning a dispute with Biancani on the altitude of the lunar mountains, but there was no discussion with the Jesuits of Clavius's views.[61]

In a long letter dated 16 July 1611, Galileo discussed Colombe's theory of the lunar markings and observations. He noted that Colombe cited Clavius's opinion from the letter to Bellarmine as if the opinion of the aged astronomer were an argument for Colombe's own. Galileo then pointed out that if Colombe liked the opinion of Clavius, then he should have disliked the fact that the other three Jesuit astronomers favored Galileo's theory— yet Colombe is silent on this embarrassing circumstance. Galileo continues,

> And Colombe does not know how easy a thing it would have been, while I was in Rome, to persuade and bring Father Clavius around to my opinion, if only his great age and continual infirmity would have allowed us to consider these matters together and make the required observations. But it would have been little less than disgraceful to disturb and burden with discussions and observations an old man, who, owing to his age, learning, and goodness, is so venerated, and who has earned immortal fame by such great and illustrious work. It is of little significance to his glory that in this particular matter he has come to hold a false opinion.[62]

How easy it would have been for the ever self-confident Galileo to convince Clavius is a question whose answer is beyond the historian's grasp. Was Clavius's health a significant factor in their discussions? We might reasonably suspect Galileo of exaggerating Clavius's poor health in an attempt to mitigate the influence of his unfavorable opinion. On the other hand, Clavius was indeed old (seventy-three years) by the time of Galileo's visit and had definitely fallen ill either before or soon after Galileo's departure from Rome. This illness is likely the reason why he was unable to write or publish his opinion on the lunar surface markings.

Galileo's description of Clavius brings into question the real extent of the discussions that took place between them during April and May 1611. Galileo suggests that Clavius's health precluded their meeting as often and as extensively as the visitor would have liked. In fact, the evidence of Galileo's correspondence records only one discussion between them, the one of 30 March, almost immediately after Galileo's arrival in Rome. Perhaps Clavius's health was in decline during the months while Galileo was in Rome, which could mean that the aged Jesuit's participation in the events of those months was minimal. That would include the exchange with Bellarmine, Galileo's telescope demonstrations, and the ceremonies at the Collegio Romano.

We have at best a rough idea of Clavius's opinion of the meaning of the markings on the lunar surface as revealed by the telescope. Unfortunately, we know even less about his thoughts on the cosmological significance of the phases of Venus. Neither the letter to Bellarmine nor Maelcote's *Nuntius sidereus Collegii Romani* reveal any difference of opinion over the meaning of those observations. The letter, of course, does not present any interpretation of this particular discovery. Yet in the case of the lunar markings, the difference of opinion between the younger astronomers and the older one was at least acknowledged. Does this mean that there was no difference of opinion concerning Venus? Maelcote's *Nuntius* reveals his opinion that the phases of Venus suggest that it receives its light from and revolves around the sun. There must have been some controversy on the matter—it is not reasonable to suppose that all the scholars of the Collegio Romano should have capitulated on such an important point. Moreover, as mentioned above, Gregory St. Vincent's account of the Galileo *festa* noted that the students of mathematics demonstrated "to the scandal of the philosophers, that Venus circles about the sun." This confirms at least a certain amount of controversy. Where did Clavius come down on this issue? Could he possibly have conceded, without protest or comment, such a cosmologically important point, namely, that the sun, not the earth, could be the center of motion of one (or more) of the major planets?

It is time to return to Clavius's only published words on this entire story, namely, those he added to the last edition of his *Sphaera*.

I do not want to hide from the reader that not long ago a certain instrument was brought from Belgium. It has the form of a long tube in the bases of which are set two glasses, or rather lenses, by which objects far away from us appear very much closer, and indeed considerably larger, than the things themselves are. This instrument shows many more stars in the firmament than can be seen in any way without it, especially in the Pleiades, around the nebulas of Cancer and Orion, in the Milky Way, and other places . . . and when the moon is a crescent or half full, it appears so remarkably fractured and rough that I cannot marvel enough that there is such unevenness in the lunar body. Consult the reliable little book by Galileo Galilei, printed at Venice in 1610 and called *Sidereal Messenger,* which describes various observations of the stars first made by him.

Far from the least important of the things seen with this instrument is that Venus receives its light from the sun as does the moon, so that sometimes it appears to be more like a crescent, sometimes less, according to its distance from the sun. At Rome I have observed this in the presence of others more than once. Saturn has joined to it two smaller stars, one on the east, the other on the west. Finally, Jupiter has four roving stars, which vary their places in a remarkable way both among themselves and with respect to Jupiter—as Galileo Galilei carefully and accurately describes.

Since things are thus, astronomers ought to consider how the celestial orbs may be arranged in order to save these phenomena.[63]

The content of Clavius's summary allows us to date roughly when he must have written this passage. Of the discoveries Clavius mentioned, the last that Galileo had announced was the observation of the phases of Venus, which was made public on New Year's Day 1611.[64] Galileo wrote to Clavius on 30 December 1610 about his discovery of the phases, thus in the strict sense giving Clavius advance notice of the news. The composition of Clavius's passage must have taken place after that date. On the other hand, Clavius makes no mention of Galileo's observations of sunspots, which the latter had shown to various people in Rome in April 1611—including Maelcote and therefore probably other Collegio Romano Jesuits as well.[65] Assuming that Clavius knew of the sunspots and that he considered them significant, we can figure that the passage above must have been written before Galileo's visit to Rome in April and May. That second assumption may not be sound, however. Drake notes that in April and May 1611 Galileo had not yet begun to consider sunspots as particularly important and was showing them off as curiosities.[66] So perhaps Clavius would not have seen fit to mention sunspots if Galileo himself did not place much importance on them.

It is unlikely that Clavius's last revisions to his *Sphaera* were finished much after the middle of 1611, because it was about that time, at the latest, that his illness set in. And the illness must have been fairly serious because it lasted at least a couple of months if Clavius was already ill when Galileo was in Rome. Grienberger made a small astronomical joke in reference to Clavius's health in a letter to Galileo of 26 June 1611 where he says "Father Clavius till now has remained immovable where last you greeted him, though now he begins to rise and set on occasion."[67] So the aged Jesuit must have been ill or at best weak through most of June. The text would have had to be on its way to Mainz by midyear in any case, so as to be printed near the end of 1611.

On balance, it seems likely that Clavius would have mentioned the sunspots, as did Maelcote, if he had known of them, and therefore that the passage was written sometime before Galileo's arrival in Rome at the end of March 1611. Furthermore, in Clavius's description of the appearance of the lunar surface, there is no hint of his difference of opinion with Galileo and the other Jesuit astronomers on that matter. Since rumor had it that Clavius later intended to write something about the debate, it seems likely that he would have alluded to it in the *Sphaera*. But if he composed the passage before the debate emerged (that is, before he had formed a clear opinion on the issue in the wake of the Bellarmine letter in April), then he would have had no reason to bring it up. In the quotation from the *Sphaera* above, Clavius says that the moon "*appears* so remarkably fractured and rough that I cannot marvel enough that there is such unevenness in the lunar body." It sounds as if he has not formed an opinion and is still trying to understand what the observation means.

We know so little about Clavius's final position on the question of the lunar irregularities that we are reduced to guesswork almost immediately. As mentioned above, we cannot conclude that his explanation of the observations coincided with Colombe's. What we can conclude is that Clavius was very reluctant to accept the lunar mountains and valleys—which seemed obvious to Galileo and many others, including the Jesuits Maelcote, Grienberger, Lembo, and Borro—probably because he refused to surrender the concept of the celestial bodies as perfect spheres. This was Colombe's starting point, too. Clavius had long before (after the nova of

1572) relinquished the Aristotelian principle of absolute celestial incorruptibility. So why would he cling to the idea that celestial bodies must be perfect spheres? Perhaps he believed, as some have interpreted Copernicus, that "a spherical shape, the most perfect geometrical form, and the one that all natural bodies endeavour to assume because of this very perfection, is not only more suitable for motion . . . but is sufficient cause thereof, and naturally engenders the most perfect and most natural of motions, namely circular motion."[68] Such an important dynamical principle, if Copernicus or Clavius held it, would make it difficult to surrender the sphericity of the planets, because there is a direct link between the shape of a celestial body and the all-important principle of uniform circular motion. There was an additional requirement in the case of the moon: Clavius had stated in the *Sphaera* that the body of the moon must rotate so that as it was carried around by its epicycle the same side would always face Earth.[69] How could a body with the jagged surface claimed by Galileo rotate smoothly within its orb?

No explicit statement of the dynamical significance of sphericity can be found in Clavius's *Sphaera*. But he does at least suggest a relationship between the sphericity of celestial bodies and their circular motion. After giving several reasons for the superior nobility and excellence of the sphere, befitting its presence in the heavens, Clavius concludes, "Who then could doubt or be hesitant that a heaven should be endowed with such a figure? Especially since a heaven . . . is revolved continually by circular motion, to which motion the spherical body, among all the others, is best suited on account of the continual and uniform succession of its parts. Thus no extremity can be an impediment, because it is moved around the center always according to the same boundaries, whence it is easily moved."[70] For Clavius, as for all theoretical astronomers up to (and perhaps even including) Kepler, the naturalness of circular motion in the heavens was a given. If Clavius saw the perfect sphericity of celestial bodies as a cause, or even a necessary condition, of celestial circular motion, then his reluctance to accept the less-than-perfect sphericity of the moon is understandable.

If, perhaps, the significance of the features on the lunar surface had escaped Clavius until the time of Galileo's visit, could the same be true of the phases of Venus? Galileo had certainly made no secret of his opinion that the obvious explanation was the correct one: Venus, and probably Mercury as well, revolve around the sun. And he expressed that opinion in the same letter of 30 December 1610, in which he also told Clavius about his discovery of Saturn's peculiar shape. But Venus's motion around the sun was only one of two conclusions that Galileo drew from his observations, both of which were later quoted by Maelcote. The other conclusion was that "we are certain that the planets are intrinsically dark and only shine [by being] illuminated by the sun."[71] It was this conclusion—and not the more cosmologically disturbing one—to which Clavius chose to refer when he wrote in the *Sphaera,* as quoted above, that "Venus receives its light from the sun as does the moon." It seems that Clavius simply ignored the indisputable conclusion that the phases of Venus proved the Ptolemaic arrangement of the planets wrong.

The other two planetary discoveries, the "companions" of Saturn and the satellites of Jupiter, may have been of less concern to Clavius—especially the former. Saturn's companions had not shown any motions with respect to Saturn itself, and no new motions would have meant no new planetary spheres and no new problems. Jupiter's satellites, while clearly requiring additional spheres centered on Jupiter, might have been neutralized as a cosmological problem by arguing perhaps that minor celestial bodies need not go around the earth, just the major ones. There is no evidence that Clavius made such an argument—I mention the point merely to indicate how the significance of the Jovian satellites might have been mitigated or rationalized in a way not possible in the case of Venus. Certainly other ad hoc arguments had been used for similar purposes.[72]

When Clavius wrote that "astronomers ought to consider how the celestial orbs may be arranged in order to save these phenomena," it is likely that he had in mind the problems of the Jovian satellites and the phases of Venus, but especially the latter, which demanded a significant cosmological alteration. Clavius would never have accepted the Copernican theory because (as he stated himself) the physical implications of a moving earth and the conflicts with Scripture made it untenable. On the other hand, he also rejected the Tychonic system because it necessarily did away with the solid celestial spheres. Perhaps Clavius hoped for a third alternative, a modification in the traditional Ptolemaic scheme, which, like the replacement of Alfonsine trepidation with Copernican librations, would save the phenomena without introducing cosmological absurdities.

That Clavius hoped for some modification of the Ptolemaic framework is strongly supported by a statement in a letter of Clavius's student and close associate Grienberger. Writing to Giuseppe Biancani, another Jesuit astronomer who seems to have studied mathematics, though perhaps only briefly, at Rome during Clavius's lifetime, Grienberger referred to the very passage from Clavius's *Sphaera,* "I know, as does anyone who was familiar with Clavius, that up to the end of his life he abhorred the fluidity of the heavens, and that he constantly sought arguments to explain the phenomena by ordinary means. Concerning the incorruptibility of the heavens he was not so much concerned. Thus when he advised that other spheres should be considered, it seems he hoped more for an explanation of the new observations by the old theory than for a complete replacement."[73] We have already seen that Clavius was willing to concede the corruptibility of the heavens, as Grienberger notes. But otherwise Clavius insisted on ordinary means (*via ordinaria*) of explanation: no fluid heavens, no motion of the earth. In his earlier days, a younger Clavius was willing to accept the corruptibility of the heavens on the evidence of the nova of 1572. But the older Clavius of 1610 was not capable (if, indeed, he had ever been) of yielding the solidity of the spheres to Tycho or the motion of the earth to Copernicus on the evidence of Galileo's observations.

Just what kind of task did Clavius set before the community of astronomers? He said "astronomers ought to consider how the celestial orbs may be arranged." The original, *videant Astronomi, quo pacto orbes coelestes constituendi sint*, uses the verb *constituere*, which carries the sense of setting up, arranging, or rearranging. It does not have the sense of remaking, renovation, or replacement that Kepler and Wilkins read into it. Grienberger, however, intended this latter sense when he employed the verb *immutare*, in his letter to describe what Clavius did not want. Had Clavius meant to indicate that the old system had to be entirely replaced, he would surely have used a word conveying the idea of replacement or substitution, if not outright overthrow. We should be careful not to read more into his statement. Clavius admitted that Ptolemaic cosmology, as traditionally formulated, could no longer be maintained with regard to its teachings about the centers of motions of at least some celestial bodies. Thus, he recognized that some changes were necessary: a concession, surely—a call to revolution, not at all.

Copernicans in the Collegio Romano?

In the March 1616 pronouncement, the Congregation of the Index condemned Foscarini's pro-Copernican book and suspended "until corrected" Copernicus's own *De revolutionibus*. These actions must have made life much more difficult for Jesuit scientists.[74] Before the condemnation, they would have had to be somewhat cautious about expressing Copernican sympathies in part so as not to offend collegial sensibilities in the Collegio Romano. But the automatic and obligatory anti-Copernican prejudice after the condemnation effectively forced them not to consider that alternative at all—at least not openly. Some expressed displeasure at the turn of events. In October 1616, Federico Cesi, the Roman aristocrat and founder of the Accademia dei Lincei, wrote to Galileo that "some time ago, Fathers Grienberger and Guldin came to me expressing great affection for you and disgust for the recent turn of events, and especially Father Guldin."[75] But that statement seems to be as strong as the dissent got. The problem is, what was the cause of the dissent? Were Grienberger and Guldin disturbed at the affront to Galileo or at the termination of debate on the Copernican alternative or was their protest against the infringement of what we might call freedom of inquiry? Such are the difficulties of interpreting Jesuit and indeed, Catholic, attitudes toward Copernicanism in the period after 1616—the scientific considerations had suddenly become more tangled in the concerns of the ecclesiastical machinery.[76]

Although, as we have seen, Clavius acknowledged shortcomings in Ptolemaic cosmology, there is no good evidence that he ever considered the Copernican theory likely, much less true. Despite this, at least one Jesuit apologist has asserted that Clavius and Grienberger appreciated the truth ("intuivano la verità") of the heliocentric theory, but that at the same time they wanted an apodictic proof, which was not available. So, as a practical matter, they concerned themselves only with the new observations.[77] There is no support whatever for this line of reasoning in Clavius's case. He never acknowledged even the possibility that the Copernican theory might be true. Further, the demand for a scientific proof of Copernicanism was not made by Clavius. It was Bellarmine, after Clavius's death, who required such a proof from Galileo if Galileo were to continue teaching the Copernican system as if it truly represented reality.

There is one document that seems to suggest that Clavius, despite his public stance, secretly stood with Copernicus, but its late date and weak links to Clavius himself put its value in serious doubt. The document is a letter of September 1633 from the mathematician Nicolas-Claude Fabri de Peiresc to Pierre Gassendi. Peiresc's letter quoted the polymath Athanasius Kircher, saying that Clavius and another Jesuit, Charles Malapert, agreed with the opinion of Copernicus but could not say so because they were obligated to defend the opinions of Aristotle.[78] But Kircher was an eleven-year-old schoolboy in Germany when Clavius died in Rome, so his testimony, even if accurately related by Peiresc, must have had its origin in some unnamed source. Malapert assented to the heliocentric oribts of Mercury and Venus and had an open mind on the orbits of the other planets, yet this is far from being a Copernican.[79] Clavius consistently rejected the Copernican cosmological position, and without more reliable testimony than Peiresc's we must accept Clavius's own word on the matter. As discussed earlier, however, Clavius did praise Copernicus's work in some areas and even draw from it, as in the replacement of Alphonsine trepidation theory. This attitude, perhaps misinterpreted by some of his successors, might account for this possible echo of a rumor that Clavius was a "closet Copernican."

There are, however, hints of Copernican sympathies in the younger generation of Jesuit astronomers—Grienberger and his contemporaries. For instance, in a letter from Piero Dini to Galileo, dated 16 May 1615 (ten months before Galileo's first troubles with the Inquisition), Dini wrote, "As far as Copernicus is concerned, there can no longer be any doubt; and as for your opinion . . . we detect not even the slightest movement against you [in Rome], and if God wills that you should be able to come here before long, I am sure that it would please everyone greatly, because I understand that many Jesuits are secretly of the same opinion, although they keep quiet."[80] A couple of months earlier Cesi, too, had written to Galileo about Jesuit sympathy for Copernicanism. The Carmelite Paolo Antonio Foscarini had just published a book attempting to show that there were no conflicts between Scripture and Copernicanism. He had come to Rome, where his book was stimulating controversy, early in 1615. Cesi wrote: "[Foscarini] claims that all [Carmelites] are Copernicans, although this may not be. [Rather] they commonly profess only the freedom to philosophize in matters of nature. He is now preaching in Rome. I will talk with Msgr. Dini and with [Foscarini] and with Father Torquato de Cuppis, a Jesuit and noble Roman, who is of the same opinion."[81] This is very interesting, because it could mean that Copernican sympathies among Jesuits went well beyond the astronomers, for de Cuppis taught at the Collegio Romano for twenty-eight years, 1609–37, in a wide range of subjects, none mathematical.[82]

On the other hand, a "Copernican sympathy" can be a pretty slippery thing, especially when held secretly. How could some other party know that the secret sympathy was truly for Copernicus? It is conceivable that a moderately well informed outside observer could hear that certain Roman Jesuits believed, for example, that all the planets orbited the sun. If this hypothetical observer were not well enough informed to know that heliocentric planetary orbits were also a feature of the Tychonic system, he might mistakenly conclude that the rumor implied Copernican sympathies. Cesi, at least, deserves the benefit of the doubt that he was well informed enough to know of the Tychonic system and distinguish it from the Copernican. But in general, without knowing how well informed the nonscientific aristocracy (like Dini) were, we must be careful in evaluating reports of sympathies—especially secretly held ones.

An exchange of letters between the Jesuit Wenceslas Kirwitzer in Graz and Grienberger in Rome also gives hints of Jesuit Copernicans. Only

Kirwitzer's letters to Grienberger survive.[83] On 26 December 1614, Kirwitzer wrote to Grienberger that he believed the Ptolemaic theory to be false and that, while finding Copernicus's theory hard to believe, he found it intriguing. In Kirwitzer's second letter to Grienberger, on 7 June 1615, he declared himself a convinced Copernican and took strong exception to the Jesuit astronomer Christoph Scheiner, who favored a version of the Tychonic system. Kirwitzer promised to write a defense of Copernicanism and answer Scheiner's criticisms of that theory. D'Elia reported that this treatise does not seem to survive, though Kirwitzer promised to send a copy to Grienberger. It is, of course, interesting in itself that the Jesuit in Graz became a convinced Copernican during the first half of 1615. But D'Elia suggests that this partial exchange also tells us something about Grienberger, because Kirwitzer's second letter, D'Elia says, contains nothing to suggest that Grienberger had differed with or questioned Kirwitzer on the Copernican sympathies he had expressed in the first letter. On this basis, D'Elia suggests that Grienberger was a Copernican too.

D'Elia's argument alone might be insufficient to conclude that Grienberger was a Copernican. However, D'Elia's contention is consistent with other circumstances, including the disgust with the 1616 condemnation that Grienberger and Guldin expressed, the multiple rumors that some Jesuits were Copernicans, and the likelihood, supported by the sentiments of de Cuppis, that by 1615 Copernicanism among the Roman Jesuits had already spread well beyond its natural constituency, namely, the mathematical astronomers.

The 1616 condemnation of Copernicanism by the Congregation of the Index quickly wiped out whatever growth there was of Copernican sympathy among the Jesuits. The few traces of Copernicanism that survived (if that is what they were), were weak and scattered for two reasons. First, the growth of Copernicanism among the Roman Jesuits had probably only begun in 1611 and was not sufficiently established before the 1616 debacle. But more important, when the 1616 condemnation came, there was a ready and convenient alternative to both the Ptolemaic and Copernican cosmologies. That, of course, was the Tychonic theory.

Tychonic Cosmology at the Collegio Romano

Clavius and the Danish astronomer Tycho Brahe were not acquainted with each other and did not really correspond. Only one letter between them, from Tycho to Clavius, is known, and apparently Clavius never responded, despite being urged to do so in a later letter from Francis Tengnagel, Tycho's son-in-law.[84] However, they could learn of each other

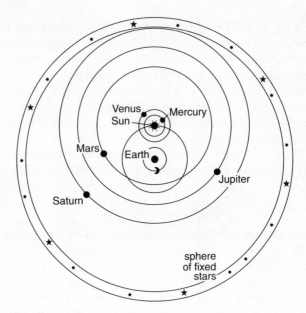

Figure 23. Tychonic cosmology. Cartography Laboratory, University of Wisconsin–Madison.

through intermediaries. Both, for instance, were correspondents of Magini, who served as a conduit between the other two for reports of activities and opinions. Tycho, of course, had proposed a cosmological arrangement of his own as an alternative to the Ptolemaic and Copernican. Tycho's scheme, like the Ptolemaic cosmology, established the earth in the center of the cosmos, where it was stationary (see fig. 23). The sun would orbit the earth, again as in the Ptolemaic conception, but the planets would orbit the sun—as they did in the Copernican system. By keeping the earth central and stationary and allowing the sun to move, Tycho's suggestion had the effect of neutralizing the most serious objections against the Copernican theory while at the same time preserving, in purely geometrical terms, the arrangement created by Copernicus. That meant that Copernicus's attractive mathematical harmonies could be retained without sacrificing the assumptions of traditional physics.

In the later editions of his *Sphaera,* Clavius cited and praised Tycho's observational work on stellar mapping, calling him "a distinguished astronomer of our time."[85] Yet Clavius nowhere in print referred to Tycho's cosmological speculations. This is especially curious, given that the Tychonic system (or variations on it) became the favored cosmological choice among the next generation of Jesuit scientists. Clavius could not have been ignorant of Tycho's theory. Knowledge of the Tychonic system was widespread, having been published first in *De mundi aetherei recentioribus phaenomenis* (1588) and also in his posthumous publications. Even had it otherwise escaped Clavius's notice, Tycho was careful to call his theory to the Jesuit's attention in some detail in the letter of January 1600. But Clavius knew about the Tychonic system well before that. In a letter to Magini dated 27 January 1595, after asking Magini to send him the information about observations, Clavius reassures Magini, "He [Tycho] confuses the whole of astronomy, since he would have it that Mars can be closer [to the earth] than the sun."[86]

The confusion, as Clavius would have it, arose from the fact that Tycho claimed he had successfully measured the parallax of Mars at opposition and found it to be larger than the solar parallax.[87] This would mean that Mars, at its closest, can be closer than the sun. (Note how the paths of Mars and the sun intersect in fig. 23.) This is not a problem for bodies orbiting freely in space (or, more aptly, for bodies moving on their own through a fluid medium), but it is a problem if the planets are supposed to be moved on rigid celestial spheres. The relationship of these parallax measurements was one of the circumstances that led Tycho to conclude that there were no solid celestial spheres, because then the spheres of the sun and Mars would perforce protrude into one another. But that very consequence led Clavius to reject Tycho's system. It appears that, for Clavius, the rigid celestial spheres were too fundamental to be surrendered on the basis of the evidence that Tycho could provide. Furthermore, the dissolution of the spheres was tantamount to an endorsement of the principal concepts of the fluid-heaven theory, which, as we have seen, Clavius consistently and emphatically rejected throughout his career.

A brief account of a lunar eclipse provides a hint of the mixed attitude toward Tycho at the Collegio Romano in early 1609. The Roman Jesuit astronomers observed the lunar eclipse of 19 January 1609, and on 31 January, Clavius's student Vremann sent Magini a written account of their observations. Though Vremann does not make clear what significance the observations have for him, it seems that he thought them supportive of Tycho's eclipse tables, for he concluded his report by stating, "And so that any critic of astronomy or of Tycho shall not think that I have concocted the stated observations, Your Excellency should know that our college's professor of mathematics was present the whole time as well as another student of Father Clavius's. But almost equally important to those two was the presence of Father Clavius himself, who in various respects is hardly a friend of Tycho."[88] At least one of those respects (and perhaps the only one) in which Clavius was "hardly a friend" of Tycho must have been on the question of cosmology.

The next generation of Jesuit astronomers did not find Tycho's cosmological system as untenable as Clavius did. We have seen that Clavius's younger colleagues seemed inclined to accept Galileo's evidence that Venus and probably Mercury as well orbited the sun. The Jesuits Giuseppe Biancani and Charles Malapert both accepted the heliocentric orbits of these planets early in the 1600s.[89] In fact, Tycho's cosmology began to find favor much earlier among the Roman Jesuits—very soon, in fact, after Galileo's discoveries and his visit to the Collegio Romano, if not before. Just as Grienberger, for example, may have had Copernican sympathies in the years preceeding 1616, there was, at the same time, support for the Tychonic system, which could only grow even stronger after the 1616 inquisitorial condemnation.

The early Tychonic influence is clear from a very interesting manuscript treatise by the Roman Jesuit Cristoforo Borro. We know very little about Borro. Sommervogel (who notes also the forms Borri and Burro) states that he was from Milan, entered the Society of Jesus in 1601, and received his education at the Collegio Romano. After some time spent as a missionary in the Indies, Borro seems to have proven unsuited to missionary work and returned to the West to teach mathematics at Coimbra. In 1632 he returned to Rome and took leave of the Jesuits, made an abortive attempt to join another religious order, and died soon thereafter.[90]

In 1610 and 1611, when the stir over Galileo's astronomical discoveries was beginning, Borro was probably a fairly advanced student at the Collegio Romano. He may still have been studying theology while teaching mathematics courses. He does not, however, appear ever to have held the chair of mathematics itself. His manuscript treatise of 1612 is entitled *De astrologia universa tractatus*. The title page accords him the title *Collegio mathematicarum scientiarum Doctore praestantissimo* and notes that he left Rome for the Indies in 1615. It is not possible to treat this tract in detail here, but a few observations on its contents will help illuminate the status of the various cosmological alternatives as seen by Borro in the Collegio Romano.[91]

Borro's general plan is evidently a series of introductory lectures on astronomy. The first set covers *astrologia contemplativa* (theoretical astronomy) and the second, *astrologia practica* (practical astronomy). As part of the theoretical section, Borro reviews the three "hypotheses" concerning the celestial orbs: first, the Copernican, then the Ptolemaic, and finally, the Tychonic.[92] Borro's description and dismissal of the Copernican system are very brief. He notes, in apparent agreement, Clavius's opinion that Copernicus's third terrestrial motion is incomprehensible.[93] But he particularly emphasizes that the Copernican theory contradicts the words of Scripture, and he gives some examples.

Borro's account of the Ptolemaic system is somewhat more detailed than the Copernican section. He explains that Peurbach is the foremost Ptolemaic author and thus he will explain Peurbach's theories—not because they are true, but because they are easier to understand.[94] Not surprisingly, Borro follows Clavius closely in his brief review of Ptolemaic astronomy.

Finally comes what he grandly calls *Nova mundani orbis hypothesis,* or "new theory of the cosmic orb." At this point Borro falls into a first-person account relating how, seven years ago (around 1605, presumably) when he was first beginning his studies, there was much confusion over astronomical theories.[95] He did not like the Ptolemaic system (but does not say why) and could not accept the Copernican (because it conflicts with Scripture), but then came along the admirable system of Tycho. Returning to his third-person account, Borro describes the Tychonic system and provides a diagram in which he is careful to show the sphere of Mars intersecting the sphere of the sun. Borro agrees with Tycho "that there are no hard and solid heavens in which the stars are embedded; rather the fabric of the universe is none other than a very fluid ethereal atmosphere and evidently a most simple thing."[96] He makes an important point of the compatibility of this theory with the words of Scripture and also mentions its agreement with the opinions of the church fathers.

Borro's text demonstrates that the Tychonic system, which would become widely accepted by Jesuit astronomers in the succeeding decades of the seventeenth century, had a very early start at the Collegio Romano— considerably earlier than Galileo's discoveries if 1605 is when Borro first encountered it. Borro's account does not make clear, however, exactly when the Tychonic system began to appear to be a solution to his dissatisfaction with the other two. Since Borro's treatise and his enthusiasm for the Tychonic system date from well before there existed any formal ecclesiastical prejudice against the Copernican system, it is safe to suggest that Borro accepted and supported Tycho's theory on its own merits and not because of any external ecclesiastical pressure. Ecclesiastical pressure, I say, because religious considerations are certainly there; Borro made no secret of the important role that agreement with scriptural authority played in his decision against Copernicus and for Tycho. However, the censorial machinery of the church would not be brought to bear on cosmological questions until after Borro had already taken ship to the East.

The other point to be made here on the basis of Borro's manuscript is that almost from its first appearance in Jesuit cosmological speculation, Tycho's theory was explicitly conflated with the fluid-heaven concept that Clavius had energetically refuted for nearly half a century but that Bellarmine had always quietly maintained. Another indication of early Jesuit interest in fluid-heaven cosmologies is Guldin's letter of 1611 to Johann Lanz about Galileo's discoveries. Guldin asked rhetorically "Where shall we locate these new planets? Which and how many orbs and epicycles shall we attribute to them? Do not Tycho's views encompass them? Shall we have it that the stars move freely like the fish in the sea?"[97] What is the relationship between these last two possibilities? Is the second a consequence of the first, or an alternative to it? The close association of the two ideas suggests a connection. But in Borro's 1612 treatise there is no doubt. By that early date he had rejected both the Ptolemaic and Copernican cosmologies and assimilated Bellarmine's ideas (though he need not have gotten them directly from the cardinal) with the "geoheliocentric" framework to arrive at a fluid Tychonic cosmology.

Borro would go on to write more on the fluid Tychonic system, and the signs of Bellarmine's cosmological influence are easy to pick out. Borro's *Collecta astronomica* (1631) contains a work with the title *Doctrina de tribus coelis, aereo, sydereo, et empireo.*[98] This title reflects Bellarmine's view that Scripture and the senses admit to only three heavens, the airy, the stellar, and the empyrean.[99] Borro also placed great emphasis on the spiral nature of the motions of heavenly bodies.[100] This, too, is an echo of Bellarmine, who had maintained, for instance, that the real motion of the sun was the complex spiral that it traces in its motion around the earth, and that astronomers had no warrant to resolve this real motion into arbitrary components such as an annual and a diurnal motion.[101]

By the 1630s Borro was far from the only writer advocating fluid Tychonic heavens—in fact, acceptance of the fluid heavens was becoming common.[102] Christoph Scheiner (a German Jesuit with strong connections to the Collegio Romano who disputed with Galileo over sunspots) taught that the heavens were a single fluid and that the planets moved according to the Tychonic scheme. He concluded, "This opinion is now the commonly accepted teaching of all astronomers. Those believers who admire anything to the contrary would reject [this theory] without adequate cause, because its evident truth is consistent with Holy Scripture and the holy fathers, is established by the soundest arguments, and taught by irresistible observations."[103] Baldini and Coyne see the final flowering of Bellarmine's cosmological influence in Riccioli's planetary theory. In his *Astronomia reformata* (1665) Riccioli's solar theory has the sun guided along its spiral path through a fluid heaven, "just as hypothesized by Bellarmine. Riccioli was the last to discuss this spiral motion, and after the middle of the century the ideas of Bellarmine were abandoned."[104]

Fusion of the Fluid-Heaven and Tychonic Cosmologies

Galileo's telescopic discoveries were the catalyst for many changes among the Collegio Romano astronomers, including Clavius's concession on the necessity to alter the astronomical system that he had defended for so many years. But perhaps the most remarkable change was the resurgence of the fluid cosmology and its melding with the Tychonic planetary arrangement. Given that for decades Clavius had been criticizing and refuting the idea that the planets move through the heavens like "fish in water or birds in air," it would seem that such an idea would not be so readily presented as a solution to the puzzles posed by Galileo. Yet Borro's treatise shows that this solution was invoked almost immediately. There are two likely reasons for this. One is, ironically, the continuing notoriety that Clavius's critique in the *Sphaera* gave to the fluid-heaven concept. It was part of the student's astronomical education to study the theory and learn to refute it. The second reason is the continual reservoir of good will toward the fluid-heaven idea in the opinions of Cardinal Bellarmine and like-minded Italian philosophers, such as Patrizi.

Galileo's discoveries were the immediate and most powerful reasons why Clavius's students gave up on Ptolemaic cosmology and began to choose from among the alternatives. But there were other factors favoring a choice of the fluid heavens, the most important of which was the general recognition that substantial change could occur in the heavens, as was agreed to by Clavius himself (among others) after the determination that the nova of 1572, followed by those of 1600 and 1604, were located in the firmament. The same conclusion was supported by the demonstrations of Tycho and Maestlin, for example, that the comet of 1577 was not an atmospheric phenomenon. Some degree of celestial corruptibility is, in itself, not disproof of the reality of celestial spheres; Clavius, for one, never drew that conclusion despite his admission of a degree of corruptibility. Nevertheless, it is easier to imagine a rearrangement of celestial matter into a nova or comet if that matter is not rigid, like the spheres, but fluid.

Tycho's system would have been a logical candidate for adaptation to a fluid cosmology because it preserves cosmological assumptions (like geocentricity) that were deeply intuitive and because it raises no conflicts with Scripture. But furthermore, it rules out rigid celestial spheres in its structure as a result of the intersection of the spheres of Mars and the sun.

It would have had further appeal to those Jesuits who, though trained in the tradition of Ptolemaic mathematical astronomy, might nevertheless have been influenced by Bellarmine's epistemological standards. We see this combination of inclinations in Borro, who stressed Bellarmine's principle of founding cosmology in scriptural and patristic writings (for example the scriptural justification for counting only three heavens) as well as the cardinal's emphasis on the primacy of spiral motions. But at the same time, Borro would have seen the Tychonic system as necessary to explain the heliocentric planetary orbits and to provide the scheme of circular motions then considered to be the only possible basis for a mathematical astronomy. In this sense, the Tychonic system would have been a true synthesis constructed from the demands of telescopic observations, the advantages of the fluid heavens, and the capabilities of a mathematical astronomy.

Christine Schofield has suggested that the reason for the prolonged success of the Tychonic system in competition with the Copernican was primarily a result of religious constraints in the beliefs of Catholics.[105] After 1616 this was certainly a factor, but the evidence from the Collegio Romano before 1616 suggests that there is more to it. It is important to realize that both before and after 1616, the problems caused by the hypothesis of the earth's motion were powerful forces preventing the acceptance of Copernicanism. The great difficulties of putting aside one's beliefs in the earth's centrality and stability, concepts so intuitively congenial and traditionally approved, cannot be underestimated. These prejudices could be overcome, of course—there were, after all, actual Copernicans in the late sixteenth century, but they were rare.

Once we acknowledge the inertia of cosmological biases, we can ask about the role of religious belief before 1616 in the preference for the Tychonic system. We have seen that even before the declaration of ecclesiastical prejudice in 1616, Borro, for one, freely chose the Tychonic system, despite what many have seen as its drawbacks: being awkward, asymmetric, and devoid of observable astronomical advantages. I suggest that a major influence in the success of the Tychonic system was not ecclesiastical constraint (which only became real in 1616) but what I would call confessional constraint. A constraint, in other words, that is self-imposed and derives more from an individual religious decision than from a doctrinal decision of the church hierarchy. Clavius, Bellarmine, Borro, and the Protestant Tycho, for that matter, all rejected Copernicanism in part because of a decision to interpret Scripture in a certain sense. For most of Clavius's life, the rejection of Copernicanism led to no dilemma because there were no compelling reasons to question the Ptolemaic system. When that was no longer the case (after Galileo's celestial discoveries), Clavius was probably too old to confront the problem creatively. Accepting the full implications of the discoveries would place him on the horns of a true dilemma: to accept the Tychonic theory would have been to surrender the celestial spheres, the cosmological constructions that made the architecture and dynamics of the cosmos comprehensible and mathematically tractable. But to accept Copernicus would have been to surrender the equally important tenets of terrestrial centrality and stability along with the congeniality between Scripture and cosmology. No other alternatives, such as homocentrics, would do either; Clavius had spent his career demolishing them. The aged professor was not up to the task of escaping that trap and could only hope that some rearrangement of the Ptolemaic orbs would fix things up.

As seen by Borro and his generation, the problem was quite different. With the Ptolemaic theory rendered obsolete by Galileo, the situation still boiled down to a choice between the Copernican and Tychonic alternatives in some form. But unlike Clavius, the younger Jesuit astronomers were not wedded to the solid celestial spheres. Borro and his fellow Jesuits were guided to choose the fluid Tychonic alternative by the constraints coming from Scripture and the opinions of the church fathers—accompanied, of course, by that theory's adaptability to a fluid cosmos and its conformity with Galileo's discoveries. The fluidity in the fluid Tychonic cosmology was vital because it seemed to provide a foundation for planetary dynamics in the absence of the rigid celestial spheres. A fluid cosmos also allowed for explanations of celestial changes like comets and novas in terms of condensation and rarefaction, coalescence and dissipation, and the like.

Finally, we should take note of the significance of Bellarmine's cosmological views with respect to the Inquisition's condemnation of 1616. For it was only then that the guardians of orthodoxy imposed ecclesiastical constraints on the cosmological choices of the Jesuit astronomers and, for that matter, all Catholics. Though Bellarmine and Clavius were associates and colleagues for years, and despite the fact that Bellarmine was not a trained mathematician, Bellarmine remained skeptical of the most fundamental methodological concept of pre-Keplerian astronomy: that the observed motions of the celestial bodies could be resolved into components of uniform circular motions. He appears to have been particularly suspicious of a system in which any newly discovered celestial motion necessarily implied major structural adaptations to the cosmology.[106] A relevant example (though not used by Bellarmine, to my knowledge) would be the way in which Clavius and Magini, in response to the observed weaknesses of trepidation theory, ended up adding yet another suprafirmamental sphere and creating an eleven-sphere universe where before there had been ten. And all three of the major spheres above the firmament were, according to the mathematicians, invisible. It is not hard to see why Bellarmine could find questionable a procedure in which an arbitrary methodology led to unobservable consequences.

Most of the evidence for Bellarmine's views comes from his *Louvain Lectures,* which were composed rather early in his career and probably before 1572.[107] But he apparently held them quietly all his life. The cardinal's much younger contempory, Federico Cesi—who was also interested in fluid cosmology—wrote that Bellarmine had never published his cosmological ideas because the schools commonly opposed his views with alleged mathematical demonstrations to the contrary. In particular, the schools, he said, hold that it is totally impossible to save the appearances without the solid orbs and their motions.[108] This is a clear reference to Clavius's teachings.

Additional evidence that Bellarmine continued to hold his unconventional cosmological views comes from a document preserved in the Jesuit archives that dates from 1616. The document, written by Grienberger, is an examination of a book by his fellow Jesuit astronomer Biancani, who taught at Parma. In discussing the idea of a single fluid heaven, Grienberger cited as one of its supporters Bellarmine, "who moreover is not at all against admitting that the planets move independently of material spheres and even admits that heavenly objects might be corruptible, maintaining that this point of view is more in keeping with Sacred Scripture and those who interpret it." Grienberger went on to note that there is nothing necessary about the resolution of the motion of celestial bodies into uniform circular motions. "No argument seems to make plausible, much less to prove, that the heavenly bodies could not, by their own power, or by the intervention of heavenly intelligences, travel those same orbits in heaven which they would in fact trace out under the hypothesis that they were moved by the rotation of more than one sphere."[109] Bellarmine's skepticism of analysis through uniform circular motions is clearly present in this statement, and free movement of the celestial bodies through a fluid heaven emerges once again as a plausible enough theory.

In addition to questioning the validity of the mathematical methodology by which the homocentric, Ptolemaic, and Copernican traditions demonstrated the great multiplicity of celestial spheres, Bellarmine denied the Aristotelian physical distinction between the terrestrial elements and the heavens and denied also the supposed unique characteristics of celestial matter—eternally uniform circular motion and incorruptibility.[110] Although Bellarmine dissociated himself from the traditional Aristotelian and Ptolemaic approach to understanding the structure of the heavens, he remained nonetheless convinced of the earth's centrality and immobility because they are attested by the two trustworthy sources of knowledge, the senses and Scripture.[111]

Now, perhaps, we can understand somewhat better Bellarmine's reaction to Galileo's advocacy of the Copernican cosmology. So long as the question was of a philosophical or mathematical nature (the adequacy, for example, of solid orbs as an explanation of celestial motions or the possibility of contrary motions in a fluid or the actual number of spheres in addition to the three attested in sacred sources), Bellarmine was satisfied to disagree tacitly and leave the mathematicians and philosophers to their disputations. But Galileo's advocacy of the Copernican system presented an alternative in which Bellarmine could see no shred of merit because it shared all the methodological unsoundness of the Ptolemaic approach, namely, a priori analysis into circular motions and spherical constructions. Moreover, according to Clavius's critique, Copernicus had derived his theory from the Ptolemaic by deducing true conclusions from false premises. Even worse than its poor methodology, Galileo's promotion of Copernicus contradicted the testimony of the Scriptures, the church fathers, and the senses by asserting that the earth moves and is not at the center of the cosmos.

Thus Bellarmine's opposition to the Copernican theory has a dual aspect, both methodological and epistemological. As Baldini has put it, "The fundamentalist reading of Genesis and other biblical texts . . . imposed on the cardinal the denial of the most radical among the astronomical innovations, the heliocentric principle."[112] In addition, Galileo could offer no satisfactory proof that would either lay to rest the cardinal's doubts about the logic of mathematical astronomy or convince him that the evidence of the Scriptures, the church fathers, and the senses should be reinterpreted. Finally, the Tychonic system, which was capable of fitting all of Galileo's discoveries without any of the problems of the Copernican, was probably as familiar to Bellarmine as it was to Borro. His knowledge of an acceptable geostatic alternative—acceptable even by the standards of the mathematicians—would have robbed Galileo's arguments for the Copernican system of any necessity. Thus it would be far from the truth to conclude that Bellarmine (the dilettante cosmologist who embraced a non-Aristotelian fluid cosmology) rejected Copernicanism because of any dogmatic Aristotelianism or ignorance of the astronomical arguments.[113]

Thus the final chapter of Clavius's astronomical career ended in the midst of controversies over issues similar to those that had occupied him for decades: the possibilty of determining the real mechanisms in the heavens and of evaluating the status of alternatives—particularly the Copernican and the fluid-heaven cosmologies. But the nature of the debates over those issues had been transformed since he had published his first *Sphaera* in 1570. Galileo and the telescope had revolutionized the investigation of the heavens, making available information about the nature and motions of the celestial bodies that was previously inconceivable. Though the homocentric sphere theories had become completely irrelevant, Galileo's work had revitalized the other rival cosmologies that Clavius had battled for decades. The recrudescence of the fluid heavens—aided perhaps by Bellarmine's influential opinions but awakened by Galileo's work and almost required by the Tychonic planetary arrangement—was in dramatic contrast to events in Clavius's earlier years. The Copernican theory was truly energized by Galileo's advocacy but not proven by him. Ptolemaic cosmology had been weakened by novas, by comets, and perhaps by its need to borrow new ideas from Copernicus's models as in the replacement of trepidation theory. After 1610 not even Clavius could repair the damage done by Galileo's discovery of the phases of Venus, and the fifteen-hundred-year reign of Ptolemaic cosmology ended, followed in 1612 by its stalwart defender.

Conclusion

This study has focused on Clavius the astronomer, but it is helpful to remember that as astronomy in his time was seen as only one branch of mathematics, so the astronomer was only one aspect of Clavius the mathematician. Much of Clavius's work remains unexamined by modern scholars except in cases of specific concepts or disputes. Yet as with his *Sphaera,* his mathematics textbooks were widely printed and used in sixteenth- and early seventeenth-century Europe. In particular, Clavius's pivotal role in the Gregorian calendar reform, and especially his vigorous defense against its critics, has yet to be fully examined. In the study of a book such as Clavius's *Sphaera,* it is important to attempt to understand the work as a whole. There is a temptation to plunge in just long enough to retrieve a choice phrase or pithy analysis on some matter of interest. But such an approach will fail to detect themes, arguments, and connections that can only emerge from a broader view. I hope I have maintained that broader view in this study. There are a great many other neglected books, now centuries old, still awaiting appreciation.

Clavius the astronomer was the kind of figure who was very different from the more famous names of his time, and it is not surprising that his astronomical activity was of a different character from theirs. Clavius was not a particularly enthusiastic or careful observer. However, he made skillful use of the observations of others, as in his "geographical parallax" argument for the location of the nova of 1572 in the firmament. He was more inclined to theoretical work, as exemplified by his repudiation of medieval trepidation theory and adoption of a modified Copernican model for the precessional motions of the stars. Yet what would have been his primary theoretical contribution, the *Theorica planetarum,* remained unpublished and, in fact, unknown until only recently. Clavius's primary role was that of an educator, and his activities in observational and theoretical astronomy were subordinate to his pedagogical priorities. His most sig-

nificant contributions were the publication of authoritative textbooks, the
establishment of mathematical subjects in the standard curriculum, and
his own teaching and training of teachers at the Collegio Romano. The
activities and priorities of Clavius's career are more comparable to an
academic astronomer such as Kepler's teacher Michael Maestlin (with
whom Clavius debated the merits of the Gregorian calendar) than to extraordinary figures such as Tycho, Galileo, or Kepler.

It is mainly the content of Clavius's astronomical teachings, and the
changes therein, that make him interesting. In the course of his decades
of teaching and publishing textbooks, we have seen a steady development
in Clavius's thought that mirrors the changes taking place in the early
period of the scientific revolution. We saw him backing away from doctri-
naire Aristotelianism much earlier than many Peripatetic philosophers of
his day. Clavius took a historical view of Aristotle, recognizing, for example, that the Greek philosopher lived long before eccentrics and epicycles
were invented as theoretical tools. Clavius was ready to abandon certain
Aristotelian doctrines when experience contradicted them, most notably
the corruptibility of the heavens on the evidence of the nova of 1572. On
a related matter, the nature of comets, Clavius seemed to back away
from the idea that they were a strictly meteorological phenomenon and
interpreted novas as comets in the firmament. He backed away from some
of the traditional Ptolemaic constructions, dispensing with trepidation theory in particular and eventually agreeing, in the face of Galileo's observations, that Ptolemaic cosmology required serious reexamination.

Clavius stood firm through the years on certain other issues. He never
questioned the basic principles of a geocentric and geostatic universe—
that view was too deeply rooted in Aristotelian physics, common sense,
and Scripture. He also held firm to the solidity of the heavens, refuting
energetically all attempts to free the planets from the celestial spheres.
Associated with that idea was the perfect spherical shape of the celestial
bodies, which he maintained in the face of Galileo's demonstrations of
the irregularities of the lunar surface. And like almost all of his predecessors and contemporaries, he never questioned the principle of uniform
circular celestial motions. Finally, in the face of his era's fashionable
skepticism, he held firm to the belief that human reason, through mathematical study, could come to understand the real structure and operation of
the celestial machinery—a position he shared with the likes of Copernicus,
Tycho, Galileo, and Kepler.

The parties involved in the cosmological debates changed greatly between 1570, when Clavius published his first *Sphaera,* and 1611, when
he published his last. Clavius's defense of Ptolemaic cosmology had been

largely concerned on one front with responding to skeptical and conservative Aristotelian philosophers like Averroës and Nifo, who represented a tradition based in the natural philosophy of the universities. The skeptical view had some following among the Jesuits—Pereira at least. The other front in the defense of Ptolemaic cosmology, the front facing the alternative cosmologies, responded to the astronomical community, to Fracastoro, Copernicus, and the like, none of whom belonged to the Jesuit order. But by 1612, Clavius's defense had been outflanked—the Roman Jesuit scholars were nearly as diverse in their cosmological views as was the astronomical community in general. Bellarmine—whose opinion carried great authority in the Society of Jesus—represented a viewpoint that combined skeptical elements with concepts drawn from the fluid-heaven cosmologies. Clavius himself continued to hold the Ptolemaic orthodoxy. Those like Borro favored a fusion of the fluid heavens and the Tychonic planetary arrangements. And there may even have been those, such as Grienberger, who favored the Copernican alternative until it became untenable for them in 1616.

The failure of the Ptolemaic cosmology to measure up to the new scientific demands of early seventeenth-century astronomy must have bewildered Clavius. But if his theoretical program had failed, his pedagogical program had not, for his decades of teaching and writing had succeeded in producing within the Jesuit order an institutional basis for mathematical astronomy. That foundation proved itself capable of generating and supporting astronomical thought of high caliber. The Jesuit mathematical program produced scholars such as Grassi, Scheiner, Biancani, Riccioli, and later Bošković and Grimaldi, to name a few. Perhaps more important, the Society of Jesus brought new energy to the practice of astronomy. During the 1600s Jesuit authors produced scores of general astronomy textbooks and many others on more specialized aspects of astronomy. Jesuit astronomical publication activity grew steadily during the seventeenth century and exploded in the first half of the eighteenth. The Society was also vigorous in establishing observatories at many of its schools scattered across Europe, providing, in effect, institutional support for astronomy on an international scale. The theories, arguments, and questions in the new astronomy of the early 1600s would have been all but unimaginable to Clavius in 1570. But the stature and support Jesuit astronomy achieved in the course of the seventeenth century were far more than he could have hoped for.

Notes

Works referred to by short title in the notes are cited in full in the bibliography.
The following abbreviations are used in the notes and in the bibliography:

APUG   Archivo della Pontificia Università Gregoriana (Archive of the Pontifical Gregorian University, Rome).

ARSI   Archivum Romanum Societatis Iesu (Roman Archive of the Society of Jesus).

BAV   Biblioteca Apostolica Vaticana (Vatican Library).

BNVI   Biblioteca Nazionale Vittorio Emmanuele II (Italian National Library, Rome).

Corrispondenza   Christoph Clavius, *Corrispondenza*. Edited by U. Baldini and P. D. Napolitani. Bibliography, s.v. Clavius.

DSB   *Dictionary of Scientific Biography*. Edited by Chas. C. Gillispie. New York: Charles Scribner's Sons, 1970.

Op. Gal. A. Favaro, ed.   *Opere di Galileo Galilei,* s.v. Galileo.

Chapter One

1. Sommervogel, *Bibliothèque,* s.v. Clavius. Sommervogel cites one edition of Clavius's *Elements* in print as late as 1738.

2. Examples of the current state of research on the historical issues of calendar reform can be found in Coyne, Hoskin, and Pedersen, eds., *Gregorian Reform of the Calendar.*

3. See, e.g., Clavius, *Apologiae calendarii novi adversus Michaelem Maestlinum,* in *Opera mathematica* 5:16 (hereafter *Op. math.*). Thomas Kuhn mistakenly suggests a more fundamental relationship between Copernican innovation and the Gregorian reform. While Pope Gregory's commission adopted Copernicus's measurement of the length of the year and his theory of precession of the equinoxes (though not his physical explanation), the heliocentric hypothesis itself was not involved. See Kuhn, *Copernican Revolution,* 126, 196. The mistaken idea of Copernican cosmology being somehow essential to the Gregorian reform goes back to the seventeenth century. Schofield (*Tychonic World Systems,* 231–32) quotes Tommaso Capanella: "The Church allows the Copernican system to be

studied simply as an hypothesis, having herself made great use of it in connection with the reform of the calendar.''

4. Dear, "Mersenne and the Learning of the Schools," 65ff; Knobloch, "Marin Mersennes Beiträge zur Kombinatorik," 360–63. Gassendi, in *Tychonis Brahei . . . vita,* cites Clavius frequently.

5. Clavius's texts were common in, e.g., the libraries of English scholars of the late sixteenth and early seventeenth centuries. See Feingold, *Mathematician's Apprenticeship.*

6. D'Elia, *Galileo in Cina,* 30–31; Casanovas, "Il P. C. Clavio professore di matematica del P. M. Ricci nel Collegio Romano.''

7. In the quotation, Sagredo is referring to Clavius's treatise on isoperimetric figures, which appeared as part of the *Sphaera* in the 1581 and all later editions. See also Galileo, *Two New Sciences,* trans. Drake, 62.

8. On Galileo's use of Clavius, see Crombie, "Sources of Galileo's Early Natural Philosophy," 162–66. Wallace (*Galileo and His Sources,* esp. 255–61, *Galileo's Early Notebooks,* 262–66, and "Galileo's early arguments for geocentrism," esp. 39–40) devotes considerable effort to showing Galileo's use of Clavius's *Sphaera.*

9. Wallace, "Galileo and the *Doctores Parisienses,*" 227. See also n. 25 below.

10. Dreyer, *Tycho Brahe,* 13; Thoren, *Lord of Uraniborg,* 11–12.

11. See Blackwell, *Galileo, Bellarmine, and the Bible* for a good discussion of the theological context (chap. 1) and later Jesuit institutional developments of the cosmological debates (chap. 6).

12. Schmitt, *Critical Survey,* 132.

13. Ibid.

14. Villoslada, *Storia del Collegio Romano,* 112.

15. Taxil, *L'Astrologie et physiognomie en leur splendeur* (Tournon, 1614), cited in Thorndike, *History of Magic and Experimental Science* 6:97. We should bear in mind that in generations past the terms "astronomy" and "astrology" did not carry such distinctly different meanings as we give them today and, in fact, were often used interchangeably.

16. Bullart, *Académie des sciences* 2:117–19.

17. Wilkins, *The Discovery of a World in the Moone,* 16; Ross, *New Planet No Planet,* 9; Froidmondt, *Ant-Aristarchus* (Antwerp, 1631), 3. Schofield (*Tychonic World Systems,* 277–78) discusses the exchange. Wilkins's treatment and Ross's response are discussed more extensively later in this chapter.

18. "Guadagnata una fama immortale." *Op. Gal.* 11:151.

19. Donne, *Ignatius His Conclave,* 17–19 (emphasis in original). On Donne's knowledge of astronomy and use of it in his satirical writings, see Gossin, "Poetic Resolutions," esp. 97–117.

20. Villoslada, *Storia;* Casanovas, "Il P. C. Clavio," and "L'astronomia nel Collegio Romano"; Homann, "Christopher Clavius and the Isoperimetric Problem," and "Christopher Clavius and the Renaissance of Euclidean Geometry."

21. D'Elia's book has also been translated into English as *Galileo in China.*

Though generally reliable, D'Elia comes to the remarkably mistaken conclusion that Clavius eventually advocated the adoption of the Copernican theory. That entire issue is considered fully in chap. 7.

22. Carrara, "I gesuiti e Galileo."

23. Naux, "Le père Christophore Clavius (1537–1612)," pt. 2, 182. In addition to many serious factual errors, Naux completely misreads Clavius's opinions on major cosmological questions.

24. Aufgebauer, "Christoph Clavius," 230.

25. See, e.g., Carugo and Crombie, "The Jesuits and Galileo's Ideas of Science and of Nature"; Crombie, "Mathematics and Platonism in the sixteenth-century Italian universities," and "Sources of Galileo's Natural Philosophy"; Wallace, *Galileo and His Sources*, various essays in *Prelude to Galileo*, esp. "Galileo and Reasoning *Ex Suppositione*," and, more recently, *Galileo's Logical Treatises* and *Galileo's Logic of Discovery and Proof*.

26. See, e.g., Dear, "Mersenne"; Donahue, *Dissolution of the Celestial Spheres;* Jardine, "Forging of Modern Realism."

27. Grant, "In Defense of the Earth's Centrality and Immobility."

28. See, e.g., Homann, "Clavius and the Isoperimetric Problem," and "Clavius and the Renaissance"; Knobloch, "Sur la vie et l'oeuvre de Christophore Clavius (1538–1612)," and "Christoph Clavius—Ein Astronom zwischen Antike und Kopernikus."

29. Baldini, *Legem impone subactis;* Clavius, *Corrispondenza*. The latter work, the first and only edition of the extensive Clavius correspondence, is a monumental effort providing important and much-needed resources for Jesuit studies and the history of astronomy and mathematics. The introductory volumes contain historiographical tools otherwise unavailable, and the entire work deserves to be published in a more widely accessible form. As it became available very shortly before my study went to press, the present work has benefited only partially from it. Examination of the correspondence has revealed nothing that substantially alters the conclusions of the research described here. I wish to thank Ugo Baldini for copies of his works and for his generous cooperation.

30. Twain, "Thirty Thousand Killed a Million," 53.

31. One census puts the number at no more than ten between 1543 and 1600. Westman, "Astronomer's Role," 106.

32. "Mais ses commentaires mêmes prouvent que si Clavius était plus savant et plus mathématicien que le moine anglais, il n'était pas plus réellement astronome; il n'a fait aucune observation, aucune recherche théorique." Delambre, *Astronomie du moyen age,* 242.

33. "Analyses of historical data suggest Sun is shrinking," *Physics Today* 3, no. 9 (September 1979): 17–19.

34. For some comments on the perils of using antique astronomical data, see O'Dell and Van Helden, "How Accurate Were Seventeenth-Century Measurements of Solar Diameter?"

35. An exhaustive chronology and listing of the biographical source materials on Clavius appears in Clavius, *Corrispondenza* 1, pt. 1:33–58.

36. Baldi's biographical sketch of Clavius, which was written as part of a series on the lives of great mathematicians, is printed in Zaccagnini, *Bernardino Baldi*. Clavius's birthdate appears there on p. 334.

37. Knobloch, "Christoph Clavius—Ein Astronom zwischen," 117.

38. Bruhns, "Christoph Clavius," *Allgemeine Deutsche Biographie* 4:298–99. This article abounds in serious errors and must be used with extreme caution. Among the worst errors, it places Clavius in Coimbra in 1596, states that he only taught for fourteen years at the Collegio Romano, that he eventually became a cardinal, and proclaims that his *Sphaera* "contains little or nothing from Sacrobosco," while, in fact, it contains the entire text of Sacrobosco's *Sphere*.

39. Zinner (*Astronomische Instrumente*, 280) noted a study finding that in Bamberg the name Schlüssel occurred only once, while the names Nagel and Klau occurred more often. This study, which I have not seen, was presumably that of Konrad Arneth, "Der Deutsche Name des Christoph Clavius," *Fränkisches Land in Kunst, Geschichte und Volkstum* 1 (1953): 37, which is cited by Knobloch, "Christoph Clavius—Ein Astronom zwischen," 116, n. 7.

40. Clavius, *Corrispondenza* 1, pt. 1:33, n. 3.

41. Pachtler, *Ratio studiorum* 3:xii.

42. Bangert, *History*, 25.

43. Ibid., 71ff.

44. Kleiser and Lamalle, comps., Chronology of Clavius, ARSI (1968).

45. Scaduto, *L'epoca di Giacomo Lainez*, 50–52; O'Malley, *First Jesuits*, 232–34.

46. Bangert, *History*, 25.

47. Farrell, *Jesuit Code of Liberal Education*, 434.

48. Rashdall, *Universities of Europe* 2:114, n. 1.

49. Baldi gave the latter's name as Cipriano Rose, which must be some sort of misunderstanding of Soares. See Clavius, *Corrispondenza* 1, pt. 1:35, n. 6.

50. Zaccagnini, *Baldi*, 335.

51. Clavius, *Corrispondenza* 1, pt. 1:36.

52. "Tiene muy buena abilidad. Es bien dispuesto pero anda muy flaco." Quoted in Kleiser and Lamalle, comps., Chronology of Clavius, ARSI.

53. Clavius, *Sphaera* (1611), 295. Up to the 5th ed. (1606), he mistakenly dated this eclipse to 1559. See Baldini, *Scientific Scene*, 145; Clavius, *Corrispondenza*, no. 159 (Brahe to Clavius), and no. 255 (Ziegler to Clavius).

54. Codina, "Sant Ignasi a Montserrat," 115. On Ignatius at Montserrat see, e.g., O'Malley, *First Jesuits*, 24–25.

55. Kleiser and Lamalle (Chronology of Clavius, ARSI) list Clavius as professor in 1567. The list of professors of the Collegio Romano, compiled by Ignazio Iparraguirre and appended to Villoslada's *Storia*, gives 1564 as Clavius's first year of teaching. Baldini and Napolitani, however, cite a letter from J. Polanco to J. Nadal indicating that 1563 was Clavius's first year of teaching. See Clavius, *Corrispondenza* 1, pt. 1:42.

56. Loria, *Storia delle matematiche*, 270.

57. *DSB*, Busard, s.v. "Christoph Clavius."

58. Zinner, *Astronomische Instrumente*, 154.

59. Zinner, *Geschichte und Bibliographie*.

60. Zaccagnini, *Baldi*, 344.

61. Ibid., 335.

62. It would be prudent to view with some caution this story of Clavius's being inspired by the *Posterior Analytics* to study mathematics. Clavius had motivations, rooted in his pedagogical agenda, to link mathematics and Aristotelian logic. The story thus makes as much sense as a rhetorical device as it does as a factual account. Moreover, Paul L. Rose (*Italian Renaissance of Mathematics*, 261) notes that Baldi himself had an interest in stressing the mathematical importance of Aristotle over Plato.

63. "Egli desideroso di ben intendergli si pose per sè stesso senz'altro aiuto di maestri ad affaticarvisi di maniera che in queste professione egli afferma d'essere, come dicono i greci, autodidascalo." Zaccagnini, *Baldi*, 335.

64. Baldini, "Scientific Scene," 141, on La Sapienza. Perhaps the lack of higher mathematical studies in the city should not surprise us; as Grendler (*Schooling in Renaissance Italy*, 84) has remarked, "Rome was a curial rather than a commercial city."

65. Though Jerónimo (not the same as the Jesuit Balthazar Torres, who taught the public mathematics course at the Collegio Romano before Clavius's arrival) never appears on the lists of Iparraguirre or Baldini and Napolitani, he seems to have been in Rome long enough to collaborate with Clavius on planning a mathematical curriculum, for their names appear together on one undated manuscript document of Roman provenance (see Lukács, *Monumenta* 1:478, item 37). Since Jerónimo seems to have had mathematical interests, Clavius could have studied informally with him in Rome (see Clavius, *Corrispondenza* 1, pt. 1:41, n. 12, and pp. 66ff. for the list of professors of the public mathematics lectures).

66. Clavius's own account of this solar eclipse is in *Sphaera* (1611), 295. See also Clavius, *Corrispondenza*, no. 255 (Ziegler to Clavius). Kepler's discussion is in *Astronomiae pars optica*, in *Opera omnia* 2:316–17. Modern calculations, which are difficult in this case because of the complexity of lunar motions, indicate that this eclipse was in the transition region between clearly annular and clearly total. Thus it is hard to say whether Clavius did, indeed, see a true but very thin solar annulus or, as some have suggested, perceived the bright inner corona of the sun. For some numerical details see "The Solar Eclipse of 1567," *Nature* 15 (1877): 342.

67. See Paul L. Rose, *Italian Renaissance of Mathematics*, esp. chap. 9.

68. Homann, "Clavius and the Renaissance," 233; Zaccagnini, *Baldi*, 344.

69. Clavius, *Sphaericorum libri III*.

70. Homann, "Clavius and the Renaissance," 235; Maierù, "Il quinto postulato."

71. Scaduto, *Maurolico*, 129–30.

72. Ibid., 134.

73. Ibid., 138–39; Clavius, *Corrispondenza* 1, pt. 1:46.

74. Paul L. Rose, *Italian Renaissance of Mathematics*, 175.

75. Scaduto, *Maurolico*, 139. Maurolico's optical treatises were three: *De erroribus speculorum, Diaphaneon seu transparentium libellus*, and *Photismi de lumine et umbra*, listed in Lindberg, *Catalogue*, 66.

76. Ziggelaar, "Papal Bull," 205. Bullart (*Académie des sciences* 2:118), also states that the reform was a work of ten years.

77. See Moyer, "Aloisius Lilius"; Ziggelaar, "Papal Bull." A good account of Clavius's contributions to the calendar reform process can be found in Baldini, "Scientific Scene."

78. Sommervogel (*Bibliothèque*) lists these editions: *Novi calendarii romani apologia* (1588); *Iosephi Scaligeri elenchus et castigatio calendarii gregoriani . . . castigata* (1595); *Romani calendarii . . . restituti explicatio* (1603); *Responsio ad convicia et calumnias Iosephi Scaligeri, in calendarium gregorianum* (1609); *Confutatio calendarii Georgii . . . Borussi* (1610); *Appendix ad novi calendarii romani apologiam continens defensionem* (1612); *Refutatio cyclometriae Josephi Scaligeri* (1612).

79. Villoslada, *Storia*, 65–66; Boero, *Menologio*, 117.

80. Boero, *Menologio*, 117.

81. Zaccagnini, *Baldi*, 344–45.

82. *New Grove Dictionary of Music and Musicians* (London: Macmillan, 1980), s.v. "Clavius."

83. "De toto orbe terrarum detegendo Hispaniae regibus sciscitantibus respondit." Von Murr, *Merkwürdigkeiten*, 124.

84. Clavius, *Corrispondenza* 1, pt. 1:51–53.

85. "Vix non mortuus," ibid., no. 119.

86. Ibid., 1, pt. 1:66.

87. On Clavius's "academy," see ibid., 1, pt. 1:68–89.

88. For example, in 1602 he was listed as "inter eos qui scribunt" Kleiser and Lamalle, comps., Chronology, ARSI.

89. Kelly, *Calendar of Documents*, 185, no. 403. The letter describing the meeting is dated 31 March 1603.

90. Of the three, only Filliucci appears in Sommervogel's bibliography. Most of his listed works are on moral theology, though one does appear to have an astronomical theme: *Stanze sopra le stelle e macchie solari, scoperte col nuovo occhiale, con una breve dichiarazione* (Rome, 1615). I have not located a copy of this work.

91. The etching by E. de Boulonois, copied closely from Villamena's portrait, was printed in Bullart, *Académie des sciences* 2:117. Villamena's work appears also in the rare compendium by Alfred Hamy, *Galerie illustrée* 2:40, where Hamy includes a brief biographical essay on Clavius, but his treatment, like Bullart's, is preoccupied with Clavius's work on the calendar reform.

92. Orbaan, *Documenti*, 169, n. 2.

93. On the Collegio Romano and the nova of 1604, see Baldini, "La nova del 1604." Clavius's participation in the lunar eclipse observations is mentioned in a letter of J. Vreman to Magini printed in Favaro, *Carteggio*, 327.

94. "Risaluto molto caramente il Padre Clavio, et mi dispiace che egli sia in letto," *Op. Gal.* 11:127.

95. Orbaan, *Documenti,* 284, which cites BAV Urbinates Latines 1080, c.107B.

96. Von Murr repeated the story but expressed some doubt about it. The compiler of the *Bibliografia universale antica e moderna* (Venice, 1823), 138–39, dismissed the story completely. Bruhns also found a denial necessary in the *Allgemeine Deutsche Biographie.*

97. Wilkins, *A Discourse Concerning a New Planet,* 21–22.

98. On Wilkins's *Discourse,* see Shapiro, *John Wilkins,* 38; more recently Moss, *Novelties in the Heavens,* 301–29; and Deason, "John Wilkins and Galileo Galilei."

99. Clavius, *Sphaera* (1611), *Op. math.* 3:75.

100. Kepler, *Opera omnia* 6:117.

101. Drake, *Discoveries and Opinions of Galileo,* 153.

102. Ross, *New Planet No Planet,* 9.

103. "Vernum nemo credet, Clavio morienti mentem adeo motam, ut terram movere voluerit, qui Clavium novit: & aliud longe est, existimare aliquid circa eccentricos & epicyclos Ptolemaei turbandum esse, atque integro systemate veteri prorsus diffracto, ad Copernicum voluisse transfugere." Froidmond, *Ant-Aristarchus,* 3.

Chapter Two

1. On humanistic currents in early Jesuit education, see Farrell, *Jesuit Code of Liberal Education,* 177ff.; Grendler, *Schooling in Renaissance Italy,* 377–80; Scaglione, *Liberal Arts,* 51–60; O'Malley, *First Jesuits,* Chaps. 6–7.

2. Harris, "Jesuit Ideology," esp. 167–211.

3. Ibid., 200.

4. Dear ("Mersenne," 28) sees Clavius's focus on utility as a reflection of "the practical emphasis of Ramus." See also Grafton and Jardine (*From Humanism to the Humanities,* 197), who discuss the breadth of the pragmatic tradition. Citing Ramus as their main example, they go on to generalize, "We find this same instrumental and pragmatic brand of humanism in every flourishing school of the later sixteenth century, Protestant or Catholic."

5. D'Elia, *Galileo in Cina,* 83, n. 1, and Sommervogel, *Bibliothèque,* s.v. Clavius.

6. Monachino, "La Fortuna di Galileo in Oriente," 818.

7. From "Gubernatorio Collegii Romani ac primo in literis et spiritualibus. Anno 1566," in Pachtler, *Ratio studiorum* 1:196.

8. One example of Clavius's curriculum documents appears in the appendix to Baldini, "La Nova del 1604" and another in Lattis, "Christopher Clavius," app. 4.

9. Ignatius Loyola, *Constitutions,* pt. 4:214.

10. "Modus quo disciplinae mathematicae in scholis Societatis possent promoveri." *Monumenta Paedagogica*, 471–73.

11. See, e.g., O'Malley, *First Jesuits*, 221–23.

12. On the social position of mathematics, see Westman, "Astronomer's Role"; Biagioli, "Social Status of Italian Mathematicians."

13. See Paul L. Rose, *Italian Renaissance of Mathematics*, esp. chaps. 8, 9; Gilbert, *Renaissance Concepts of Method*, chap. 3; Grendler, *Schooling in Renaissance Italy*, 309–11. See also the good survey of related issues by Gascoigne, "A Reappraisal."

14. Aristotle, *Posterior Analytics* 72a30–72b5.

15. Robertus Anglicus made a similar argument, though not stated so formally, in his commentary on Sacrobosco's *Sphere* (Thorndike, *Sphere of Sacrobosco*, 245).

16. See Baldini, "L'astronomia del Bellarmino"; Cosentino, "L'insegnamento delle matematiche"; Crombie, "Mathematics and Platonism"; Giacobbe, "Il *Commentarium de certitudine mathematicarum*," and "Epigoni nel seicento"; Wallace, "Certitude of Science," and *Galileo and His Sources*.

17. Gilbert, *Renaissance Concepts of Method*, 90–91; De Pace, *Matematiche e il mondo*, chaps. 1–2.

18. Benedict Pereira, *De communibus omnium rerum naturalium principiis et affectionibus, libri quindecim* (Cologne, 1595), 120, quoted in Gilbert, *Method*, 91. Pereira's book was first published in 1562. On Pereira and the status of mathematics see also Giacobbe, "Un gesuita progressista"; and De Pace, *Matematiche e il mondo*, 75ff.

19. "Non aliter resolvi poterunt omnes aliae propositiones non solum Euclidis, verum etiam caeterorum Mathematicorum. Negligunt tamen Mathematici resolutionem istam in suis demonstrationibus, eo quod brevius ac facilius sine ea demonstrent id quod proponitur, ut perspicuum esse potest ex superiore demonstratione." Clavius, *Op. math.* 1:28. This same statement appeared in his very first edition of the *Elements* (1574).

20. For an illuminating discussion of the debate over the status of mathematical sciences as interpreted by a generation of Jesuit scholars younger than Clavius, see Wallace, *Galileo's Logic of Discovery and Proof*, esp. 111–14. Another valuable study is Biagioli, "Social Status of Italian Mathematicians."

21. "Theoremata enim Euclidis, caeterorumque Mathematicorum, eandem hodie, quam ante tot annos, in scholis retinent veritatis puritatem, rerum certitudinem, demonstrationum robur, ac firmitatem. . . . Cum igitur disciplinae Mathematicae veritatem adeo expetant, adament, excolantque, ut non solum nihil, quod fit falsum, verum etiam nihil, quod tantum probabile existat, nihil denique admittant, quod certissimis demonstrationibus non confirment, corroborentque, dubium esse non potest, quin eis primus locus inter alias scientias omnes fit concedendus." Clavius, *Op. math.* 1:5.

22. "Procedunt enim semper ex praecognitis quibusdam principiis ad conclusiones demonstrandas, quod proprium est munus, atque officium doctrinae sive

disciplinae, ut et Aristoteles I. posteriorum testatur; neque unquam aliquid non probatum assumunt Mathematici." Ibid., 1:3.

23. "Scientiarum Mathematicarum quasdam in intellectilibus duntaxat ab omni materia separatis, quasdam verro in sensilibus, ita ut attingant materiam sensibus obnoxiam, versari." Ibid.

24. For a more elaborately articulated but less metaphysical scheme of mathematical disciplines, cf. John Dee's 1570 preface to Euclid's *Elements*. The modern edition of the preface, with an introduction by Allen Debus, appears under the title *The Mathematicall Praeface*.

25. "Ad has omnes utilitates accedit maxima iucunditas, atque voluptas, qua cuiusque animus his artibus colendis, exercendisque perfunditur. Sunt enim hae praecipuae ex septem artibus liberalibus, in quibus non solum ingenui adolescentes, verum etiam nobiles viri, principes, reges ac imperatores ad honestissimum, maximeque liberalem oblectationem animi . . . pariunt." Clavius, *Op. math.* 1:6.

26. "Christophorus Clavius, cuius commentarii caeteris omnibus praeferendi sunt" (Barozzi, *Cosmographia,* 9); "Et infinitorum eius commentatorum interque eos doctissimi et copiosissimi Christophori Clavii" (Kepler, *Opera omnia* 6:306).

27. "C'est le meilleur commentaire de l'astronomie (de *Sphaera*) de Sacrobosco." Houzeau and Lancaster, *General Bibliography of Astronomy* 1, pt. 1:596.

28. Johnson, "Astronomical Text-books in the Sixteenth Century," 301.

29. The standard modern edition and translation is Thorndike, *Sphere of Sacrobosco.* For a detailed analysis of Sacrobosco and his *Sphere,* see Pedersen, "In Quest of Sacrobosco." On the preference for Sacrobosco's work over the *Spheres* of other authors, see Thorndike, 23, and Pedersen, 175.

30. There were, e.g., those of Robert Grosseteste, John Peckham, and Campanus of Novarra, noted in Thorndike, *Sphere of Sacrobosco,* 23.

31. Ibid., 24–40.

32. Johnson, "Astronomical Text-books," 293.

33. See Pedersen's work on this subject, esp. "Astronomy," 303–37, "Corpus Astronomicum," and "Theorica Planetarum Literature."

34. Thorndike, *Sphere of Sacrobosco,* 48. Another useful survey of *Sphere* commentaries is McMenomy, "Discipline of Astronomy."

35. Duhem (*Le Système du monde* 3:239, n. 4) cites a printing as late as 1656.

36. Flamsteed, *An account,* 26.

37. The discussion here involves only introductory textbooks and does not include highly technical, innovative, or controversial works. A fuller treatment can be found in Johnson, "Astronomical Text-books."

38. Johnson cites several editions of each of the following (for brevity's sake only the date of the first edition is noted here); Oronce Finé, *De mundi sphaera sive cosmographia* (1542); Gemma Frisius, *De principiis astronomiae et cosmographiae* (1530); Francesco Maurolico, *Cosmographia* (1543); Francesco Barozzi, *Cosmographia* (1585).

39. Important examples cited by Johnson, all of which enjoyed multiple edi-

tions, are Alessandro Piccolomini, *De la sfera del mondo* (1540); Cornelius Valerius, *De sphaera et primis astronomiae rudimenta* (1558); Caspar Peucer, *Elementa doctrinae de circulis coelestibus et primo motu* (1551).

40. Johnson, "Astronomical text-books," 300–301.

41. A partial summary of the editions of Clavius's *Sphaera* is contained in Lattis, "Christopher Clavius," app. 2. The earliest extant version of Clavius's work is a manuscript preserved in BAV, Urb. Lat. 1303, 1304. The text bears the heading "Expositio in Sphaera Ioan. de Sacro Bosco." References in the text strongly suggest it was composed around 1565. It is not an autograph, nor is it the source text of the first printed edition, but it seems to have been copied and illuminated by a professional scribe perhaps working from Clavius's lecture notes. I am preparing a study of this prototype of the published *Sphaera*.

42. Johnson, "Astronomical Text-books," 296.

43. Dicks, *Early Greek Astronomy,* 117.

44. The figures cited in this discussion correspond to the sidereal periods of the superior planets but the synodic periods of the inferior.

45. The complex history of this problem cannot even be sketched here in any detail. For a general introduction to the history of ancient astronomy, see Lindberg, *Beginnings of Western Science,* esp. chap. 5. More detailed discussions can be found in, e.g., Dreyer, *History of Astronomy;* Pannekoek, *History of Astronomy;* Kuhn, *Copernican Revolution;* or Dicks, *Early Greek Astronomy.*

46. See Dicks, chap. 6, for a discussion of what we really know about Eudoxus's theories.

47. Hargreave, "Reconstructing Planetary Motions," 344–45.

48. A concise discussion of many of the characteristics of Ptolemy's planetary theories and his cosmological concepts can be found s.v., "Ptolemaic Planetary Theory" and "Ptolemy's Cosmology" in *Encyclopedia of Cosmology* (New York: Garland, 1993).

49. The fundamental source for the early history of this problem is Van Helden, *Measuring the Universe.*

Chapter Three

1. Clavius frequently does not name his adversaries. Depending on the context they can be Averroës and his later followers—e.g., Agostino Nifo—but they can also be advocates of rival cosmologies like Copernicus and Fracastoro. Their identities sometimes emerge as his responses to them develop.

2. Sixteenth-century cosmological diversity has not gone completely unappreciated. In particular, see Barker, "Stoic Contributions"; Donahue, *Dissolution of the Celestial Spheres,* "Kepler's Cosmology"; Grant, "New Look at Medieval Cosmology"; Schofield, *Tychonic World Systems.*

3. Westfall, "Rise of Science," 219.

4. Kuhn, *Copernican Revolution,* 175 (emphasis in original).

5. Some earlier standard histories are far worse. Take an older text still in common use: Dreyer (*History of Astronomy,* 338) on the same point says, "In thus giving the relative dimensions of the whole system Copernicus scored heavily

over Ptolemy, as no geocentric system can give the smallest clue to the distances of the planets.'' His point is technically true only in the sense that with the geocentric astronomy the planetary order and spacing is not implicit in the geometry. But the problem is that Dreyer implies this to be a failing of the pre-Copernican astronomy, which it is not—unless we persist in judging medieval arguments by modern standards. Another attempt to cleave Renaissance astronomy from cosmology is made in Price, ''Contra-Copernicus.'' Once again, see Van Helden, *Measuring the Universe,* for a balanced approach to the problems.

6. Kuhn, *Structure of Scientific Revolutions,* 69.

7. There is good reason to doubt whether the statement attributed to Alfonso X has anything to do with astronomical theory and indeed, whether it is any more than a myth. See Goldstein, ''Blasphemy of Alfonso X.''

8. The many reprintings of the *Sphaera,* in sometimes unauthorized and often obsolete versions, make comparisons and citations a complex business. My convention for distinguishing unambiguously between the various revisions of Clavius's *Sphaera* is to denote a particular revision level by the year of that version's first appearance followed by the part of the title indicating how many times it had been revised. Thus *Sphaera* 1593 (*nunc quarto*) refers to the fourth revision (and fifth version), which first appeared in 1593. The only exceptions are the first edition and the final revision, which I have taken as my standard in this work. The former came out in 1570 and the latter appeared in vol. 3 of Clavius's *Opera mathematica* (1611). These two versions may be recognized merely by their years of publication. Because some of the editions were reprinted, citations by page will note the place and date of publication if there is any possibility of ambiguity. Thus: *Sphaera* (Venice, 1596), p. 80.

9. This *disputatio* (a complete translation of which I am preparing) and other passages attributed to the 1581 version could have appeared as early as 1575 if, as some sources suggest, Clavius published a now lost *Sphaera* edition in that year. The bibliographies of Houzeau and Lancaster, Sommervogel, and Scheibel all cite a 1575 edition. Further, the words ''nunc *iterum* ab ipso auctore'' on the title page of the 1581 edition suggest that the work had been revised again, thus implying a revision between the 1570 and 1581 editions. Yet no copies of the book, if it existed, seem to have survived. It is unfortunate that the extinction of sixteenth-century titles is not a rare phenomenon. See Hirsch, *Printing, Selling, and Reading.*

10. The term ''Ptolemaic system'' is frequently used to refer imprecisely to at least two different ideas. One is the set of abstract mathematical models presented by Ptolemy in the *Almagest,* the other is the medieval and Renaissance conception of those models as embodied in material celestial spheres (see, e.g., Goldstein, ''Arabic Version of Ptolemy's *Planetary Hypotheses,*'' 3). The latter concept constitutes, and is thus better called, a cosmology, and I will use that term to distinguish the cosmological conception from the merely mathematical theories or ''system.''

11. A good survey of late medieval and Renaissance opinions on these issues is Grant, *In Defense of the Earth's Centrality.*

12. On this question in medieval *Sphere* commentaries, see Thorndike, *Sphere of Sacrobosco*, 49–52. See McMenomy ("Discipline of Astronomy," 246–50) on Capuanus de Manfredonis and Lefèvre d'Étaples, and on others passim. Schofield (*Tychonic World Systems*, 4–5) asserts that this acceptance of Ptolemaic constructs in the sixteenth century was quite general. For a comprehensive discussion see Grant, "Eccentrics and Epicycles."

13. Peurbach died in 1461 without publishing his *Theoricae novae planetarum*. His student and colleague Regiomontanus put it into print years later. See Pedersen (*Survey of the* Almagest, chap. 13) for a general discussion of the *Planetary Hypotheses;* for the text, see Goldstein, "Arabic Version of Ptolemy's *Planetary Hypotheses.*" On Peurbach, see Aiton, "Peurbach's *Theoricae novae planetarum.*"

14. "Eruditus quidam vir & religiosus vitam degens in provincia Peru . . . altera experientia consistit in partibus coeli rarioribus, cuiusmodi non paucae cernuntur . . . prope polum antarcticum, ita ut nigror quidam plerisque in locis coeli appareat, ac si coelum quodammodo esset perforatum." Clavius, *Sphaera* (1611), 47. The identity of this fellow is unknown. However, David Dearborn suggests (personal communication) that Clavius's source might have been José de Acosta, a Jesuit who, according to Sommervogel, went to the West Indies as a missionary in 1571 and eventually served as the Society's provincial of Peru. He returned to Europe, visited Rome, and later published some of the first accounts of native American sky lore in his *Historia natural y moral de las Indias* (Seville, 1590). Though Clavius's comment predates Acosta's published work, the reports concerning the appearance of the southern sky could have come to Rome through correspondence or other channels.

15. The question of the nature of the celestial spheres in medieval and early Renaissance cosmology has, in recent years, been the crux of a debate among some historians of astronomy. Clavius's opinions on the matter, though important in their own right, are rather too late to shed much light on the historiographical dispute, which tends to focus on Copernicus's concept of a celestial sphere. Nevertheless, the writings stimulated by the debate provide the interested reader with a spirited if occasionally confusing introduction to the issues. The following listing is not complete but follows the main line of the discussion. See Donahue, "Solid Planetary Spheres" (1975); Swerdlow, "Pseudodoxia Copernicana" (1976); Aiton, "Celestial Spheres and Circles" (1981); Jardine, "Significance of the Copernican Orbs" (1982); Rosen, "Dissolution of the Solid Celestial Spheres" (1985); Grant, "Celestial Orbs" (1987); and Barker, "Copernicus" (1990).

16. "Illud certe consequens videtur, secundum communem astronomorum sententiam, duritiem et hanc coelorum constitutionem stare non posse." Christoph Scheiner, *De maculis solaribus*, in *Op. Gal.* 5:69.

17. Roger Bacon, *Opus tertium*, quoted in Grant, "Eccentrics and epicycles," 190.

18. See Goldstein, "Arabic Version of Ptolemy's *Planetary Hypotheses*," 3.

19. It is difficult to translate *eccentricus secundum quid* succinctly because the term *secundum quid* carries significant baggage from its origins in medieval philosophical discussions. "Virtual eccentric" conveys the sense most important

here, namely, the device's function of constituting part of an eccentric in effect but not in fact. E. J. Aiton ("Peurbach's *Theorica*") has used the terms "relative eccentric" and "absolute eccentric." The terms *simpliciter* and *secundum quid* are not original with Clavius. They appear at least as early as the late fourteenth-century *Sphere* commentary of Pierre d'Ailly. See Grant, "Eccentrics and Epicycles," 193. In the *Theorica planetarum* literature the nonuniform orbs are frequently called *orbes partiales,* or partial orbs, in recognition of their contribution to the *sphaera tota,* or complete sphere.

20. "Solus circulus est." Clavius, *Sphaera* (1611), 310.

21. "Sed iam ad phaenomena explicanda accedamus, quibus maxime astronomi sunt impulsi, ut eccentricos orbes atque epicyclos in sphaeris coelestibus invenerint," ibid., 291.

22. This general arrangement was established by Ptolemy, who concluded that the eighth sphere completed a revolution in thirty-six thousand years. Later astronomers, on the basis of other observations, modified this figure. See Ptolemy, *Almagest,* 327–29.

23. Among them Thābit Ibn Qurra.

24. "Nunc ratum sit & certum, nonum orbem motu isto tardissimo ab occidente in orientem trahere secum 8 inferiores sphaeras coelestes, nullo vero pacto supremam sphaeram. Iuxta enim sententiam astronomorum, quicumque orbis superior suo motu circumfert inferiorem sibi contiguum & concentricum, non autem superiorem." Clavius, *Sphaera* (1611), 29.

25. "Nullus tamen planeta inferior movetur ad motum proprium planetae superioris, eo quod non circa idem centrum propriis lationibus feruntur." Ibid. 30.

26. "Videmus enim sphaeras omnium planetarum, simul cum firmamento, et nono coelo, spatio 24 horarum ad motum diurnum primi mobilis rapi ab ortu in occasum. Rursus experimur, easdem sphaeras planetarum, una cum firmamento ad motum nonae sphaerae trahi ab occasu in ortum. . . . Denique animadversum est, omnes coelos planetarum paulatim etiam moveri ad motum trepidationis, seu accessus, et reccessus octavae sphaerae. . . . Cum igitur maxima singularitas motuum in planetis reperiatur, ita ut nullius motus proprius inferiori planetae communicetur, ut cuius . . . notum esse potest, et a nimine negatur, (Iuppiter enim nihil prorsus habet ex motu 30 annorum Saturni: itemque Marti nihil communicatur ex motu 12 annorum Iovis, et sic de caeteris, ut omnes affirmant.) perspicuum esse videtur, orbes planetarum vectores non esse concentricos. Alioquin motus cuiuslibet superioris omnibus inferioribus planetis communicaretur, quemadmodum id contingere videmus in sphaeris totalibus, ut diximus." Ibid. 298.

27. Schofield (*Tychonic World Systems,* 4–5) suggests, on the basis of her survey, that the Ptolemaic order of the planets, like the Ptolemaic planetary constructions, was widely accepted in the sixteenth century.

28. For Sacrobosco's statement on the order of the planets and for a comparison to other commentaries, see Thorndike, *Sphere of Sacrobosco,* 119 and 51, respectively.

29. Ptolemy, *Almagest,* 419–20. On Ptolemy's lunar parallax measurement, see Henderson, *On the Distances between Sun, Moon, and Earth,* 36–37.

30. The term *diversitas aspectus,* for parallax, was old even when Clavius

used it; it can be found in medieval *Sphere* commentaries. Clavius's students abandoned the venerable Latin term and took up the more fashionable *parallaxin* from Greek, whence we derive the modern term. See, e.g., Maelcote's apposition of the old and new terms in his discussion of the parallax of the nova of 1604 in Baldini, "La nova del 1604," 73.

31. Clavius, *Sphaera* (1611), 43.

32. Van Helden, *Measuring the Universe*, 16–19.

33. Maurolico argues along similar lines in his *Cosmographia,* fols. 19v–20r.

34. Clavius, *Sphaera* (1611), 45.

35. Ibid., 45–46.

36. "Quamvis enim nulla earum sufficienter hunc ordinem colligat, omnes tamen simul sumptae confirmant, coelos eo ordine collocatos esse. Nam ex diversitate aspectus infallibiliter colligitur ordo Lunae, Mercurii, Veneris, & Solis. Ex velocitate vero & tarditate motus convenienter supra hos quatuor planetas collocatur Mars, deinde Iupiter, postremo Saturnus, supra omnes vero planetas firmamentum. . . . Ex eclipsibus denique licet non omnium planetarum ordo firmiter possit colligi, tamen Lunam cogimur infimo loco ponere & omnes planetas sub firmamento." Ibid., 44.

37. "Solem convenienter statui in medio planetarum." Ibid.

38. Ibid., 47.

39. "Constat igitur ex omnibus iis, quae diximus, ordinem a nostro auctore praescriptum inter planetas esse verum, et magis conformem astronomis peritis, alios autem minime." Clavius, *Sphaera* (1570), 95. "Constat igitur ex omnibus iis, quae diximus, ordinem a nostro auctore praescriptum inter planetas esse veriorem, et magis conformem astronomis peritis." Clavius, Sphaera (1581, *nunc iterum*), 70. Wallace points out the shift from *verum* to *veriorem* in "Galileo's Early Arguments for Geocentrism," 33.

40. Wallace, "Galileo's early arguments for geocentrism," 33.

41. Here is one of the cases in which the phantom 1575 edition, if it existed, might shed some light on this situation.

42. Van Helden, *Measuring the Universe*, 29–33.

43. "Et hoc iudicio praecepimus, quod nulla vacuitas est inter circulos." Al-Farghānī, *Rudimenta astronomica, differentia* xxi.

44. Ptolemy in Goldstein, "Arabic Version of Ptolemy's *Planetary Hypotheses,*" 8.

45. Van Helden gives a thorough discussion of the development of this scheme from Ptolemy to the Middle Ages in *Measuring the Universe,* esp. chaps. 3 and 4.

46. On Maurolico, see Paul L. Rose, *Italian Renaissance of Mathematics,* 160–76 passim.

47. Maurolico's table appears in an appendix to his *Cosmographia* (1543), and al-Farghānī's numbers appear in *Rudimenta astronomica* (1537), *differentia* xxi. On the relationship of these tabulations, see Van Helden, "Galileo on the Sizes and Distances of the Planets," 70. Van Helden reproduces al-Farghānī's numbers in *Measuring the Universe,* 30, and again in "Telescope and Cosmic Dimensions," 107.

48. Van Helden, *Measuring the Universe*, 53.

49. Grant, "Eccentrics and Epicycles," 189–90. This article gives a good survey of the objections raised against the Ptolemaic constructions and some of the solutions proposed by medieval philosophers.

50. Ibid., 204.

51. "Quod autem hoc motu nunc ad terram magis accedant, nunc longius ab ea demoveantur, hoc non est absurdum: quia hic accessus et recessus non fit per lineam rectam, quem solum a corporibus coelestibus Aristoteles exclusit, cum solis elementis conveniat, quae gravia sunt, ac levia. Quod si quis contendat, Aristotelem contrarium putasse, condonandum ei hoc erit. Locutus est enim de illis duntaxat motibus, qui suo tempore cogniti erant, quales sunt a medio, et ad medium per lineam rectam, et circa medium mundi. Quod si motus eccentricorum et epicyclorum suo tempore noti fuissent, non dubito, quin aliter de motu circa medium locutus fuisset." Clavius, *Sphaera* (1611), 302.

52. "Secundam obiectionem solvemus, si dicamus, omnes orbes eccentricos, etiam illos secundum quid, atque epicyclos, perfectissime esse sphaericos, quoad propria centra. Superficies enim extimae omnium horum orbium secundum omnes partes aequaliter a suis centris absunt. Neque vero obstat, quod orbes eccentrici secundum quid, crassiores sunt una parte, quam alia: quia nulla ratio naturalis persuadere potest, omnes orbes coelestes debere esse uniformis et aequalis crassitiei. Si vero Aristoteles contrarium docuit, nos ei hac in parte non credimus." Ibid.

53. "Ad obiectionem septimam negandum est, terram quiescentem necessariam esse in quolibet centro, ut circa illam orbes coelestes moveantur. Quamvis enim Deus Opt. Max. terram hanc vel omnino auferret, vel alio impelleret extra centrum mundi, adhuc tamen coeli motu diurno veherentur circa medium mundi." Ibid., 303.

54. There are even strong theological overtones in Ptolemy, bk. 1, which would thus have provided adequate precedent for the later Christian authors to address concerns of their own faith in their astronomy texts (see Ptolemy, *Almagest*, 35–37; Taub, *Ptolemy's Universe*, chap. 5).

55. "Supra hos undecim coelos mobiles theologi, ut Strabus, Venerabilis Beda & omnis iam theologorum coetus, aliud coelum esse affirmant, immobile quidem & nulla praeditum stella, sed felicem angelorum & beatorum sedem ac patriam, quod vocant coelum empyreum." Clavius, *Sphaera* (1611), 24. On earlier ideas of the number and order of the spheres and the existence of the empyrean, see Grant, "Cosmology," 275–80. Grant traces the concept of the empyrean heaven to biblical commentaries near the beginning of the twelfth century or possibly even earlier. If Clavius is correct in attributing the concept to Walafrid Strabo (d. 849), then it goes back to the ninth century at least.

56. "Statuunt ergo astronomi huius temporis, in universum esse duodecim coelos, undecim quidem mobiles, unum vero, ex sententia theologorum, immobile prorsus." Clavius, *Sphaera* (1611), 24.

57. Van Helden, *Measuring the Universe*, 53.

58. For example, Agostino Ricci, *De motu octave sphaere* (1513); and Johann Werner, *De motu octavae sphaerae* (Nuremberg, 1522). On this topic in other

texts, see Johnson, "Astronomical Text-Books," 301–2. The standard title harks back to the much earlier work of the same title by Thābit Ibn Qurra on trepidation.

59. Johnson, "Astronomical Text-Books," 298–99.

60. Swerdlow and Neugebauer, *Mathematical Astronomy*, 127–28.

61. Thanks to Owen Gingerich, who helped me track down the first appearance of this passage.

62. "Caeterum et hoc observandum diligenter est, distantias, crassities, magnitudinesque coelorum ac stellarum eo modo inventas ut praescriptum est a nobis, quamvis immensae sint et fidem humanam superare quodammodo videantur, esse tamen minimas, quae esse possint: propterea quod astronomi ponunt eccentricum orbem cuiusque orbis coelestis tangere convexum et concavum ipsius coeli in uno tantum puncto. . . . Credibile autem est, Deum Opt. Max. orbes illos coelestes condidisse densiores. . . . Quo posito, certum est, distantias, crassities, magnitudinesque coelorum, ac stellarum longe esse maiores, quam ab astronomis sunt repertae. Solum igitur demonstratum est a nobis, quo pacto omnia haec ex ipsis motibus colligi possint. Nam etsi fortasse maior illa crassities, ac distantia condita est a Deo, per motus tamen illam cognoscere nullo modo possumus." Clavius, *Sphaera* (1611), 119.

63. The history of these and other related questions has been treated recently in Grant, *Much Ado about Nothing;* and Dick, *Plurality of Worlds.*

64. "Extra hunc vero mundum, seu extra coelum empyreum, nullum prorsus corpus existit, sed est spacium quoddam infinitum (si ita loqui fas sit) in quo etiam toto Deus existit sua essentia, in quo infinitos alios mundos perfectiores etiam hoc, fabricare posset, si vellet, ut theologi asserunt." Clavius, *Sphaera* (1611), 47. This comment appeared in Clavius's first *Sphaera* (1570) and every edition thereafter.

65. An extract of the official Jesuit philosophical doctrine on this point is published in Grant, *Source Book in Medieval Science*, 560–63.

66. Oresme, *Le livre du ciel et du monde*, 177.

Chapter Four

1. Berry (*Short History*, 121) does bring up Fracastoro's system but incorrectly identifies it as Ptolemaic.

2. Grant ("New Look at Medieval Cosmology," 417–19) has also pointed out the historiographical neglect of sixteenth-century cosmological diversity and traced its roots, in part, to Galileo. Vasoli ("Patrizi sull'infinità dell'universo," 297) has also pointed out the many cosmological options of this era.

3. On the varieties of Renaissance Aristotelianism, see Schmitt, *Aristotle and the Renaissance*, 10–33, and on Achillini and Nifo, ibid., 99–101.

4. Like his later nemesis Clavius, Fracastoro, too, was an advocate of, and dabbler in, calendar reform schemes. See Pellegrini, *Un veronese precursore*, 136.

5. On homocentrics in antiquity, see Dreyer, *History of Astronomy*, chaps. 4, 5.

6. On the failings of homocentric theories, see Hargreave, "Reconstructing Planetary Motions," 335–45.

7. Goldstein, "Status of Models," 138–39; Kellner, "On the Status of the Astronomy in Maimonides."

8. Dreyer (*History of Astronomy*, 264–67) gives a very brief treatment of al-Bitrūjī. For a fuller, more technical treatment, see Goldstein, *Al-Bitrūjī: On the Principles of Astronomy*. For the thirteenth-century Latin translation of al-Bitrūjī by Michael Scot, see Carmody, *Al-Bitrūjī*.

9. Goldstein, "On the Theory of Trepidation," 244–46.

10. An excellent review of the history of homocentric astronomy in the Middle Ages and Renaissance is contained in Di Bono, *Le sfere omocentriche*, 11–74.

11. Thomas Aquinas, *Exposition of Aristotle's Treatise*, 33.

12. Roger Bacon, *Liber secundus communium naturalium de celestibus*, 430–32.

13. Kren, "Homocentric Astronomy."

14. On Bernard of Verdun and Jean Buridan see Grant, *Source Book in Medieval Science*, 520–29, and "Eccentrics and Epicycles."

15. Duhem, *Système du monde* 10: 361–62. Duhem attributes this report to Gassendi.

16. See Shank, " 'Notes on al-Bitrūjī.' "

17. Swerdlow, "Aristotelian Planetary Theory," 36.

18. Giovanni Battista Amico, *De motibus corporum coelestium iuxta principia peripatetica sine eccentricis et epicyclis* (Venice, 1536; republished, with extensive and very valuable commentary by Di Bono, as *Le sfere omocentriche*). See also Swerdlow, "Aristotelian Planetary Theory," 36.

19. Di Bono, *Le sfere omocentriche*, 62–63.

20. Swerdlow gives a technical account of this mechanism, which is generally known today as the "Tusi couple," in "Aristotelian Planetary Theory," 38–41.

21. Fracastoro, *Homocentrica*, fol. 10v. On the infirmities of Fracastoro's mathematics and Clavius's criticisms of them, see Lattis, "Homocentrics, Eccentrics, and Clavius's Refutation of Fracastoro."

22. *Aristotelis opera cum Averrois commentariis* 5, *De caelo*, lib. 2, quaes. 2, 3.

23. Nifo, *In quattuor libros de celo et mundo et Aristote. et Avero. expositio*, bk. 2, fols. 23–26. On Nifo, see also Duhem, *To Save the Phenomena*, 48.

24. Clavius's occasionally repetitive reaction to homocentrics occurs in various places in his textbook. The most concentrated segment, however, is found in the *disputatio* on eccentrics and epicycles. See esp. Clavius, *Sphaera* (1611), 290–304.

25. "Subtile sane, sed omnino futile figmentum. . . . Si esset illa densitas, eaedem stallae fixae in zodiaco existentes uno tempore maiores nobis apparerent, quando nimirum illis supponuntur partes illae densiores, quam alio tempore, quod cum experientia pugnat." Ibid., 293. Cf. Fracastoro, *Homocentrica*, fol. 58.

26. "Scio auctores orbium concentricorum confingere intra singulorum planetarum orbes, singulos orbes restituentes, quos Fracastorius circitores appellat, quorum officium sit, ut quantum superiores planetae inferiores trahunt suis motibus, tantum ipsi inferiores planetas in contrariam partem restituant. Verum hoc figmento simile esse videtur. Praeterquam enim, quod hac ratione maxima confusio in moti-

bus introducitur, non video, quo pacto primum mobile omnibus inferioribus sphaeris motum diurnum possit communicare, cum in medio positi sint circitores illi, qui inferiores sphaeras omnino prohibent, ne a superioribus rapiantur." Ibid., 299.

27. "Ponunt enim, ut apud Fracastorium est manifestum, orbes, seu spheras mobiles 77 vel 79, octo quidem stellatas, reliquas vero omnes stellis privatas, quarum sex supra firmamentum collocant. Quod non solum maiori parti astronomorum adversatur, qui hactenus tres tantum sphaeras coelestes non stellatas supra firmamentum invenerunt; verum etiam pugnat cum omnibus peripateticis, qui, ex Aristotelis sententia, ne unum quidem orbem supra firmamentum admittere volunt." Ibid., 299–300.

28. Thorndike, *Sphere of Sacrobosco,* 202–3.

29. McMenomy, *Discipline of Astronomy,* 188. McMenomy also discusses Robertus Anglicus.

30. Andalo's more famous close contemporary Jean Buridan (1295–1358) also employs the phrase. See Grant, "Cosmology," 298, n. 55.

31. Dales, "De-Animation of the Heavens," 546, n. 58. Pietro discusses the motions of the planets in *differentia* 5 of his *Lucidator.*

32. Thorndike, *Sphere of Sacrobosco,* 203.

33. Barker, "Stoic contributions," 141.

34. Ibid., 138–39.

35. Ibid., 143–47.

36. Giovanni Pontano, *Opera,* vol. 3 (Venice, 1519), fols. 144v–145r, cited in Jardine, "Scepticism," 95. Lerner ("Problème de la matière céleste," 263–66) mentions Pontano and some other early sixteenth-century advocates of the fluid heavens.

37. Rossi, "Francesco Patrizi," 370, "La negazione delle sfere," and "Sfere celesti e branchi di gru"; see also Brickman, "Introduction to Patrizi's 'Nova de universis philosophia' "; Saitta, *Il pensiero italiano nell'umanesimo e nel Rinascimento* 2:567–68; Vasoli, "Patrizi sull'infinità dell'universo," 284–87, 291–96. Rosen made an unconvincing attempt to trace Patrizi's views to Tycho in "Francesco Patrizi and the Celestial Spheres."

38. "Si novatione delectarer, forsan aliquid comminisci possem simile Fracastorianis aut Patricianis conceptionibus." Kepler, *Opera omnia* 6:306.

39. Jardine, *Birth of History and Philosophy of Science,* 154–55; Kepler, *New Astronomy,* 117.

40. Kristeller, *Eight Philosophers,* 113. Lerner (*Tre saggi,* 98–99) also names Jean Pena and Christoph Rothmann as advocates of fluid heavens.

41. On the cosmological views of this influential Jesuit, see Baldini, "L'Astronomia del Cardinale Bellarmino"; Baldini and Coyne, *Louvain Lectures.* For a brief summary of Bellarmine's views, see Blackwell, *Galileo, Bellarmine, and the Bible,* 40–45.

42. Baldini and Coyne, *Louvain Lectures,* 16. The general view of three heavens was not original with Bellarmine and can be found, e.g., in Thomas Aquinas, though the two theologians had different ideas of the physical natures involved. See, e.g., Grant, "Cosmology," 275.

NOTES TO PAGES 97–101 **239**

43. Baldini and Coyne, *Louvain Lectures,* 12.

44. Ibid., 20–22. The translators also comment (33, n. 39) on Bellarmine's skepticism about the Ptolemaic constructions.

45. Baldini, "L'Astronomia del Bellarmino," 302 passim; Baldini and Coyne, *Louvain Lectures,* 33, n. 39.

46. "Motu proprio sicut aves per aerem et pisces per aquam." Baldini and Coyne, *Louvain Lectures,* 18–19.

47. Correspondence between Federico Cesi and Johann Faber and a note by Grienberger show that Bellarmine made no secret of his views. See *Op. Gal.* 13:429; Baldini and Coyne, *Louvain Lectures,* 42–43.

48. Clavius, *Sphaera* (1611), 22.

49. "Omnis scientia nostra secundum philosophorum dogmata, a sensu oritur." Ibid.

50. "Atqui in astris reperiuntur diversi motus & oppositi. Cum ergo astra non per se moveantur, ut pisces in aqua, vel aves in aere, ut Aristoteles vult cum philosophis & nos paulo post demonstrabimus, sed ad motum orbis, in quo sunt, sicuti nodus in tabula ad motum tabulae, vel clavus infixus in rota aliqua ad motum rotae; oportebit concedere plures coelos, quam unum." Ibid.

51. "Sententia eorum, qui dicunt coelum quiescere & stellas per se moveri" and "Sententia eorum qui dicunt coelum moveri ab ortu in occasum, stellas vero per se ab occasu in ortum." Ibid., 24.

52. "Quidam vero asserunt, non solum coelum, verum etiam terram quiescere, stellas vero per sese moveri, ut aves in aere, seu pisces in mari, ab oriente in occidentem: Sed quoniam hac ratione non possent planetae duobus ferri motibus, quod pugnat cum experientia, cum non solum planetas videamus ab ortu in occasum moveri, sed etiam ab occasu in ortum." Ibid.

53. "Idcirco alii coelum moveri ab oriente in occidentem, secumque stellas circumducere, singulas vero stellas, singulos etiam habere motus ab occidente in orientem, affirmant. Quam ob rem, inquiunt, efficitur, ut omnia astra eodem tempore videantur motum diurnum absolvere, in temporibus vero inaequalibus ea moveri ab occasu in ortum deprehendamus." Ibid.

54. "Ut in sequentibus demonstrabimus, impossibile est stellas per sese moveri . . . sed necesse est, eas ad motum duntaxat orbis, in quo sunt, circumduci." Ibid.

55. "Sumatur quaevis stella, sive fixa sit, sive erratica, quam aliquis dicat per sese moveri. Haec stella movetur motibus quodammodo oppositis, ut supra diximus. Movetur enim simpliciter & continue ab oriente in occidentem & simul eodem tempore secundum quid & continue ab occidente in orientem, quemadmodum supra expositum fuit, atque demonstratum. At vero nullum corpus idem numero cieri potest diversis motibus, atque adeo oppositis, eodem tempore: Implicat enim contradictionem unum & idem corpus simul procedere ab oriente in occidentem & eodem instanti ab occidente in orientem, ita ut neuter motus alterum interrumpat, sed uterque sine ulla intermissione uniformiter progrediatur, nisi altero motu moveatur tanquam ad vehiculum alterius. . . . Confirmatur hoc ipsum multo magis in planetis. Moventur enim adhuc pluribus motibus, quam duobus

illis ab ortu in occasum & ab occasu in ortum, & nunc velocius videntur moveri ab occidente in orientem, nunc tardius: Videntur interdum stare, interdum retrocedere in occidentem, &c. ut in Theoricis planetarum explicatur. Si igitur stellae per sese moverentur, non posset sufficiens ratio huiusce varietatis afferi: Si autem ad motum coeli moveri dicantur, facili negocio omnes apparentiae locum habent ut in Theoricis planetarum explicabitur.'' Ibid., 48.

56. ''Immo si ita moverentur, et non potius ad motum orbium, in quibus sunt, nullam certam scientiam de illorum motibus habere possemus. Cum enim, ut in superioribus apparentiis dictum est, planetae aliquando magis, aliquando minus a terra absint: interdum velocius moveatur, interdum quasi cursum inhibeant; nunc stare videantur, nunc progredi sub zodiaco ab occasu in ortum, nunc retrogredi; quis est, qui non videat, planetas, si moventur ut pisces, seu aves, aliquando suos circulos, quos ab occasu in ortum describunt debere relinquere, ut magis possint a terra recedere, et ad eandem accedere, aliquando autem proprium cursum negligere, rursusque in oppositam partem retrocedendo niti; aliquando denique cursum omnino sistere in coelo, ut penitus non moveantur? Quae si fierent, quonam modo, obsecro, eorum periodi definiri poterunt, qua item ratione cognosci, quanam in parte coeli altius a terra digressuri sint planetae, et iterum ad terram reversuri, &c.'' Ibid., 299.

57. Giannettasio, *Universalis cosmographiae elementa*, 116.

58. A variant of this idea has been found by Michael Shank in a fifteenth-century manuscript by ''G. Marchio,'' who may be, according to Shank, identical with the Franciscan Guy de la Marche. See Shank, '' 'Notes on al-Bitrūjī,' '' 18–22. The treatise describes the planets as moving though toroidal circles (*circuli columnpnares*) between which is some sort of fluid medium.

59. ''Dicunt enim, unicum tantum esse coelum, atque hoc ipsum unico motu moveri ab oriente in occidentem, una cum omnibus stellis; stellas vero propriis motibus ab occidente in orientem ferri, ut aiunt, solutas ab orbibus coelestibus; non quidem tanquam pisces in mari, vel aves in aere, ne detur penetratio corporum, aut scissio coeli, sed per canales quosdam.'' Clavius, *Sphaera* (1611), 48.

60. ''Itaque secundum hos auctores totum coelum erit refertum istis canalibus pro multitudine stellarum ad instar animalis, quod repletum est variis ac multiplicibus venis.'' Ibid.

61. ''Absurda quidem, quoniam sine ulle necessitate, aut ratione probabili, ponit corpus coeleste perforatum tot canalibus & refertum undique corpore illo fluxibili, quod nemo philosophorum hactenus concedere visus est. Insufficiens vero, quia impossibile est defendere iuxta hanc sententiam omnia phenomena, quae astronomi diligentissime observarunt in motibus coelestibus . . . tamen nullo modo plures motus, praeter hos duos, stella quaevis habere potest . . . cum igitur in Luna plures sint deprehensi motus, nempre sex, ut minimum . . . planetae, ut ex *Theoricis planetarum* liquet, non semper aequaliter distant a centro terrae, sed nunc propiores, nunc vero remotiores apparent, quod nullatenus fieri posset, si stellae per sese in dictis canalibus moverentur, nisi dicatur illos canales esse eccentricos cum mundo. . . . Nam cum canales illi sint infixi corpori coelesti, necessario efficeretur, ut planeta quicunque in eadem semper parte coeli maxime a terra

distaret, etc., quod est falsissimum. . . . Omitto apparentias de variatione latitudi-
num . . . necnon de retrogradatione, etc. quas nullo pacto praedicta opinio tueri
potest. . . . Constat igitur stellas non per se moveri, sed ad motum coelorum, in
quibus sunt infixae.'' Ibid.

Chapter Five

1. On many important aspects of the reception of Copernicus's work, see
Westman, ''Three Responses to the Copernican Theory,'' 285–345, and ''The
Copernicans,'' 76–113.
2. For Nuñez's reference to Copernicus, see Swerdlow, ''Pseudodoxia Copernicana,'' 141–42. On Maurolico, see Dreyer, *History of Astronomy,* 356–57; and
also Rosen, ''Maurolico's Attitude toward Copernicus.''
3. Thorndike, *History of Magic and Experimental Science* 6:6–7; Johnson,
''Astronomical Text-books,'' 285–86.
4. ''L'auteur a évité de parler du système de Copernic.'' Houzeau and Lancaster, *General Bibliography* 1, pt. 1:596.
5. Johnson, *Astronomical Text-books,* 302.
6. Grant (''Celestial Orbs'') provides a survey of medieval opinions on this
subject and acknowledges there that the solidity of the celestial orbs was generally
accepted in the sixteenth century. On this matter, as in many other cases, some
of Galileo's early writings echo Clavius's teachings. See Wallace, *Galileo's Early
Notebooks,* 147.
7. Pereira, *De communibus omnium rerum naturalium principiis et affectionibus* (Rome, 1562). Pereira's teachings on celestial matter are cited and summarized
in Donahue, *Dissolution of the Celestial Spheres,* 28–30.
8. This general view is not original with Pereira but can be traced at least to
Thomas Aquinas. See Jardine, ''Significance of the Copernican Orbs,'' 175. On
Aquinas's views, see Litt, *Corps célestes.*
9. Pereira, *De communibus* 9:179, 10:181–83. Cited in Donahue, *Dissolution
of the Celestial Spheres,* 29.
10. On Ramus, see Jardine, ''Scepticism in Renaissance Astronomy,'' 95,
and *Birth of History of Science,* 234–36. On Bellarmine's views, see Baldini,
''L'astronomia del Bellarmino,'' 298–302, and *Legem impone subactis,* 285–344.
11. ''Philosophus siquidem praecipue naturam ac substantiam coeli conatur
investigare . . . astrologus vero de eodem corpore coelesti agit hac praecisa ratione,
qua circa medium universi est mobile, ut videlicet assignet periodos et varietates
omnium motuum.'' Clavius, *Sphaera* (1611), 6.
12. Carmody, *Al-Bitrūjī,* esp. 42–46; Goldstein, *Al-Bitrūjī,* 5.
13. ''Quoniam cum hi motus, ut aiunt, sint contrarii, necesse est alterum eorum
esse violentum, quod fieri non potest, immo absurdum videtur concedere vio-
lentiam in corporibus coelestibus, tum quia nullum violentium est perpetuum;
motus autem coeli perpetuus est, ex Aristotelis sententia.'' Clavius, *Sphaera*
(1611), 25.
14. Carmody, *Al-Bitrūjī,* 46–47; Goldstein, *Al-Bitrūjī,* 64–65.

15. "Quod nulla proportio in hac retardatione cernatur: Octava enim sphaera absolvit, secundum Ptolemaeum, suum circuitum spatio 36,000 annorum: Saturnus 30 annis: Iuppiter 12, Mars 2, Sol uno anno, Venus ac Mercurius eodem fere tempore: Luna denique 27 diebus and 8 horis: ubi manifeste vides, nullam certam proportionem inveneri." Clavius, *Sphaera* (1611), 26.

16. "Immo si ita moverentur . . . nullam certam scientiam de illorum motibus habere possemus. . . . Quis est qui non videat, planetas, si moventur ut pisces, seu aves, aliquando suos circulos, quos ab occasu in ortum describunt debere relinquere, ut magis possint a terra recedere, et ad eandem accedere; aliquando autem proprium cursum negligere, rursusque in oppositam partem retrocedendo niti; aliquando denique cursum omnino sistere in coelo, ut penitus non moveantur? Quae si fierent, quonam modo, obsecro, eorum periodi definiri poterunt, qua item ratione cognosci, quanam in parte coeli altius a terra digressuri sint planetae, et iterum ad terram reversuri, etc." Ibid., 299.

17. "Ex orbibus eccentricis et epicyclis, non solum apparentiae iam olim cognitae defenduntur, sed etiam futurae praedicuntur quarum tempus omnino ignoratur; ita ut si ego dubitem, an v.g. in plenilunio Septembris anni 1587 futura sit eclipsis Lunae, certus omnino reddar ex motibus orbium eccentricorum et epicyclorum, futuram esse eclipsim, ita ut amplius non dubitem. Immo ex eisdem motibus cognosco, quo hora illa eclipsis inceptura sit, et quanta pars Lunae sit obscuranda. Eodemque modo omnis eclipses tam Solares quam Lunares praedici possunt, earumque tempus, et magnitudines." Ibid., 301.

18. For more detail, see the explanation of trepidation in Pedersen and Pihl, *Early Physics and Astronomy*, 183–85.

19. "Ponunt enim, ut apud Fracastorium est manifestum, orbes, seu spheras mobiles 77 vel 79 octo quidem stellatas, reliquas vero omnes stellis privatas, quarum sex supra firmamentum collocant. Quod non solum maiori parti astronomorum adversatur, qui hactenus tres tantum sphaeras coelestes non stellatas supra firmamentum invenerunt; verum etiam pugnat cum omnibus peripateticis, qui, ex Aristotelis sententia, ne unum quidem orbem supra firmamentum admittere volunt." Clavius, *Sphaera* (1611), 299–300.

20. Clavius says here thirty-three rather than thirty-four, apparently forgetting to update this passage with the additional sphere added after the rejection of trepidation. He listed the number of orbs in each planetary model in the tabular *Theorica* at the end of the *Sphaera*. His numbers, excluding epicycles, are Sun, three; Moon, four; Mercury, five; Venus, Mars, Jupiter, and Saturn, three each. Except for the sun, each has an epicycle, which accounts for six more. Finally the firmament and three suprafirmamental spheres brings the total to thirty-four.

21. "Tantum confusionem vitant ii, qui eccentricos orbes ponunt in coelis; quia in universum orbes duntaxat 33 concedunt, ambientes quidem terram 27 sex vero epicyclos, qui toti extra terram extant. Unde non erit tanta motuum multitudo. . . . Itaque cum, secundum celeberrimum philosophorum axioma, frustra fiat per plura, quod fieri potest aeque bene per pauciora; ponantur autem a nobis triplo fere pauciores eccentrici, quam ab adversariis concentrici; et non solum aeque bene, sed multo melius omnia φαινόμενα per eccentricos defendantur quam

per concentricos, cum sexcentarum apparentiarum ratio per concentricos dari
nequeat, ut ex dictis perspicuum est; quis dubitabit, potius in coelis esse orbes
eccentricos et epicyclos constituendos, quam concentricos, praesertim cum naturali
philosophiae eccentrici nihil omnino repugnent, ut ex solutionibus argumentorum
Averrois, eiusque sectatorum constabit?'' Clavius, *Sphaera* (1611), 300.

22. ''Aliquid supervacaneum et otiosum.'' Ibid., 302.

23. Ibid., 303.

24. ''Ex antiquis igitur nonnulli, quorum dux fuit Aristarchus Samius 400
annis ante Ptolemaeum, quem ex recentioribus secutus est Nicolaus Copernicus in
opere de revolutionibus coelestibus, hunc ordinem inter corpora totius universi
confinxerunt: ut Sol in centro, seu medio mundi immobilis sit collocatus; circa
quem orbis Mercurii; deinde orbis Veneris; circa hunc orbis magnus, terram una
cum elementis & Luna continens; circa quem orbis Martis; deinde coelum Iovis;
postea globus Saturni; ultimo tandem stellarum fixarum sphaera sequatur. Verum
haec opinio multis experimentis refragatur & communi omnium philosophorum
astrologorumque sententiae. Debet enim terra consistere in medio totius mundi,
ut postea demonstrabimus plurimis experientiis ac phaenomenis.'' Ibid., 42.

25. Ibid., 67.

26. Ibid., 68. Ptolemy, *Almagest,* 41-42.

27. Clavius, *Sphaera* (1611), 69-70.

28. ''Ac profecto natura iure optimo terram in medio mundi collocasse videtur,
ut tam vile ac rude corpus ab omnibus partibus coeli, quod est corpus praestantissi-
mum, aequiliter semoveretur.'' Ibid., 70.

29. Wilkins, *A Discourse concerning a New Planet, Tending to Prove, that
(It Is Probable) Our Earth is One of the Planets,* in *Mathematical and Philosophical Works,* 190.

30. Lovejoy, *Great Chain of Being,* 102. C. S. Lewis expanded eloquently
(as always) on the role of this idea in medieval cosmology in his *Discarded Image.*

31. ''Cum in hac rerum universitate innumerabilia pene sunt, oculis hominum
sentibusque proposita, quibus humi strata mortalitas ad summi illius opificis con-
templationem possit exsurgere; tum coelestium orbium syderumque species &
pulchritudo ac moles ipsa, est euismodi ut intuentium animos a terrenis impurisque
cogitationibus ad sublimes easdemque iucundissimas facillime traducat ac rapiat.''
[Because there are nearly innumerable things in this universe put before the eyes
and senses of man, by means of which (denizens of) the mortal level may rise to
the contemplation of its highest creator, the types and beauty as well as the matter
of the celestial orbs and of the stars easily transfers and carries off the souls of
those looking (at the heavens) from earthly and impure to sublime and most happy
thoughts.] Thanks to Peter Sobol for help with the translation. Dedicatory letter
to Prince William, Duke of Bavaria, Clavius, *Sphaera* (Rome, 1570).

32. ''Quo concesso, quis non videt, minus gravia, cuiusmodi sunt arborum
folia, paleae et reliqua omnia corpora, post ipsam in aere debere relinqui, cum
eius motum celerrimum consequi nequeant?'' Clavius, *Sphaera* (1611), 106.

33. ''Si super axem mundi moveri dicatur, efficitur, ut nubes, aves, et omnia,
quae in aere existunt, in contrariam partem cernantur moveri . . . quoniam videlicet

consequi non possent motum terrae rapidissimum, ut pote qui in spatio 24 horarum absolvitur. Neque vero dici potest, aerem eadem celeritate cum terra circumduci, quoniam constat, ipsum modo huc, modo illuc fluctuare, prout nimirum in hanc, vel illam partem a variis ventis agitatur, ut quotidiana experientia nos docet." Ibid.

34. Ptolemy, *Almagest,* 45.

35. "Omnia aedificia corruerent et nulla ratione diu consistere possent." Clavius, *Sphaera* (1611), 106.

36. "Neque aqua in vase posita, quod circumvolvatur quantumvis velociter, si orificium eius ad partes exteriores vergat." Ibid. Grant, who discusses this argument briefly ("In Defense of the Earth's Centrality," 54), suggests that Clavius is ambiguous about whether the vessel rotates on its own axis or is swung about on the end of a cord. However, the description of the orientation of the opening of the vessel makes it clear that Clavius intended the latter interpretation, since "inward" or "outward" would have no meaning in the case of rotation (as opposed to revolution).

37. "Pari ratione efficeretur, lapidem, seu sagittam aliquam magna vi sursum directe proiectam, non in eundem locum recidere, veluti in navi aliqua celerrime mota accidere conspicimus." Clavius, *Sphaera* (1611), 106.

38. Galileo, *Dialogue,* 190.

39. Grant suggests Clavius as the source of Bartolomeo Amico's arguments and Amico, in turn, as Galileo's source. See Grant, "In Defense of the Earth's Centrality," 42, n. 154, and 54. Grant cites Bartolomeo Amico, *In Aristotelis libros De caelo et mundo dilucida textus explicatio* (Naples, 1626). However, Amico is not a necessary intermediary, since Wallace has shown that Galileo was very familiar with Clavius's *Sphaera* (see Wallace, *Galileo and his Sources,* 255ff).

40. "Neque etiam circumvertetur circulariter praeter naturam, nempe ad motum coeli, quoniam hac ratione semper eadem coeli pars vertici nostro immineret; unde neque astra orirentur, neque occiderent: quod absurdum est." Clavius, *Sphaera* (1611), 106.

41. Ibid.

42. "Concludamus igitur cum communi astronomorum atque philosophorum sententia, terram esse omnis motus localis tam recti, quam circularis, expertem. Coelos ipsos continue circa ipsam circumagi, praesertim quia hoc concesso, multo facilius omnia phaenomena defenduntur, nullumque inconveniens inde consequitur." Ibid.

43. "Favent huic quoque sententiae sacrae literae, quae plurimis in locis terram esse immobilem affirmant, Solemque ac caetera astra moveri testantur: Legimus enim Psalmo 103 *Qui fundasti terram super stabilitatem suam, non inclinabitur in seculum seculi.* Item in Ecclesiaste cap. 1 *Terra in aeternum stat, oritur Sol & occidit & ad locum suum revertitur, ibique renascens gyrat per meridiem & flectitur ad aquilonem.* Quid clarius dici poterat? Clarissimum quoque testimonium quod Sol moveatur perhibet nobis Psalmus 18 in quo ita legitur *In Sole posuit*

*tabernaculum suum & ipse tanquam sponsus procedens de thalamo suo, exultavit
ut Gigas ad currendam viam a summo coelo egressio eius, et occursus eius usque
ad summum eius, nec est, qui se abscondat a calore eius.* Rursus inter miracula
refertur, quod Deus aliquando Solem aut retroduxit, aut prorsus, ut consisteret,
effecit.'' Ibid. The translation preserves the early seventeenth-century understanding of the scriptural passages by quoting from the Douay translation of the Vulgate,
which was translated during Clavius's lifetime and published during his last years
as *The Second Tome of the Holie Bible,* 189, 317–18, 42–43. The allusion to the
sun moving backward would seem to be referring to the biblical story of King
Hezekiah, 2 Kings 20.1–11.

44. Clavius, *Sphaera* (1611), 42.

45. Tolosani's treatise has been published by Garin, ''Alle origini della polemica anticopernicana.''

46. Clavius, *Sphaera* (1611), 107.

47. ''Dicendum est igitur nullam aliam esse causam, propter quam terra in
medio mundi quiescat, quam ipsius gravitatem. Hinc enim fit, ut semper quaerat
esse in infimo loco, qui est remotissimus a coelo, centrum videlicet totius universi,
quod cum semel possederit, naturaliter ab eo divelli non potest, quia contra suam
naturam, ac inclinationem ascenderet.'' Ibid.

48. ''Ita ut si esset tota terra ab una parte ad alteram perforata & grave aliquod
incederet in foramen illud, perveniret solum maximo impetu ad centrum, non
autem ad alteram partem, quia tunc ascenderet, licet in principio, ob motus im-
petum, huc, illucque fluctuaret aliquantisper, donec, paulatim remisso motus impetu, in medio quiesceret.'' Ibid.

49. Hoskin and Molland, ''Swineshead on Falling Bodies,'' 150.

50. Ibid., 152.

51. See ''Textual Traditions in Medieval Astronomy,'' in chap. 2, above.

52. ''Disputat nunc in ultimo huius operis capite de motu aliorum coelorum,
que fit ab occasu in ortum: ac praecipue de motu Solis ac Lunae, ut nobis aperiat
rationes eclipsium Lunarium, et Solarium. At quoniam haec omnia brevissime ab
auctore perstringuntur, propterea et nos brevissimi hac in parte erimus, praesertim
quia tractatio haec, si pro dignitate tractari debet, longiorem expostulat sermonem,
pertinetque ad Theoricas planetarum, quas, favente Deo, brevi in lucem edemus.''
Clavius, *Sphaera* (1611), 290.

53. ''Operae pretium me facturum arbitror, si breviter hoc loco (ut illis, qui
enixe id a me flagitarunt, satisfaciam) adducam experientias varias, quibus Ptolemaeus, Alphraganus, Thebit, et alii fere Astronomi omnes maxime permoti fuer-
unt, ut in coelis orbes eccentricos, et epicyclos esse crederent. Deinde vero propo-
nam potissimas rationes Averrois, sectatorumque ipsius, quibus huiusmodi orbes
impetunt, et omnino destruere conantur. Tertio denique easdem dissolvam, et
frivolas esse ostendam, ut quilibet intelligat, astronomos non sine ratione, sed
magna industria, et incredibili felicitate hosce orbes in coelis invenisse, philosophos autem, qui Averroem sequuntur, temere tanto impetu in eosdem insultare.''
Ibid.

54. The *disputatio* is not part of the first *Sphaera* (1570). If it appeared in the phantom 1575 edition, then the earlier date would make it even more likely that the petition came from Clavius's immediate associates.

55. Clavius had a great many students, of course, over his decades of teaching, most of whom never pursued mathematical studies beyond the elementary level. I have mentioned Grienberger, Maelcote, and Grassi as examples of the much smaller group who spent many years studying and teaching with him at the Collegio Romano. One could cite others, e.g., Giuseppe Biancani, who were eminent mathematicians but who spent little or no time studying directly with Clavius. In the case of Biancani, for instance, there is only documentation for a single year that he spent, presumably working with Clavius, at the Collegio Romano. Many more could legitimately consider themselves "students" of Clavius in the sense of having learned from his texts. On Biancani see Clavius, *Corrispondenza* 1, pt. 2:18–19.

56. Jardine has translated and quoted some of Clavius's methodological discussion in his "Forging of Modern Realism." Jardine's article uses Clavius's arguments as the background to a discussion of similar arguments found in Kepler, *Apologia Tychonis contra Ursum*. All translations used here, unless otherwise noted, are my own.

57. The three principles necessary for change are a substrate (i.e., prime matter) and a pair of contraries. Aristotle, *Physics* 1.7. An alternate formulation of the three would be matter, form, and privation.

58. "Postremo ita licebit propositum concludere. Sicut in philosophia naturali per effectus devenimus in cognitionem causarum, ita etiam in astronomia, quae de corporibus coelestibus a nobis remotissimis agit, necesse est, ut in cognitionem ipsorum, coordinationem, constitutionemque perveniamus ex effectibus, hoc est, ex motibus stellarum per sensus nostros perceptis. Quemadmodum enim ex generatione et corruptione mutua rerum naturalium, philosophi naturales cum Aristotele materiam primam cum aliis duobus principiis transmutationis naturalis, et multa alia collegerunt: sic etiam astronomi per motus coelorum in genere varios ab ortu in occasum, et ab occasu in ortum investigarunt certum numerum sphaerarum coelestium; alii quidem octo, quod octo tantum diversos motus in genere cognoverint, alii autem decem ex decem motibus diversis in genere notatis." Clavius, *Sphaera* (1611), 300.

59. "Item eadem ratione per alia φαινόμενα ordinem inter coelestes sphaeras constituerunt, ut cap. 1 copiose a nobis est expositum. Quamobrem conveniens est, et rationi maxime consentaneum, ut ex motibus planetarum particularibus, et variis apparentiis astronomi inquirant numerum partialium orbium, qui planetas tam variis motibus circumducunt, eorumque constitutionem, ac figuras: ea tamen lege, ac conditione, ut omnium motuum, apparentiarumque causae possint commode assignari, nullumque inde absurdum, quod philosophiae naturali repugnet, inferri possit." Ibid.

60. "Quocirca cum eccentrici orbes et epicycli sint eiusmodi, ut per illos astronomi nullo labore omnia φαινόμενα tueantur, ut partim ex dictis liquet, partim ex Theoricis planius intelligetur, nullumque ex ipsis absurdum, aut incom-

modum sequatur in naturali philosophia, ut mox ex solutione argumentorum, quae
contra huiusmodi orbes ab adversariis afferri solent, constabit merito decreverunt
astronomi, planetas in orbibus eccentricis, atque epicyclis vehi, non autem in
concentricis, cum per hos tueri non possimus tam multiplicem varietatem in motibus planetarum.'' Ibid.

61. ''Verum hanc rationem enervare conantur adversarii dicentes se concedere,
positis orbibus eccentricis et epicyclis, omnia φαινόμενα posse defendi, non
tamen ex hoc sequi, dictos orbes in rerum natura reperiri, sed esse omnino fictitios:
tum quia fortassis omnes apparentiae possunt commodiore via defendi, licet ea
nobis adhuc sit ignota, tum etiam, quia fieri potest, ut per dictios orbes vero
apparentiae defendantur, quamvis ipsi omnino fictitii sint et nullo modo vera causa
illarum apparentiarum: quemadmodum etiam ex falso verum colligere licet, ut ex
Dialectica Aristotelis constat.'' Ibid.

62. ''His possumus addere confirmationem hoc modo. Nicolaus Copernicus in
opere *De revolutionibus orbium caelestium* tuetur omnia phenomena alia via, po-
nendo scilicet firmamentum immobile et fixum, Solem quoque fixum in centro
universi, tribuendoque terrae existenti in tertio coelo triplicem motum, etc. Quare
necessarii non sunt eccentrici et epicycli ad φαινόμενα tuenda in planetis. Rursus
Ptolemaeus per epicyclum reddit omnium apparentiarum causam in Sole, quas per
eccentricum defendit: Non ergo colligi potest ex tertio nostro argumento, Solem
in eccentrico moveri, cum fortassis in epicyclo vehatur.'' Ibid.

63. ''Dicendum nihilominus est, tertium nostrum argumentum suum robur re-
tinere, responsionemque adversariorum nihil concludere. Primum enim, si commodiorem viam habent, exhibeant illam nobis, contentique erimus et illis maximas age-
mus gratias. Nihil enim aliud contendunt astronomi, quam ut omnia φαινόμενα
in coelo quam commodissime tueantur, sive hoc fiat per eccentricos orbes et
epicyclos, sive alio modo. Et quia nulla via hactenus commodior inventa est, quam
ea, quae per eccentricos et epicyclos omnia defendit, credibile valde est, sphaeras
coelestes ex orbibus eiusmodi constare. Quod si commodiorem viam nobis non
possunt exhibere, certe acquiescere deberent huic viae ex tam variis φαινόμενοις
collectae: si prorsus destruere nolunt non tantum philosophiam naturalem, quae in
scholis praelegitur, sed etiam intercludere aditum ad omnes alias artes, quae per
effectus causas investigant.'' Ibid.

64. ''Quotiescunque enim quispiam per effectus manifestos causam aliquam
collegerit, dicam idem prorsus, quod ipsi, nimirum aliam fortasse causam nobis
ignotam dari posse illorum effectuum. . . . Si propterea non recte colligitur ex
apparentiis, eccentricos et epicyclos in coelis reperiri, quia ex falso colligi potest
verum, ruet universa philosophia naturalis. Nam eodem pacto, quando aliquis ex
effectu noto concludet, hanc vel illam esse illius causam, dicam ego, verum id
non esse quia ex falso licet colligere verum: atque ita omnia principia naturalia a
philosophis inventa destruentur. Quod cum sit absurdum, non recte enervari videtur
nostri argumenti vis ac robur ab adversariis.'' Ibid.

65. The seminal historical work in this area is Duhem, *To Save the Phenomena*,
which analyzes sixteenth-century opinions on this issue using the realist/instrumen-
talist dichotomy. A more sophisticated and satisfactory approach can be found in

several studies by Jardine in which he distinguishes between realists, skeptical realists, and skeptics. See esp. Jardine, *Birth of History of Science,* chaps. 6, 7, "Forging of Modern Realism," and "Scepticism in Renaissance Astronomy."

66. Jardine, *Birth of History of Science,* 231–32, and "Scepticism in Renaissance Astronomy," 85–86.

67. Jardine, "Scepticism in Renaissance Astronomy," 88–92.

68. "Dici etiam potest, regulam illam Dialecticorum *Ex falso sequitur verum,* non esse ad rem; quia aliter ex falso infertur verum, et aliter per eccentricos et epicyclos defenduntur φαινόμενα. Ibi enim ex vi formae syllogisticae verum ex falso colligitur. Unde cognita veritate alicuius propositionis, possunt disponi praemissae falsae in tali forma, ut necessario ex vi syllogismi propositio illa vera concludatur. Ut quia ego scio, animal esse sensitivum, possum conficere talem syllogismum. Omnis planta est sensitiva: Omne animal est planta. Igitur omne animal est sensitivum. Quod si de conclusione aliqua dubitem, nunquam ex falsis praemissis acquiram certitudinem illius, etiamsi ex vi syllogismi recte colligatur: quia alioquin omnia facile hoc modo concluderem. Ut si ambigam, num omnis stella sit rotunda, licet ex vi huius syllogismi. *Omnis lapis est rotundus: Omnis stella est lapis. Igitur omnis stella est rotunda,* recte illud inferam ex falsis praemissis, numquam tamen certus reddar de praedicta conclusione mihi dubia." Clavius, *Sphaera* (1611), 300–301.

69. "At ex orbibus eccentricis et epicyclis, non solum apparentiae iam olim cognitae defenduntur, sed etiam futurae praedicuntur, quarum tempus omnino ignoratur: ita ut si ego dubitem, an v.g. in plenilunio Septembris anni 1587 futura sit eclipsis Lunae, certus omnino reddar ex motibus orbium eccentricorum et epicyclorum, futuram esse eclipsim, ita ut amplius non dubitem. Immo ex eisdem motibus cognosco, qua hora illa eclipsis inceptura sit, et quanta pars Lunae sit obscuranda. Eodemque modo omnes eclipses tam Solares, quam Lunares praedici possunt, earumque tempus et magnitudines, cum tamen nullum certum inter se ordinem servent, ita ut determinatum temporis intervallum inter duas proximas interiiciatur; sed aliquando in uno anno duae contingant, aliquando una, et aliquando nulla." Ibid., 301.

70. "Non est autem credibile, quod nos cogamus coelos (cogere autem videmur, si eccentrici et epicycli sint figmenta, ut adversarii volunt) ut nostris obediant figmentis, moveanturque uti nos volumus, vel uti nostris principiis congruit." Ibid.

71. "Quod vero attinet ad Nicolaum Copernicum, dicimus, eum non respuere eccentricos et epicyclos tanquam fictitios et philosophia repugnantes. Ponit enim ipse idem terram, tanquam epicyclum; et in Luna statuit epicycli epicyclum: Sed hoc solum conari, ut periodos motuum planetarum emendet, quas iam claudicare invenerat. Difficile enim admodum est, periodos motuum ita definire, ut multis annorum seculis a vero non devient, cum nullus unquam mortalium unius planetae potuerit periodum ita determinare, ut non supersint aut desint aliquae minutiae, quae in magno annorum intervallo notabilem errorem inducant." Ibid.

72. Baldini, "L'astronomia del Bellarmino," 295–96.

73. "Si può dunque concludere che se Bellarmino e l'ambiente del Collegio

Romano forse non avvertirono la questione *storica* della paternità della prefazione di Osiander, ciò non implica che sfuggì loro quella *epistemologica* della sua congruenza con la struttura logica dell'opera." Ibid., 296.

74. "Itaque quod alia via Copernicus φαινόμενα tueatur, mirum non est. Quia enim ex motibus eccentricorum et epicyclorum cognovit tempus, quantitatem et qualitatem apparentiarum tam futurarum, quam praeteritarum, potuit ut erat ingeniosissimus, novam viam excogitare, qua illae apparentiae commodius (ut ipse putabat) defendi possent, et periodi motuum aliqua ex parte emendari, quas iam animadverterat claudicare, quod praecipuum videtur fuisse studium Copernici, ut diximus, quemadmodum etiam cognitam aliquam conclusionem possumus in pluribus syllogismis, etiam ex falsis praemissis inferre." Clavius, *Sphaera* (1611), 301.

75. "Tantum autem abest, ut propter doctrinam Copernici tollantur eccentrici et epicycli, ut multo magis propterea ponendi sint. Idcirco enim astronomi hos orbes excogitarunt, quia certo certius ex variis phaenomenis deprehenderunt, planetas non ferri, semper aequali distantia a terra. Quod quidem libenter Copernicus admittit, cum secundum eius doctrinam planetae semper inaequalem a terra habeant distantiam, ut patet expositione terrae extra centrum mundi in tertio coelo. Solum hoc ex eius positione colligitur, non esse certum omnino, talem esse constitutionem eccentricorum et epicyclorum, qualem Ptolemaeus facit: quandoquidem multa φαινόμενα possunt alia via defendi. Neque vero nos in hac quaestione aliud contendimus lectori persuadere, quam planetas non ferri aequali semper distantia a terra: atque adeo vel esse in coelis orbes eccentricos et epicyclos eo ordine, quo eos posuit Ptolemaeus, vel certe aliquam horum effectuum ponendam esse causam aequivalentem eccentricis et epicyclis." Ibid.

76. "Quod si positio Copernici nihil falsi et absurdi involueret, dubium sane esset, utri opinioni, Ptolemaeine, an Copernici potius, (quod attinet ad huiusmodi φαινόμενα tuenda) adhaerendum esset. Sed quoniam multa absurda et erronea in Copernici positione continentur, ut quod terra non sit in medio firmamenti, moveaturque triplici motu, quod qua ratione fieri possit, vix intelligo, cum secundum philosophos uni corpori simplici unus debeatur motus: et quod Sol in centro mundi statuatur, sitque omnis motus expers. Quae omnia cum communi doctrina philosophorum, et astronomorum pugnant, et videntur iis, quae sacrae literae plerisque locis docent, contradicere, ut copiosius cap. 1 pertractavimus." Ibid.

77. Duhem, *To Save the Phenomena,* 95.

78. Langford, *Galileo, Science and the Church,* 87–90.

79. "Idcirco anteponenda videtur opinio Ptolemaei huic Copernici inventioni. Ex quibus omnibus liquet, tam esse probabile, dari eccentricos orbes, et epicyclos, quam probabile est, dari octo, aut decem, vel etiam undecim coelos mobiles, cum tam coelorum numerus, quam dicti orbes ex φαινόμενοιζ et motibus inventi sint ab astronomis." Clavius, *Sphaera* (1611), 301.

80. "Iam vero ex eo, quod Ptolemaeus tam per epicyclum, quam per eccentricum φαινόμενα Solis tuetur, solum colligitur, incertum esse, an in eccentrico, an in epicyclo Sol feratur. Sed utrumvis dicatur, perspicuum est, Solem inaequaliter a terra distare, et minime in orbe concentrico ferri, quod satis nobis est, ut diximus. Potius tamen Ptolemaeus elegit eccentricum orbem in Sole, propterea

quod centrum terrae ambit et circundat. Sed proponamus iam argumenta Averrois, eiusque sectatorum, eaque refellamus, ut hinc quoque appareat, eccentricos et epicyclos non esse monstra, aut portenta, nihilque omnino philosophiae naturali repugnare, ut falso adversarii putant." Ibid.

81. A fuller discussion of the term "monster" and its implications in Clavius's argument can be found in Dear, "Jesuit Mathematical Science," 145–47.

82. Westman, "Melanchthon Circle."

83. Ibid., 166–67.

Chapter Six

1. Jardine, "Forging of Modern Realism."

2. Jervis, *Cometary Theory*, 121–23. For a brief survey of opinions on the comets and novas of the late sixteenth century, see Lerner, *Tre saggi*, chap. 3.

3. There is a difference of opinion in the secondary literature on when Clavius first published his views on the nova. Wallace ("Galileo's Early Arguments for Geocentrism," 33) and Baldini ("La nova del 1604," in *Legem impone subactis*, 155) maintain that the discussion of the nova first appeared in the 1581 edition of the *Sphaera*. On the other hand, Hellman ("Maurolyco's 'Lost' Essay," 322, n. 2) said the discussion was not present in the 1581 *Sphaera* that she examined in Vienna. My examination of two copies of the 1581 edition (held by BAV and Regenstein Library, University of Chicago) as well as the copy in Owen Gingerich's library (personal communication) confirms Hellman's finding. The discrepancy raises the interesting (if unlikely) possibility that there are, in fact, two versions of the 1581 (*nunc iterum*) edition. I have not yet been able to examine a sufficient number of copies of the 1581 edition to resolve this discrepancy, but the weight of the evidence favors the first appearance in 1585.

4. Modern astronomers classify the "new stars" of 1572 and 1604 as supernovas—catastrophic stellar explosions. The nova of 1600 was the blue supergiant star now called P Cygni (also called 34 Cygni). Astronomers have found that stars (novalike variables) such as P Cygni are prone to irregular brightness variations and occasional eruptive outbursts like that observed in 1600. Despite these modern distinctions, in this account I will use the term "nova" for all of these events.

5. Clavius, *Sphaera* (1607), 221, and *Sphaera* (1611), 105. The following discussion is based entirely on Clavius's published text. Aufgebauer ("Clavius und die Nova von 1572," 22) cites a Latin treatise on the nova in the document collection of the Hagener Volkssternwarte, but he gives no citation. He attributes that document to Clavius but makes no attempt to justify that attribution. In another article ("Christoph Clavius—Astronom, Mathematiker, Chronologe," 229), Aufgebauer states that Clavius's account of the 1572 nova is located in the Österreichische Nationalbibliothek in Vienna but again gives no citation. He may have meant MS 10689 of the *Tabulae codicum manu scriptorum*, the incipit of which corresponds to Clavius's condensation of Maurolico's account of the nova. Paul L. Rose (*Italian Renaissance of Mathematics*, 175) confirms that the Vienna manuscript is Clavius's digest of Maurolico's account and adds that a complete copy

of Maurolico's treatise is preserved in the Biblioteca Nazionale di Napoli, MS I.E.56, fols. 2–10. The complete text of Maurolico's treatise was published by Hellman in "Maurolyco's 'Lost' Essay."

6. "Medici & astronomi, qui Antuerpiae idem sidus novum contemplatus est." Clavius, *Sphaera* (1611), 105. The quotation opening this chapter comes from Maurolico's treatise on the nova of 1572. Maurolico, the descendant of Byzantine exiles, referred to Electra, the lost member of the Pleiades, who was said to have faded from sight in sorrow at the fall of Troy.

7. "Nonnulli enim, licet pauci, putarunt." Clavius, *Sphaera* (1611), 103.

8. Ptolemy's star catalog counts thirteen stars in Cassiopeia. Clavius included his own version of Ptolemy's star catalog in all editions of the *Sphaera*.

9. Dreyer, *Tycho Brahe*, 64.

10. "Cuius rei etiam testis sum ego ipse, qui Romae anno 1573 mense Decembri, praeter novum illud astrum, (diminutum tamen, ita ut stellis tertiae magnitudinis par videretur) in Cassiopeia alia tredecim conspexi: nec vero ego unus Romae, sed complures alii mecum," and later, "Quod quidem ego cum multis aliis Romae saepius observavi." Clavius, *Sphaera* (1611), 104.

11. "Periti astronomi ubique locorum notaverunt, illam stellam eundem situm habere inter stellas fixas . . . adeo ut nullam pene aspectus varietatem in ea tam variis locis deprehenderint. Quod cum ita sit, quis dubitare poterit, illam non in suprema regione aeris, ubi caeteri cometae generantur, sed supra Lunam locum esse fortitam?" Ibid.

12. "Censeo stellam illam, quaecunque illa fuerit, in firmamento, ubi stellae fixae sunt, extitisse. Nam eam in regione aetherea & non in elementari apparuisse . . . quod in ea non sit deprehensa aspectus diversitas . . . credam stellam illam novam in firmamento, non in alio quovis orbe coelesti, extitisse . . . quod neque ego, neque ullus omnino astronomus, quod quidem sciam, alium motum in ea animadverterit, praeter eum, quem in fixis sideribus observamus." Ibid.

13. Dreyer, *Tycho Brahe*, 48, 59; Hellman, *Comet of 1577*, 144; Thoren, *Lord of Uraniborg*, 67–68.

14. Thoren, *Lord of Uraniborg*, 73.

15. "Ita mihi persuadeo, stellam illam vel tunc a Deo Opt. Max. procreatam esse in coelo octavo, ut magnum aliquid portenderet (quod cuiusmodi fit, adhuc ignoratur), vel certe in ipso coelo gigni posse cometas, sicut in aere, licet rarius id contingat." Clavius, *Sphaera* (1611), 105.

16. Thoren, *Lord of Uraniborg*, 66.

17. Wilkins, *Discovery of a New World*, in *Mathematical and Philosophical Works*, 95.

18. "Hoc si verum est, videant Peripatetici, quomodo Aristotelis opinionem de materia coeli defendere possint. Dicendum enim fortasse erit, coelum non esse quintam quandam essentiam, sed mutabile corpus, licet minus corruptibile sit, quam corpora haec inferiora. . . . Quicquid tandem sit (meam enim sententiam in tanta re non interpono), mihi in praesentia satis est, paucis demonstrasse, astrum illud de quo loquimur, in firmamento sedem habuisse." Clavius, *Sphaera* (1611), 105.

19. Grant, "Were There Significant Differences?" 13.

20. "Aetherea namque regio, sive coelestis, nec alterari, nec augeri, diminuive, nec generari, corrumpive potest, secundum philosophos." Clavius, *Sphaera* (1611), 20.

21. "Aetherea namque regio, sive caelestis, nec alterari, nec augeri, diminuive, nec generari, corrumpive potest." Clavius, *Sphaera* (Rome, 1570), 53.

22. Emphasis mine; see n. 20 above for text.

23. Schmitt, *Aristotle and the Renaissance,* 7.

24. On Riccioli, Clavius, and other Jesuits on celestial corruptibility, see Grant, "Celestial Incorruptibility," 111–18; and Donahue, *Dissolution of the Celestial Spheres.*

25. Hellman, "Maurolyco's 'Lost' Essay," 334.

26. A much later document from the Collegio Romano on the nova of 1604 reports that the 1572 nova had disappeared sixteen months after its appearance. See Baldini, "La nova del 1604," 70. In that document Clavius is invoked as a witness of the truth of the account, but it is not clear whether the sixteen-month span of visibility resulted from his observations—though it seems likely that he was the source as Clavius would have been the only Collegio Romano astronomer present in both 1573 and 1604.

27. Hellman, "Maurolyco's 'Lost' Essay," 326, 333.

28. "Idem dicendum est de stella illa nova, quae (ut ex Germania ad me perscriptum est) anno 1600 in Cygno iuxta eam, quae in pectore lucet, apparuit & adhuc perseverat. Item de alia, quae primum anno 1604 in mense Octobri visa est inter gradum 17 & 18 Sagittarii, habens latitudinem borealem gr. 2 aut circiter: quamvis cum haec scriberem ita esset imminuta, ut vix appareret. [The 1607 ed., 221, is slightly different: "quamvis nunc ita (cum haec scribo) imminuta sit, ut . . ."] Idem, inquam, dicendum est. Utraque enim stella propter eadem argumenta in firmamento collocanda est. Proptera quod & ubivis locorum in eadem distantia ab aliis stellis fixis deprehensa est, ita ut nullam admiserit aspectus diversitatem & nullus alius motus, praeteream, quem in stellis fixis notamus, in ea est animadversus." Clavius, *Sphaera* (1611), 105. Here, as always, Clavius uses the older term "diversity of aspects" where we say "parallax."

29. "La stella nova si vede qui a Roma, et con istrumenti habbiamo trovato sempre la medesima distantia dalle stelle fisse, come d'Arcturo, Lyra, Cygno, et altri; si che pare che stia nel firmamento." (In referring to Lyra and Cygnus as fixed stars, Clavius presumably means the principal star of each constellation, Vega and Deneb, respectively.) Clavius to Magini in Favaro, *Carteggio,* 283; Clavius, *Corrispondenza,* no. 236.

30. Vreman to Magini in Favaro, *Carteggio,* 327.

31. Baldini, "La nova del 1604," 66.

32. "Da Cossenza in Calabria scrive un medico matematico." Clavius to Magini in Favaro, *Carteggio,* 283–84; Clavius, *Corrispondenza,* no. 236.

33. Phillips, "Correspondence," 220, item 288; Clavius, *Corrispondenza,* no. 201.

34. On comets in the Middle Ages, see, e.g., Thorndike, *Latin Treatises on Comets;* Jervis, *Cometary Theory;* Hellman, *Comet of 1577,* 1–117. On the comet

of 1577, see Thorndike, *History of Magic and Experimental Science* 6:75ff.; Hellman, *The Comet of 1577;* Thoren, "Comet of 1577"; Westman, "Comet and the Cosmos."

35. Thorndike, *Sphere of Sacrobosco,* 119.

36. Clavius, *Sphaera* (1611), 15–20.

37. Leibniz, *Philosophical Papers,* 79.

38. "[Aer] a philosophis in tres regiones distribui. In supremum scilicet, mediam, & infimam. Suprema, in qua cometas deferri conspicimus, propter motum eius continuum, quem habet a primo mobili & ignis vicinitatem & solarium radiorum continuam emissionem per eandem, calida semper existit." Clavius, *Sphaera* (1611), 20. Compare *Sphaera* (1570), 51–52.

39. Clavius, *Sphaera* (1611), 105 (see note 15 above for Latin text).

40. Hellman, *Comet of 1577,* 199ff. For publication details, see her bibliographical app., 318–430.

41. Hellman, *Comet of 1577,* chap. 4 and bibliographical app.

42. Grassi, frontispiece, *De tribus cometis.*

43. Westman, "Comet and the Cosmos."

44. Thoren, "Comet of 1577," 53.

45. Drake, *Galileo at Work,* 31, 455.

46. Favaro, *Carteggio.*

47. Magini's work in this area is summarized in Delambre, *Histoire de l'Astronomie,* 509–12; and in Favaro, *Carteggio,* 69–71, though Favaro followed Delambre's sketch to some extent. I have drawn on both accounts in the present treatment as well as on Thorndike, *History of Magic and Experimental Science* 6:56–59.

48. Ptolemy, *Almagest,* 131–72. See also Peurbach's solar theory in Aiton, "Peurbach's *Theoricae novae planetarum.*"

49. Some critics of the Ptolemaic epicycle had raised this objection, and Clavius had responded to it briefly in his fourth chapter, maintaining that there was nothing wrong with allowing the moon to rotate in its epicycle so as to keep the same face toward Earth (Clavius, *Sphaera* [1611], 302, 303–4); this is discussed more fully in Lattis, "Homocentrics." For a brief discussion of the history of the problem of lunar rotation, see Gabbey, "Innovation and Continuity," 112–20.

50. "Credami V. S. che faria una cosa fuora di modo utile e grata, se stampasse l'osservationi, per le quali sono state composte le sue teoriche . . . l'essorto caldamente a farlo. E non bisogna aspettare quello che farà Tico Dano, perchè mi pare che non finirà mai." "Et gli ricordo di nuovo l'osservationi delle theoriche. Iddio N. S. conservi V. S. nella sua santa gratia." Clavius to Magini, 27 January 1595, in Favaro, *Carteggio,* 214–16.

51. Ibid., 72.

52. It was, of course, Johannes Kepler who actually carried out the detailed data analysis that Tycho wanted, though it appeared only after Tycho's death and with conclusions that he probably would not have approved.

53. Tycho to Magini, of 28 November 1598, in Dreyer, *Tycho Brahe,* 271.

54. Copernicus, *De revolutionibus,* bk. 2, Chap. 14.

55. Baldini, *Scientific Scene,* 149. This globe, the only astronomical instru-

ment that can be directly connected to Clavius, stood for years in the old Collegio Romano (photographs attest to its presence) and exists in Rome to this day. Its connection to Clavius was first recognized by Ugo Baldini.

56. For a more complete discussion of trepidation, see Pedersen and Pihl, *Early Physics and Astronomy,* 183ff.; Goldstein, "Theory of Trepidation"; or Toomer, "Survey of the Toledan Tables," 118–22.

57. Ptolemy, *Almagest,* 327–29.

58. Clavius, *Sphaera* (1611), 29. The modern view is that the differences were spurious and the variation in the rate of precession an illusion brought about by uncritical acceptance of the accuracy of observations. Misplaced trust in observations also convinced medieval and Renaissance astronomers of the reality of some other phenomena, most notably, a variation in the maximum declination of the sun and a mismatch between the "true" equinox and the point where the sun crosses the celestial equator.

59. The changes I attribute to the 1581 *Sphaera* could conceivably have appeared in the edition of 1575, if it existed.

60. "Sed ut verum fateamur, licet propter phaenomena, seu apparentias, quas paulo post adducemus, necessario concedendus videatur huiusmodi motus in octava sphaera, vel aliquid simile, tamen valde incertum est, eum ita fieri, ut Alphonsini docent. Multa enim absurda illum consequi videntur, ut alibi docebimus." Clavius, *Sphaera* (1581), 56.

61. "Ostendendum nunc est, quae phaenomena, apparentiaeve astronomos coegerint, ut hunc motum in coelo ponerent. *Non pauci enim motum hunc omnino explodendum a scholis astronomorum, tamquam ridiculum, arbitrantur.* Primo ergo observarunt; stellas fixas inaequaliter incedere ab occidente in orientem." Ibid., 62.

62. "Quemadmodum autem certum videtur, ut vel motus trepidationis, vel aliquid simile in octava sphaera concedatur propter apparentias dictas: ita incertissimus est modus, quo eum astronomi explicant: ut nimirum principia Aries et Libra octavae sphaerae describant circulos circa initia Aries et Libra nonae sphaerae, quorum semidiametri continuant grad. 9 cum ex hac positione multa consequantur, quae cum experientia pugnare videntur, ut in theorica octavae sphaerae copiose explicabimus." Ibid., 63.

63. "Non possum hoc loco silentio praeterire duo argumenta eruditissimi cuiusdam viri, ac nobilissimi, qui non multis ab hinc annis floruit, quibus demonstare nititur in scriptis quibusdam ad hanc rem confectis, quae ego in congregatione, quae iussu summi Pontificis de calendarii correctione Romae nuper habebatur, perlegi non indiligenter." Clavius, *Sphaera* (1611), 32.

64. Dobrzycki, "Astronomical Aspects of the Calendar Reform," 123.

65. "Disputationem perutilem de quadruplici motu octavae sphaerae, secundum periodos a Nicolao Copernico inventas; ubi vanitas motus trepidationis validissimis rationibus confutatur & undecimum caelum, primum mobile astruitur." Clavius, *Sphaera* (1593, *nunc quarto*), *Ad lectorem*. The result of this development was, counting the empyrean, a cosmos of twelve heavens. However, the general idea of a twelve-sphere cosmos was not particularly new. Baldini and Coyne note, e.g., a twelve-sphere theory presented by J. Werner, *De motu octavae sphaerae*

(chap. 3, n. 58 above), which was known, they say, to the Collegio Romano astronomers (Baldini and Coyne, *Louvain Lectures,* 33).

66. "Huc accedit, auctores huiusmodi motas trepidationis non tradere praecepta, quibus maxima declinatio solis . . . possint supputari: quia videlicet intelligebant, calculum ex motu trepidationis subductum minime phaenomenis, atque experientiae respondere quae res argumento est, motum istum in rerum natura non existere, sed prorsus esse commentitium & sine ullo fundamento confictum." Clavius, *Sphaera* (1611), 35.

67. A comprehensive treatment of Copernican precession theory is in Swerdlow and Neugebauer, *Mathematical Astronomy,* chap. 3.1. A more cursory account is available in Dreyer, *History of Astronomy,* 328–31.

68. "Quemadmodum autem quadruplicem istum motum octavae sphaerae, cum eorum periodis a Copernico praescriptis libenter recipimus & amplectimur, ita modum, quo in illis explicandis utitur, omnino reiicimus. Nam ut posteriores duos motus, seu potius librationes octavae sphaerae nobis ob oculos ponat, assumit absonas admodum & absurdas hypotheses, & a communi hominum sensu remotas ne dicam temerarias, cum Solem statuat in mundi centro omnis motus expertem, terram autem multiplici praeditam motu, cum reliquis elementis ac lunari globo in tertio coelo, inter Venerem & Martem collocet." Clavius, *Sphaera* (1611), 36.

69. Copernicus, *On the Revolutions,* bk. 3, chaps. 7–8. As Clavius notes (but as Dreyer failed to note), Copernicus gives these periods in Egyptian years, which are defined as exactly 365 days each.

70. Clavius points out that the figure would be like two ellipses that are tangent to each other in such a way that their minor axes are aligned with the solstitial colure. "Ferme duae ellipses se mutuo secundum latitudinem tangentes, ita ut minores earum axes lineam rectam constituant, abscindantque ex coluro 24 minuta." Clavius, *Sphaera* (1611), 36.

71. "Sed quis non videt, haec inter sese omnino pugnare? Si namque polus per colurum sursum & deorsum versus quasi repit, qui intelligi potest, eundem eodem tempore extra colurum posse vagari? Aut si hinc atque inde evagatur, eundem posse eodem tempore per colurum sursum atque deorsum versus moveri? Ego certe ingenue fateor, me contrarietatem hanc numquam perfecte intelligere potuisse." Ibid.

72. "Quocirca prudenter Ioannes Antonius Maginus Patavinus vir doctissimus, reiectis hisce hypothesibus & retentis motuum periodis quas Copernicus constituit, quadruplicem illum motum octavae sphaerae tueri ac defendere conatur per hypotheses usitatas & ab omnibus astronomis & philosophis receptas; quippe qui terrestrem hunc globum omni carentem motu in totius universi centro, ut ratio postulat, collocet." Ibid.

73. Copernicus, *On the Revolutions,* bk. 3, chap. 6.

74. Copernicus's use of the "Tusi Couple" to achieve the rectilinear librations is in ibid., chap. 4. See also Swerdlow and Neugebauer, *Mathematical Astronomy,* 47–48.

75. Dreyer, *Tycho Brahe,* 356. Thoren judges this to be accurate in *Lord of Uraniborg,* 290.

76. On the *Theorica planetarum,* see various articles by Olaf Pedersen, esp.

"Origins of the 'Theorica Planetarum,' " and "Astronomy," 316–19. An English translation of the *Theorica* appears in Grant, *A Source Book in Medieval Science*, 451–65.

77. See Aiton, "Peurbach's *Theoricae novae planetarum*,"; and *DSB*, s.v. "Georg Peurbach."

78. "Praesertim quia tractatio haec longiorem expostulat sermonem, pertinetque ad Theoricas Planetarum, in quas, annuente Deo, brevi commentarios conscribemus." Clavius, *Sphaera* (1570), 492.

79. "Pertinatque ad Theoricas planetarum, quas favente Deo, brevi in lucem edemus." Clavius, *Sphaera* (1611), 290.

80. These documents are preserved in *Studia* 1C, fasc. 14, ARSI. Baldini published the contents of the third list as an appendix to his article "La nova del 1604," 90–95. The edited text of the *Ordo secundus* appeared in Lattis, "Christopher Clavius," app. 4.

81. Baldini, "La nova del 1604," 93, n. 27.

82. "Hanc nos edemus . . . Sed forte maioris authoritatis esset, si commentarios in Epitoma Joan. Regiom. conscriberemus." *Ordo tertio*, item 18, *Studia* 1C, fasc. 14, ARSI.

83. The discovery of the document, the nature of the manuscripts and texts, and an analysis of the solar theory appear in Baldini, *Legem impone subactis*, 124–53, and Baldini's edited text of the solar theory is included as app. 1 (469–564). The original document is APUG MS 776. Baldini's edition and analysis of the fragmental lunar theory will appear in "The Jesuits and the Scientific Revolution," ed. Mordechai Feingold (Princeton, NJ: Princeton University Press, forthcoming). The present discussion of these texts is based entirely on Baldini's work, which he has generously made available to me in advance of its publication.

84. Baldini, *Legem impone subactis*, 141.

85. Ibid., 140.

86. Goldstein, "Theory and Observation," 46–47. Goldstein notes only two exceptions to the neglect of this issue, namely, Ibn al-Shatir and Levi ben Gerson. To them we can add Henry of Langenstein, according to Zinner, *Entstehung und Ausbreitung*, 82.

87. Baldini, *Legem impone subactis*, 140.

88. On the deficiencies of Ptolemy's lunar theory, see Copernicus, *De revolutionibus*, bk. 4, chap. 2, on the Copernican theory, bk. 4, chap. 3, and on the theory's better predictions of diameter and parallax, bk. 4, chap. 22. I am indebted to Owen Gingerich for bringing this issue to my attention.

Chapter Seven

1. A thorough discussion of Tycho's cosmos is omitted here because Clavius never discussed it in any detail. The interested reader can consult a number of sources, e.g., Thoren, *Lord of Uraniborg*, chap. 8 or Dreyer, *History of Astronomy*, chap. 14.

2. A good summary of the events leading up to Galileo's publication of *Si-*

dereus nuncius is contained in the introduction to Van Helden's authoritative translation of that work, 1–24.

3. There was one more edition of the *Sphaera,* a posthumous edition of 1618. However, since Clavius could have had no control over its content, I have not used it as a source for the current study. The five volumes of his *Opera mathematica* were published at Mainz under the editorial supervision of Johann Ziegler S.J. On Ziegler's role, see Phillips, "Correspondence," 217, n. 28. Though the title page of the complete five-volume set gives the year 1612, both the general title page of vol. 3 and the individual title page of the *Sphaera* in that volume state 1611.

4. Clavius's illness is mentioned in a letter from Biancani to Grienberger dated 14 June 1611, *Op. Gal.* 11:127.

5. "Quae cum ita sint, videant Astronomi, quo pacto orbes coelestes constituendi sint, ut haec phaenomena possint salvari." Clavius, *Sphaera* (1611), 75.

6. "Potrei anco nominargli altri matematici, i quali, mossi da gli ultimi miei scoprimenti, hanno confessato esser necessario mutare la già concepita costituzione del mondo, non potendo in conto alcuno più sussistere." *Op. Gal.* 5:328. A marginal notation in the Favaro edition specifically names Clavius alongside this sentence.

7. Foscarini, *Lettera,* 11. Thanks to Irving Kelter for helping me find this. English translations of Foscarini's famous letter can be found in Blackwell, *Galileo, Bellarmine, and the Bible;* and Kelter, "In Defense of Copernicus."

8. "Christophorus Clavius . . . monet astronomos, ut sibi, propter haec tam nova et hactenus invisa phaenomena, antiquissima autem re, sine dubio de alio coelorum systemate provideant." Scheiner, *De maculis solaribus et stellis circa Jovem errantibus accuratior disquisitio* (1612), in *Op. Gal.* 5:69.

9. Westman, "Copernicans and the Churches," 95; Donahue, *Dissolution of the Celestial Spheres,* 108.

10. "Che il Clavio prima di morire avesse visto la necessità di pensare a modificare le antiche posizioni per adottare il sistema Copernicano, non sembra che possa essere revocato in dubbio." D'Elia, *Galileo in Cina,* 14–15, n. 3. See also preface, vii.

11. Langford, *Galileo, Science and the Church,* 81.

12. Carugo and Crombie, "The Jesuits and Galileo's Ideas"; Crombie, "Mathematics and Platonism," "Sources of Galileo's Natural Philosophy"; various essays in Wallace, *Prelude to Galileo,* esp. "Galileo and Reasoning *Ex Suppositione,*" *Galileo's Early Notebooks,* and *Galileo and His Sources.*

13. Drake, *Galileo at Work,* 12–13; Wallace, *Galileo and His Sources,* 223.

14. "Benchè non è degna di lei; ma lo fo per continuare l'amicitia tra noi." *Op. Gal.* 10:121.

15. Van Helden, introduction, in Galileo, *Sidereus nuncius,* 20.

16. Phillips, "Correspondence," 212, n. 17.

17. "Io so molto bene che *tarde credere est nervus sapientiae:* però non mi risolvo a nulla, ma prego V.R.za che me ne dica in confidenza liberamente la sua opinione intorno questo fatto." *Op. Gal.* 10:288.

18. "Io stato sempre ostinato a non creder gli pianeti novi, hora sono costretto di vacillare per il contenuto d'una lettera del S. or Galilei di 17 Xmbre, di questo tenore: 'Sono finalmente comparse alcune osservationi circa i Pianeti Medicei, veduti da alcuni Padri Giesuiti, scolari del P. Clavio, e dal medesimo P. Clavio' Desidero, V.R.za confermi l'aviso, in quanto tocca lei et suoi scolari, per cavarci totalmente di dubbio." Ibid. 11:14.

19. "Dalla lettera di V.R.za resto sincerato et assicurato con molto mio gusto de'miracoli trovati dal S.or Galilei circa le stelle di Giove, Saturno et Venere, perchè sin hora, non ostanti le tante sue asseverationi, ne restai sempre con qualche scrupolo, sapendo quanto facil cosa sia l'ingannare sè stesso." Ibid. 11:45. Clavius's reply is in *Corrispondenza*, no. 324 and in *Op. Gal.* 20:600-601.

20. Drake, *Galileo at Work*, 161.

21. "Questi Clavisi, che sono tutti, non credono nulla; et il Clavio fra gli altri, capo di tutti, disse a un mio amico, delle quattro stelle, che se ne rideva, et che bisognierà fare uno ochiale che le faccia e poi le mostri, et che il Galileo tengha la sua oppinione et egli terrà la sua." *Op. Gal.* 10:442.

22. Drake, *Galileo at Work*, 160.

23. Drake, *Discoveries and Opinions*, 75.

24. Villoslada, *Storia*, 195.

25. *Op. Gal.* 10:431.

26. "Inveni ex nostris unum, Ioannem Paulum Lembum, qui antequam quicquam intellexisset de tuis, perspicillis quibusdam, non tam ad imitationem alterius sed potius vi coniecturae factis, tum lunae inaequalitatem, tum stellas in Pleiadibus, Orione et aliis plurimas, observavit; Planetas tamen novos non vidit." Ibid. 11:33-34.

27. Zinner, *Entstehung und Ausbreitung*, 489, n. 12.

28. Van Helden, *Invention of the Telescope*, 26.

29. Galileo, *Sidereus nuncius*, 37.

30. "Posteo vero, non parvo cum labore ac diligentia, tantae perfectionis perspicilla fieri procuravit . . . quibus tandem novos Planetas, saltem puriore caelo, deteximus." *Op. Gal.* 11:34.

31. "Pretende aver fatte osservazioni celesti col telescopio, prima d'aver avuto notizia di quelle di G[alileo]." Ibid. 20:227, s.v. Lembo.

32. Drake ("Galileo, Kepler, and Phases of Venus," 199) estimates that Clavius received the telescope from Santini during the first third of November 1610. *Corrispondenza*, no. 324 confirms this.

33. On Galileo's arrival in Rome, see *Op. Gal.* 11:79, n. 1; on confirmation of Saturn and Venus observations, see Welser to Clavius, ibid., 45 and *Corrispondenza*, no. 324.

34. *Op. Gal.* 11:34.

35. On Galileo's patronage relationships, see Biagioli, *Galileo, Courtier,* esp. chap. 1.

36. "Fui il giorno seguente da i Padri Giesuiti, et mi trattenni lungamente col Padre Clavio e con due altri Padri intendentissimi della professione et suoi allievi. . . . Ho trovato che i nominati Padri, havendo finalmente conosciuta la verità de i nuovi Pianeti Medicei, ne hanno fatte da 2 mesi in qua continue osservazioni, le quali vanno proseguendo; et le haviamo riscontrate con le mie, et si rispondano

giustissime. Loro ancora si affaticano per ritrovare iperiodi delle loro revoluzioni; ma concorrono col Matematico dell'Imperatore in guidicare che sia per esser negozio difficilissimo et quasi impossibile." *Op. Gal.* 11:79–80.

37. Ibid. 3:863–64; Galileo, *Opere,* ed. Albéri, 5, pt. 1:37–38.

38. Orbaan, *Documenti,* 283; Gabrieli, "Verbali delle adunanze," 479.

39. On Schreck, see Sommervogel, *Bibliothèque,* s.v. Terrentius; D'Elia, *Galileo in Cina.*

40. Lagalla at least mentions him as such in *De phoenomenis in orbe lunae* (1612); see *Op. Gal.* 3, pt. 1:366. On Desmiani, see Rosen, *Naming of the Telescope,* 30–31.

41. Drake, *Galileo at Work,* 174, 187.

42. On Maelcote, see *Op. Gal.* 11:445. Maelcote referred less explicitly to the sunspot observations in his *Nuntius sidereus Collegii Romani* (n. 50, below). On Guldin, see *Op. Gal.* 5:10.

43. "So che le RR. VV. hanno notitia delle nuove osservationi celesti di un valente mathematico per mezo d'un instrumento chiamato *cannone* overo *ochiale;* et ancor io ho visto, per mezo dell'istesso instrumento, alcune cose molto maravigliose intorno alla luna et a Venere. Però desidero mi facciano piacere di dirmi sinceramente il parer loro intorno alle cose sequenti." Ibid. 11:87. Note that there was not yet a standard term for the instrument we call a telescope, though we see here a stage in the development of the modern Italian word *cannocchiale.*

44. "Questo desidero sapere, perchè ne sento parlare variamente; et le RR. VV., come essercitate nella scienze mathematiche, facilmente ne sapranno dire se queste nuove inventioni siano ben fondate, o pure siano apparenti et non vere. Et se gli piace, potranno mettere la risposta in questo istesso foglio." Ibid., 88.

45. Ibid., 92–93.

46. "Gl'altri tre Padri, che mille volte l'hanno diligentemente osservata, inclinino, anzi interamente aderischino, alla mia opinione." Galileo to Gallanzone Gallanzoni, ibid., 18:412, n. 1. An equivalent though less strongly worded version of this assertion in a different copy of the same letter appears in ibid. 11:151.

47. Cristoforo Borro, *De astrologia,* BNVI Fondo Gesuitico, s.v. Chr. Burrus, 587, fol. 13r. Borro and his astronomical treatise are considered more fully later in this chapter.

48. "Entrò Galileo nella grande aula delle accademie . . . e noi, in sua presenza, esponemmo davanti a tutta l'Università del Collegio Gregoriano i nuovi fenomeni; e dimostrammo con evidenza, sebbene con scandalo dei filosofi, che Venere gira attorno al sole." Quoted in Villoslada, *Storia del Collegio Romano,* 198. Note that, like its modern counterpart, the Collegio Romano was also known as the Gregorian after the patronage of Pope Gregory XIII to the Society.

49. "Galileo con questa publica demonstratione se ne tornarà a Firenze consolatissimo et si può dire laureato dall'universal consenso di questa università." Orbaan, *Documenti,* 284.

50. Favaro notes, "Letto alla presenza di Galileo stesso nel Collegio romano dal P. Odo van Maelcote nel maggio del 1611." *Op. Gal.* 3, pt. 1:13. The text of the address appears in *Op. Gal.* 3, pt. 1:293–98.

51. Galileo, *Sidereus nuncius,* 84. I do not know what critic of the Copernican

theory used this objection. Clavius never mentioned it in his refutations of Copernicus's system.

52. *Op. Gal.* 10:500.

53. Ibid. 3, pt. 1:298.

54. Ibid., 294–95.

55. "Lunare corpus figura nequaquam perfecte sphaerica, sed aspera admodum inaequalique superficie, circumscribi. . . . Quod si quis vestrum huius aspectus causam densitatem raritatemque variam corporis lunaris, vel quid simile, afferri posse putet, ego iudicium meum non interpono. Mihi enim, utpote Nuncio, quae vidi et e Caelo accepi de Lunae maculis, narrasse sufficiat. Vos de rerum consequentiis iudicate." Ibid., 295.

56. "En tibi iam certum, Venerem moveri circa Solem (et idem, procul dubio, dicendum de Mercurio) tanquam centrum maximarum revolutionum omnium planetarum. Sed et illud indubitatum, Planetas non nisi mutuato a Sole lumine illustratos splendescere: quod tamen non existimo verum esse in stellis fixis." Ibid., 297. Maelcote is quoting here from Galileo's letter at ibid. 10:500.

57. "An vero deinde circularis plenae instar Lunae videnda sit, et an haec varietas, ex ipsius circulari motu circa centrum Solis, an vero aliunde, proveniat, et id genus alia definire aut investigare, nec huius temporis, quod mihi iam elapsum sentio, nec mei est muneris, qui non vatem aut arbitrum tantarum rerum." Ibid. 3, pt. 1:298.

58. For a brief explanation, see Van Helden's conclusion in Galileo, *Sidereus nuncius*, 107–9.

59. L. delle Colombe to Clavius, 27 May 1611, *Op. Gal.* 11:118.

60. Ibid., 132, 158, 168.

61. See, e.g., Galileo to Grienberger, 1 September 1611, ibid., 178ff.

62. "Et non sa il S. Col[ombe]. che facil cosa mi saria stata, mentre fui in Roma, il persuadere et ridurre nella mia sentenza il Patre Clavio, se la gravissima età et la sua continua indisposizione havessero tollerato che noi insieme fussimo di queste materie stati in trattamento et fatte le necessarie osservazioni: ma saria stato poco meno che sacrilegio l'affaticare et molestare con discorsi et osservazioni un vecchio, per età, per dottrina et per bontà così venerando, il quale havendosi con tante et sì illustri fatiche guadagnata una fama immortale, poco importa alla sua gloria che egli in questo solo particolare trapassi e resti con opinione falsa." Ibid., 151.

63. "Nolo tamen hoc loco Lectorem latere, non ita pridem ex Belgio apportatum esse instrumentum quoddam instar tubi cuiusdam oblongi, in cuius basibus compacta sunt duo vitra, seu perspicilia, quo obiecta a nobis remota valde propinqua apparent & quidem longe maiore, quam re ipsa sunt. Hoc instrumento cernuntur plurimae stellae in firmamento, quae sine eo nullo modo videri possunt: praesertim in Pleiadibus; circa Nebulosam Cancri; in Orione; via lactea & alibi. . . . Luna quoque quando est corniculata, aut semiplena, mirum in modum refracta & aspera apparet, ut mirari satis non possim, in corpore Lunari tantas inesse inaequalitates. Verum hac de re consule libellum Galilaei Galilaei, quem Sidereum Nuncium inscripsit, Venetiis impressum anno 1610, in quo varias observationes

stellarum a se primo factas describit. Inter alia, quae hoc instrumento visuntur, hoc non postremum locum obtinet, nimirum Venerem recipere lumen a Sole instar Lunae, ita ut corniculata nunc magis, nunc minus, pro distantia eius a Sole appareat. Id quod non semel cum aliis hic Romae observavi. Saturnus quoque habet coniunctas duas stellas ipso minores, unam versus orientem & versus occidentem alteram. Iuppiter denique habet quatuor stellas erraticas, quae mirum in modum situm & inter se & cum Iove variant, ut diligenter & accurate Galilaeus Galilaei describit. Quae cum ita sint, videant astronomi, quo pacto orbes coelestes constituendi sint, ut haec phaenomena possint salvari." Clavius, *Sphaera* (1611), 75.

64. Drake, *Galileo at Work,* 164.

65. Ibid., 167. Maelcote mentions sunspots in his *Nuntius sidereus Collegii Romani,* 295 (n. 50, above).

66. Drake, *Galileo at Work,* 167.

67. "P. Clavius adhuc ibidem fixus est, ubi postremo salutatus est: incipit tamen quandoque oriri et occidere." *Op. Gal.* 11:131.

68. Koyré, *Astronomical Revolution,* 58.

69. See Chap. 6, n. 49, above.

70. "Quis iam dubitare, aut haesitare poterit, coelum tali esse figura praeditum? Praesertim cum coelum . . . continue volvatur motu circulari, cui quidem motui corpus sphaericum, inter reliquas, maxime est accommodatum, ob continuam & uniformem partium successionem, ita ut nihil extrinsecus esse possit impedimento, propterea quod circa centrum eisdem semper loci limitibus circumagitatur; unde & facillime movetur." Clavius, *Sphaera* (1611), 50.

71. "Siamo certi come essi pianeti sono per sè tenebrosi et solo risplendono illustrati dal sole." Galileo to Clavius, *Op. Gal.* 10:500.

72. Recall that Magini, for example, had argued that planets exhibiting retrograde motion should possess epicycles but others should not. Or consider Riccioli's planetary system in which celestial bodies with satellites (Sun, Jupiter, and Saturn) were supposed to orbit the earth, while those without (Mercury, Venus, Mars) orbited the sun.

73. "Scio enim Clavium, et sciunt qui cum ipso familiariter egerunt, ad finem usque vitae a liquiditate caelorum abhorruisse, et subinde inquisivisse rationes, quibus via ordinaria phaenomena defenderet. De incorruptibilitate tantum caelorum minus fuit sollicitus. Itaque cum de alia sphaera cogitandum monuit, optasse videtur, ut aliquis observationes novas, hypothesi veteri accommodaret potius, quam ut penitus immutaret." Grienberger to Biancani, 1618, in Baldini, *Legem impone subactis,* 237–38.

74. On the details of the 1616 condemnation, see Langford, *Galileo, Science and the Church,* 92–104. A fairly complete account told from Galileo's perspective is contained in Drake, *Galileo at Work,* 252–59. Blackwell, *Galileo, Bellarmine, and the Bible,* contains a good account and a number of important documents. The definitive documentary history of the case is now Finocchiaro, *Galileo Affair.*

75. "Il Padre Grembergero et il Padre Gulden, molti giorni sono, furno trovarmi, mostrando buon affetto verso V.S. et disgusto dell'essito de'passati negotiati, et massime il Padre Gulden." *Op. Gal.* 12:285.

76. Schofield (*Tychonic World Systems*, 282–89) discusses some examples of later Jesuit sympathy for Copernicanism.

77. Bricarelli, *Galileo*, 231.

78. "P. Clavius mesmes n'improuvoient nullement l'advis de Copernicus." *Op. Gal.* 15:254.

79. Schofield, *Tychonic World Systems*, 174.

80. "E quanto al Copernico, hormai nou se ne dubita più; e quanto all'opinione di V.S. . . . non si sente nè pure un minimo motivo contro di V.S.; e se a Dio piacessi che lei potessi venir qua fra qualche tempo, son sicuro che darebbe gran sodisfatione a tutti, perchè intendo che molti Gesuiti in segreto sono della medesima opinione, ancorchè taccino." *Op. Gal.* 12:181. Dini, who had Florentine credentials, was the nephew and courtier of Cardinal Bandini in Rome.

81. "Lo scrittore reputa per Copernicei tutti i S.ri compagni, ancorchè ciò non sia, professandosi solo communemente libertà de filosofare *in naturalibus*. Hora predica in Roma. Io trattarò con Mons.r Dini e con questo e con il P. Torquato de Cuppis, Gesuita, nobile Romano, che è del'istesso senso." Ibid., 150. On Foscarini and his impact see Kelter, "In defense of Copernicus," and "Paolo Foscarini's *Letter to Galileo*"; Blackwell, *Galileo, Bellarmine, and the Bible*.

82. Iparraguirre's list of professors in Villoslada, *Storia*, indicates that de Cuppis was professor of logic, 1609–10, 1616–17; physics, 1610–11, 1617–18; metaphysics, 1611–12, 1618–19; and moral theology, 1620–37.

83. This incomplete exchange is described by D'Elia, *Galileo in Cina*, 36–38. D'Elia cites Kirwitzer's two letters as APUG 534, fols. 83 and 90.

84. See Phillips, "Correspondence," 210, items 107, 110, and n. 14; Clavius, *Corrispondenza*, nos. 159, 163; and Norlind, "Unpublished Letter."

85. "Tycho Brahe Danus, excellens nostra aetate astronomus." Clavius, *Sphaera* (1611), 74.

86. "E che confonda tutta l'Astrologia, poichè vole, che Marte possi stare più basso che'l Sole." Clavius to Magini, 27 January 1595, in Favaro, *Carteggio*, 215.

87. The circumstances of Tycho's claim and the likely motivations for it are nicely summarized in Gingerich and Westman, *Wittich Connection*, 70–72.

88. "Et acciò qualch'uno poco amico dell'Astronomia, o poco fautore di Tichone, non pensi ch'io mi habbi finto le sudette osservationi, sappi V.E. che a tutte vi è stato presente il professor della Mathematica in questo nostro Collegio, et un altro discepolo del P. Clavio, ma alle ultime due delle quali io fo conto, et quali quasi sole importano, vi si è trovato presente l'istesso P. Clavio, il quale pure per varii rispetti è poco amico di Tichone." Favaro, *Carteggio*, 326–27. Favaro judges that the "professor of mathematics" must have been Grienberger. Yet the rosters of professors by both Iparraguire (Villoslada, *Storia*) and Baldini and Napolitani (Clavius, *Corrispondenza*) agree that the chair of mathematics at this time was held by Maelcote, who therefore seems the more likely candidate.

89. Schofield (*Tychonic World Systems*, 174) cites Biancani, *Sphaera mundi* (Mutina, 1635) and Malapert, *Austriaca sidera heliocyclia* (Douay, 1633).

90. Sommervolgel, *Bibliothèque*, s.v. Borro, Christophe.

91. Borro, *De astrologia,* (n. 47 above). A more detailed study of this document is now in progress.

92. Ibid., fols. 19v–27r.

93. By the third motion, Borro must mean the compound motion resulting from the two librations, as these are what Clavius had criticized. For Clavius's evaluation of the Copernican motions of the earth see Clavius, *Sphaera* (1611), 36.

94. "Explicabimus, non quidem illi, ut vere adherentes, sed ut auctorum faciliorem intelligentias habeamus." Borro, *De astrologia,* fol. 20v.

95. "Septem ab hinc annis, ex quo animus mathematicis scientiis applicare cepi . . . animadvertissem confusionem." Ibid., fol. 24r.

96. ". . . nam dico cum Ticone nullos dari caelos duros, et solidos, in quibus sydera infixa sint, sed universam caeli machinam nil aliud esse quam auram aeteream liquidissimum videlicet ac simplicissimum corpus." Ibid. fol. 25.

97. "Verum planetas hos novos quo in loco collacabimus, quos et quotos orbes et epicyclos illis attribuemus? Amplexandane Tychonis sententia? Astra ut pisces in mare moveri libere pervatemus?" [*sic*] Guldin to Lanz, 1 March 1611, in Zinner, *Entstehung und Ausbreitung,* 489, n. 12. My translation is based on the conjecture that "pervatemus" is a misreading of "pervolemus."

98. Cited in Sommervogel, *Bibliothèque,* s.v. Borro. One of Borro's cosmological treatises on the fluid Tychonic heavens was translated into Persian (see Sayili, "Persian Manuscript"). Borro's work is described in Schofield, *Tychonic World Systems,* 187–88, 228–29; and also in Donahue, *Dissolution of the Celestial Spheres,* 236–38.

99. Baldini and Coyne, *Louvain Lectures,* 16. Bellarmine did not deny the possibility of more than these three heavens, but the existence of more than these three would have to come through arguments less trustworthy than the evidence of the senses or the testimony of Scripture.

100. Schofield, *Tychonic World Systems,* 187.

101. Baldini and Coyne, *Louvain Lectures,* 33, n. 39, 41–42, n. 94. See also Baldini, "L'astronomia del Bellarmino," 302.

102. Baroncini, "L'insegnamento della filosofia naturale," 176–77. Also Donahue, "Planetary Spheres," esp. 269–75. John Wilkins had concluded, "There are no solid orbs. . . . I rather think that they are all of a fluid (perhaps aereous) substance." Wilkins, *Discovery of a New World* (in *Mathematical and Philosophical Works*), 27.

103. "Denique haec sententia videtur nunc communis & omnium Astronomorum recepta doctrina, qui contrarium tenentes utcunque admirantur, propterea quod veritatem, ut ipsis videtur, planam, Sacrae Scripturae & Sanctis Patribus conformem, rationibus firmissis stabilitam, experimentis irrefragabilibus instructam, absque caussa sufficiente repudient." Scheiner, *Rosa ursina,* 769.

104. Baldini and Coyne, *Louvain Lectures,* 43, n. 94; Lerner ("Problème de la matière céleste," 266–77) discusses briefly the fluid-heaven concepts of Bellarmine, Cesi, and Tycho.

105. Schofield, *Tychonic World Systems,* 318.

106. Baldini, "L'astronomia del Bellarmino," 302.

107. Baldini and Coyne, *Louvain Lectures,* 5.

108. Cesi to Giovanni Faber, 1 June 1628, *Op. Gal.* 13:429–30. This letter was first published in Scheiner's *Rosa ursina,* 731–32.

109. Quoted and translated by Baldini and Coyne, *Louvain Lectures,* 42–43. The document they cite is identified as ARSI, Fondo Gesuitico 655, 115r.

110. Baldini, "L'astronomia del Bellarmino," 302.

111. Baldini and Coyne, *Louvain Lectures,* 33, n. 39.

112. Baldini, "L'astronomia del Bellarmino," 303.

113. Westfall makes a similar point in his article emphasizing the importance of Bellarmine's theological perspectives to the 1616 condemnation of Copernicanism by the Inquisition (see Westfall, "Trial of Galileo," esp. 18).

Bibliography

Manuscript Sources

APUG. Vol. 771. *Horologiorum nova descriptio.* Fols. 66r–204v contain a fragment of Clavius's *Commentary on Sacrobosco* in Clavius's hand.

ARSI. Chronology of Clavius. Compiled by A. Kleiser and Edm. Lamalle, 1968.

———. *Studia* 1c, fasc. 14. Christopher Clavius. "Ordo secundus brevior."

BAV. Urbinates Latines 1303, 1304. *Christophori Clavii Expositionis in Sphaeram Ioannis de Sacrobosco Tom. primum.*

BNVI. Fondo Gesuitico 587. Fols. 1r–38v contain Cristoforo Borro's *De Astrologia Universa Tractatus.*

Published Sources

Aiton, E. J. "Celestial Spheres and Circles." *History of Science* 19 (1981): 75–114.

———, trans. "Peurbach's *Theoricae novae planetarum:* A Translation with Commentary." *Osiris,* 2d ser., no. 3 (1987): 5–44.

"Analysis of Historical Data Suggest Sun Is Shrinking" *Physics Today* 3, no. 9 (September 1979): 17–19.

Ariew, Roger. "Christopher Clavius and the Classification of the Sciences." *Synthese* 83 (1990): 293–300.

Aristotle. *Aristotelis Opera cum Averrois Commentariis.* Venice, 1562–74. Reprint, Frankfurt am Main: Minerva GmbH, 1962.

Ashworth, William B., Jr. "Catholicism and Early Modern Science." In *God and Nature,* 136–66. See Lindberg and Numbers.

Aufgebauer, Peter. "Christoph Clavius—Astronom, Mathematiker, Chronologe." *Die Sterne* 43, no. 11/12 (1967): 228–30.

———. "Clavius und die Nova von 1572." *Sterne und Weltraum* 7 (1968): 22.

Averroës. *On Aristotle's "De generatione et corruptione": Middle Commentary and Epitome.* Translated by Samuel Kurland. Cambridge, Mass.: Mediaeval Academy of America, 1958.

Bacon, Roger. *Liber secundus communium naturalium Fratris Rogeri de celestibus.* In *Opera hactenus inedita Rogeri Baconi.* edited by Robert Steele, Fasc. 4. Oxford: Clarendon Press, 1913.

Baldini, Ugo. "Additamenta galilaeana I: Galileo, la nuova astronomia e la critica all'Aristotelismo nel dialogo epistolare tra Guiseppi Biancani e i revisori romani della Compagnia di Gesù." *Annali dell'Istituto e Museo di Storia della Scienza di Firenze* 9, fasc. 2 (1984): 13–43.

―――. "Additamenta galilaeana II: Galileo nelle lettere dell'elettore di colonia e di Ricardo de Burgo a Christoph Grienberger." *Nuncius* 2 (1987): 3–36.

―――. "L'astronomia del Cardinale Bellarmino." In *Novità celesti e crisi del sapere: Atti del Convegno Internazionale di Studi Galileiani*, 293–305. See Galluzzi.

―――. "L'attivitá scientifica nel primo settecento." In *Storia d'Italia. Annali 3. Scienza e tecnica nella cultura e nella societá dal Rinascimento a oggi*, edited by G. Micheli, 468–545. Turin: Einaudi, 1980.

―――. "Christoph Clavius and the Scientific Scene in Rome." In *Gregorian Reform of the Calendar*, 137–69. See Coyne, Hoskin, and Pedersen.

―――. "*La filosofia e gli epistolari:* Verso una definizione storica della 'filosofia' del Galileismo. Gli epistolari come strumento interpretivo." *Rivista di Storia della Filosofia* (1987): 213–35.

―――. "Una fonte poco utilizzata per la storia intellettuale: le 'censurae librorum' e 'opinionum' nell'antica Compagnia di Gesú." *Annali dell'Istituto Storico Italo-Germanico in Trento* 11 (1985): 19–67.

―――. *Legem impone subactis: Studi su filosofia e scienza dei Gesuiti in Italia, 1540–1632. Università degli studi "G. d'Annunzio" di Chieti, collana dell'istituto di filosofia*, n.s., 3. Rome: Bulzoni Editore: 1992.

―――. "La nova del 1604 e i matematici e filosofi del Collegio Romano: Note su un testo inedito." *Annali dell'Istituto e Museo di Storia della Scienza di Firenze* 6, fasc. 2 (1981): 63–98.

Baldini, Ugo, and George V. Coyne. *The Louvain Lectures (Lectiones lovanienses) of Bellarmine and the Autograph Copy of His 1616 Declaration to Galileo.* Vatican City: Specola Vaticana, 1984.

Baldwin, Martha R. "Magnetism and the Anti-Copernican Polemic." *Journal for the History of Astronomy* 16 (1985): 155–74.

Bangert, William V. *A Bibliographical Essay on the History of the Society of Jesus.* St. Louis: Institute of Jesuit Sources, 1976.

―――. *A History of the Society of Jesus.* St. Louis: Institute of Jesuit Sources, 1972.

Barker, Peter. "Copernicus, the Orbs, and the Equant." *Synthese* 83 (1990): 317–23.

―――. "Stoic Contributions to Early Modern Science." In *Atoms, Pneuma, and Tranquillity.* See Osler.

Barker, Peter, and Roger Ariew, eds. *Revolution and Continuity: Essays in the History and Philosophy of Early Modern Science.* Studies in the History and Philosophy of Science 24. Washington, D.C.: Catholic University of America Press, 1991.

Barnickel, J. B. *Clavius Welt-Einheitskalender: Bamberger Beiträge zur Kalender-Reform.* Bamberg: J. M. Reindl, 1932.

Baroncini, Gabriele. "L'insegnamento della filosofia naturale nei collegi italiani dei gesuiti (1610–1670): Un esempio di nuovo aristotelismo." In *"Ratio studiorum,"* 163–215. See Brizzi.

Barozzi, Francesco. *Cosmographia.* Venice, 1585.

Bellarmine, Robert. *A Most Learned and Pious Treatise, Full of Divine and Humane Philosophy, Framing a Ladder, whereby our Mindes May Ascend to God, by the Steps of His Creatures.* Translated by T. B. Douai, 1616. Reprinted in vol. 22 of *English Recusant Literature, 1558–1640,* edited by D. M. Rogers. Menston, Yorkshire: Scolar Press, 1970.

Berry, Arthur. *A Short History of Astronomy from Earliest Times through the Nineteenth Century.* 1898. Reprint, New York: Dover, 1961.

Biagiolo, Mario. *Galileo, Courtier: The Practice of Science in the Culture of Absolutism.* Chicago: University of Chicago Press, 1993.

———. "The Social Status of Italian Mathematicians, 1450–1600." *History of Science* 27 (1989): 41–95.

Blackwell, Richard J. *Galileo, Bellarmine, and the Bible.* Notre Dame, Ind.: University of Notre Dame Press, 1991.

Blancanus (Biancani), Iosephus. *Sphaera Mundi seu Cosmographia.* Bologna, 1620.

Boero, Giuseppe. *Menologio di pie memorie d'alcuni religiosi della Compagna di Gesù che fiorirono in virtù e santità.* Vol. 2. Rome: Civiltà Cattolica, 1859.

Boncompagni, B. "Lettera di Francesco Borozzi al P. Christoforo Clavio." *Bullettino di Bibliografia e di Storia delle Scienze Matematiche e Fisiche* 17 (1884): 831–37.

Bonelli, M. L. R., and William Shea. *Reason, Experiment, and Mysticism in the Scientific Revolution.* New York: Science History Publications, 1975.

Bricarelli, Carlo. *Galileo Galilei: L'opera, il metodo, le peripezie.* Rome: Civiltà Cattolica, 1931.

Brickman, Benjamin. "An Introduction to Francesco Patrizi's *Nova de universis philosophia.*" Ph.D. diss., Columbia University, 1941.

Brizzi, Gian Paolo, ed. *La "Ratio Studiorum": Modelli culturali e pratiche educative dei Gesuiti in Italia tra Cinque e Seicento.* Centro studi "Europa delle Corti"/Biblioteca del Cinquecento 16. Rome: Bulzoni Editore, 1981.

Brodrick, James. *The Origin of the Jesuits.* Chicago: Loyola University Press, 1986.

———. *The Progress of the Jesuits (1556–79).* Chicago: Loyola University Press, 1986.

Bruhns [Carl Christian]. "Christoph Clavius." *Allgemeine Deutsche Biographie.* 1876. Reprint, Berlin: Duncker & Humblot, 1968.

Bullart, Isaac. *Académie des Sciences et des Arts, Contenant les Vies, & les Eloges Historiques des Hommes Illustres.* Vol. 2. Amsterdam, 1682.

Butts, R. E., and J. C. Pitt. *New Perspectives on Galileo.* Dordrecht: D. Reidel, 1978.

Calisi, Marinella. "L'origine delle collezioni scientifiche del Museo Astronomico e Copernicano in Roma alla fine del XIX secolo." *Nuncius* 4 (1989): 249–302.

————, ed. *Il Museo Astronomico e Copernicano*. Rome: Osservatorio Astronomico di Roma, 1982.

Carmody, Francis J. "The Planetary Theory of Ibn Rushd." *Osiris* 10 (1952):
556–86.

————, ed. *al-Bitrūjī, "De motibus celorum": Critical Edition of the Latin Translation of Michael Scot*. Berkeley and Los Angeles: University of California
Press, 1952.

Carrara, Bellino. "I gesuiti e Galilio." *Rivista di Apologia Cristiana* (July–August
1914).

Carugo, Adriano, and Alistair C. Crombie. "The Jesuits and Galileo's Ideas of
Science and of Nature." *Annali dell'Istituto e Museo di Storia della Scienza
de Firenze* 8, fascs. 23–68 (1983).

Casanovas, Juan. "L'astronomia del Collegio Romano nella prima metà del seicento." *Giornale di Astronomia* 10 (1984): 149–55.

————. "Il P. C. Clavio professore di matematica del P. M. Ricci nel Collegio
Romano." *Atti del Convegno Internazionale di Studi Ricciani*, 229–39. Macer-
ata: Centro Studi Ricciani, 1984.

Clavius, Christoph. *Apologiae calendarii novi adversus Michaelem Maestlinum
mathematicum Tubigensem*. Vol. 5 of *Opera mathematica*. Mainz, 1612.

————. *Commentariu[s] in Sphaeram Ioannis de Sacro Bosco*. Vol. 3 of *Opera
mathematica*. Mainz, 1612.

————. *Christoph Clavius: Corrispondenza*. Edited by U. Baldini and P. D.
Napolitani. Pisa: Università di Pisa, Dipartimento di Matematica, Sezione di
Didattica e Storia della Matematica, 1992.

————. *Euclidis Elementorum libri XV*. Frankfurt, 1654.

————. *Gnomonices libri octo, in quibus non solum horologiorum solarium, sed
aliarum quoque rerum, quae ex gnomonis umbra cognosci possunt, descrip-
tiones geometrice demonstrantur*. Rome, 1581.

————. *Horologiorum nova descriptio*. Rome, 1599.

————. *In Sphaeram Ioannis de Sacro Bosco commentarius*. Rome, 1570.

————. *In Sphaeram Ioannis de Sacro Bosco commentarius. Nunc iterum ab ipso
auctore recognitus et multis ac variis locis locupletatus*. Rome, 1581.

————. *In Sphaeram Ioannis de Sacro Bosco commentarius. Nunc tertio ab ipso
auctore recognitus, & plerisque in locis locupletatus*. Rome, 1585.

————. *In Sphaeram Ioannis de Sacro Bosco commentarius, Nunc quarto ab ipso
auctore recognitus et plerisque in locis locupletatus. Maiori item cura cor-
rectus*. Lyons, 1593.

————. *In Sphaeram Ioannis de Sacro Bosco commentarius, Nunc tertio ab ipso
auctore recognitus et plerisque in locis locupletatus. Maiori item cura cor-
rectus*. Venice, 1596.

————. *In Sphaeram Ioannis de Sacro Bosco commentarius. Nunc tertio ab ipso
auctore recognitus et plerisque in locis locupletatus*. Venice, 1601.

————. *In Sphaeram Ioannis de Sacro Bosco commentarius. Nunc quinto ab ipso
auctore hoc anno 1606 recognitus et plerisque in locis locupletatus. Accessit
geometrica atque uberrima de crepusculis tractatio*. Rome, 1607.

————. *Sphaericorum libri III a Christophoro Clavio Bambergensi SI. Perspicuis demonstrationibus, ac scholiis illustrati. Item eiusdem Christophori Clavii sinus, lineae tangentes, et secantes, triangula rectilinea, atque sphaerica.* Rome, 1586.

Codina, Arturo. "Sant Ignasi a Montserrat." *Archivum Historicum Societatis Iesu* 7 (1938): 104–17.

Copernicus, Nicholas. *On the Revolutions of the Heavenly Spheres.* Translated by Edward Rosen. Baltimore: Johns Hopkins University Press, 1978.

Cosentino, Giuseppe. "L'insegnamento delle matematiche nei Collegi Gesuitici nell'Italia settentrionale." *Physis* 13 (1971): 205–17.

Coyne George V., M. A. Hoskin, and O. Pedersen, eds. *Gregorian Reform of the Calendar: Proceedings of the Vatican Conference to Commemorate its 400th Anniversary, 1582–1982.* Vatican City: Pontificia Academia Scientiarum, Specola Vaticana, 1983.

Crombie, A. C. "Mathematics and Platonism in the Sixteenth-Century Italian Universities and in Jesuit Educational Policy." In *Prismata*, 63–94. See Maeyama and Saltzer.

————. "Sources of Galileo's Early Natural Philosophy." In *Reason, Experiment, and Mysticism*, 157–75. See Bonelli.

Dales, Richard C. "The De-Animation of the Heavens in the Middle Ages." *Journal for the History of Ideas* 41 (1980): 531–50.

Dalla Chiara, Maria Luisa. *Italian Studies in the Philosophy of Science.* Boston Studies in the Philosophy of Science 47. Dordrecht: D. Reidel, 1981.

Dear, Peter. "Jesuit Mathematical Science and the Reconstitution of Experience in the Early Seventeenth Century." *Studies in History and Philosophy of Science* 18 (1987): 133–75.

————. *Mersenne and the Learning of the Schools.* Ithaca, N.Y.: Cornell University Press, 1988.

————. "Mersenne and the Learning of the Schools: Continuity and Transformation in the Scientific Revolution." Ph.D. diss., Princeton University, 1984.

Deason, Gary B. "John Wilkins and Galileo Galilei: Copernicanism and Biblical Interpretation in the Protestant and Catholic Traditions." In *Probing the Reformed Tradition: Historical Studies in Honor of Edward A. Dowey, Jr.*, edited by Elsie Anne McKee and Brian G. Armstrong, 313–38. Louisville, Ky.: Westminster/John Knox Press, 1989.

De Dainville, François. *L'Education des Jésuites.* Paris: Éditions de Minuit, 1978.

————. *La Géographie des Humanistes.* Paris: Beauchesne & Sons, 1940.

D'Elia, Pasquale. *Galileo in China: Relations through the Roman College between Galileo and the Jesuit Scientist-Missionaries (1610–1640).* Translated by Rufus Suter and Matthew Sciascia. Cambridge, Mass.: Harvard University Press, 1960.

————. *Galileo in Cina: Relazioni attraverso il Collegio Romano tra Galileo e i gesuiti scienziati missionari in Cina (1610–1640).* Rome: Gregorian University, 1947.

Dee, John. *The Mathematicall Praeface to the Elements of Geometrie of Euclid of Megara (1570)*. With an introduction by Allen G. Debus. New York: Science History Publications, 1975.

Delambre, [Jean Baptiste Joseph]. *Histoire de l'astronomie du moyen age.* Paris, 1819. Reprint, New York: Johnson Reprint Corp., 1965.

―――. *Histoire de l'astronomie moderne.* Paris, 1821. Reprint, New York: Johnson Reprint Corp., 1969.

De Pace, Anna. *Le matematiche e il mondo: Ricerche su un dibattito in Italia nella seconda metà del Cinquecento.* Milan: Franco Angeli, 1993.

Di Bono, Mario. *Le sfere omocentriche di Giovan Battista Amico nell'astronomia del cinquecento con il testo del "De motibus corporum coelestium."* Genoa: Consiglio Nazionale delle Ricerche—Centro di Studio sulla Storia della Tecnica, 1990.

Dick, Steven J. *Plurality of Worlds: The Origins of the Extraterrestrial Life Debate from Democritus to Kant.* Cambridge: Cambridge University Press, 1982.

Dicks, D. R. *Early Greek Astronomy to Aristotle.* Ithaca, N.Y.: Cornell University Press, 1970.

Dobryzcki, Jerzy. "Astronomical Aspects of the Calendar Reform." In *Gregorian Reform of the Calendar,* 117–27. See Coyne, Hoskin, and Pedersen.

Donahue, William H. *The Dissolution of the Celestial Spheres.* New York: Arno Press, 1981.

―――. "The Solid Planetary Spheres in Post-Copernican Natural Philosophy." In *The Copernican Achievement,* 244–75. See Westman.

Donne, John. *Ignatius His Conclave.* 1611. Reprint, Oxford: Clarendon Press, 1969.

Drake, Stillman. *Discoveries and Opinions of Galileo.* Garden City, N.Y.: Doubleday Anchor, 1957.

―――. *Galileo at Work: His Scientific Biography.* Chicago: University of Chicago Press, 1978.

―――. "Galileo, Kepler, and Phases of Venus." *Journal for the History of Astronomy* 15 (1984): 198–208.

Dreyer, J. L. E. *A History of Astronomy from Thales to Kepler.* 1906. Reprint, New York: Dover, 1953.

―――. *Tycho Brahe: A Picture of Scientific Life and Work in the Sixteenth Century.* London: Adam & Charles Black, 1890. Reprint, Gloucester, Mass.: Peter Smith, 1977.

Duhem, Pierre. *Le Système du monde: Histoire des doctrines cosmologiques de Platon à Copernic.* Paris: Hermann, 1913–1959.

―――. *To Save the Phenomena: An Essay on the Idea of Physical Theory from Plato to Galileo.* Translated by Edmund Dolan and Chaninah Maschler. Chicago: University of Chicago Press, 1969.

al-Farghānī. *Rudimenta astronomica Alfragrani.* Nuremberg, 1537.

Farrell, Allan P. *The Jesuit Code of Liberal Education: Development and Scope of the "Ratio studiorum."* Milwaukee: Bruce Publishing, 1938.

Favaro, Antonio, ed. *Carteggio inedito di Ticone Brahe, Giovanni Keplero e di altri celebri astronomi e matematici dei secoli XVI e XVII con Giovanni Antonio Magini.* Bologna, 1886.

Feingold, Mordechai. *The Mathematician's Apprenticeship: Science, Universities, and Society in England, 1560–1640*. Cambridge: Cambridge University Press, 1984.

Fejér, Josephus. *Defuncti primi saeculi Societatis Jesu: 1540–1640*. Rome: Institutum Historicum Societatis Iesu, 1982.

Finocchiaro, Maurice A. *The Galileo Affair: A Documentary History*. Berkeley and Los Angeles: University of California Press, 1989.

Flamsteed, John. *An Account of the Revd. John Flamsteed, the First Astronomer-Royal*. Edited by Francis Baily. London: Dawson, 1966.

Foscarini, Paolo Antonio. *Lettera . . . sopra l'opinione de'Pittagorici e del Copernico della mobilità della terra e stabilità del sole*. Naples, 1615.

Fracastoro, Girolamo. *Homocentrica: Sive de Stellis*. Venice, 1538.

Froidmond, Libert. *Ant-Aristarchus sive orbis-terrae immobilis*. Antwerp, 1631.

Gabbey, Alan. "Innovation and Continuity in the History of Astronomy: The Case of the Rotating Moon." In *Revolution and Continuity*, 95–129. See Barker and Ariew.

Gabrieli, Giuseppe. "Verbali delle adunanze e cronaca della prima Accademia Lincea (1603–1630)." *Atti della Reale Accademia Nazionale dei Lincei*. Memorie della classe di scienze morali, storiche e filologiche, 6th ser., vol. 2 (1927–29): 461–512.

Galilei, Galileo. *Dialogue concerning the Two Chief World Systems*. Translated by Stillman Drake. Berkeley and Los Angeles: University of California Press, 1953.

——. *Le Opere di Galileo Galilei*. Edited by Eugenio Albéri. 11 vols. Florence: Società Editrice Fiorentina, 1842–48.

——. *Le Opere di Galileo Galilei*. Edited by Antonio Favaro. Edizione Nazionale. 20 vols. Florence: G. Barbèra, 1892–1904.

——. *Sidereus Nuncius or the Sidereal Messenger*. Translated by Albert Van Helden. Chicago: University of Chicago Press, 1989.

——. *Two New Sciences: Including Centers of Gravity and Force of Percussion*. Translated by Stillman Drake. Madison: University of Wisconsin Press, 1974.

——, Horatio Grassi, Mario Guiducci, and Johann Kepler. *The Controversy on the Comets of 1618*. Translated by Stillman Drake and C. D. O'Malley. Philadelphia: University of Pennsylvania Press, 1960.

Galluzzi, Paolo, ed. *Novità celesti e crisi del sapere: Atti del Convegno Internazionale di Studi Galileiani*. Florence: G.Barbèra, 1984.

Garin, Eugenio. "Alle origini della polemica anticopernicana." *Colloquia Copernicana*, vol. 2. Studia Copernicana 6:631–42. Wroclaw: Polish Academy of Sciences Press, 1973.

Gascoigne, John. "A Reappraisal of the Role of the Universities in the Scientific Revolution." In *Reappraisals of the Scientific Revolution*, 207–60. See Lindberg and Westman.

Gassendi, Pierre. *Tychonis Brahei, equitis dani, astronomorum coryphaei vita. Accessit Nicolai Copernici, Georgii Peurbachii, & Ioannis Regiomontani Astronomorum celebrium, vita*. Paris, 1654.

Giacobbe, Giulio Cesare. "Il *Commentarium de certitudine mathematicarum disciplinarum* di Alessandro Piccolomini." *Physis* 14 (1972): 162–93.

———. "Epigoni nel seicento della 'Quaestio de certitudine mathematicarum': Giuseppe Biancani." *Physis* 18 (1976): 5–40.

———. "Un gesuita progressista nella 'Quaestio de certitudine mathematicarum' rinascimentale: Benito Pereyra." *Physis* 19 (1977): 51–86.

Giannettasio, Nicola Partenio. *Universalis cosmographiae elementa.* Naples, 1688.

Gilbert, Neal W. *Renaissance Concepts of Method.* New York: Columbia University Press, 1960.

Gingerich, Owen, and Robert S. Westman. "The Wittich Connection: Conflict and Priority in Late Sixteenth-Century Cosmology" *Transactions of the American Philosophical Society* 78, pt. 7, 1988.

Goldstein, Bernard R. "The Arabic Version of Ptolemy's *Planetary Hypotheses.*" *Transactions of the American Philosophical Society* 57 (1967): 41–55.

———. *al-Bitruji On the Principles of Astronomy.* 2 vols. Yale Studies in the History of Science and Medicine, no. 7. New Haven, Conn.: Yale University Press, 1971.

———. "The Blasphemy of Alfonso X: History or Myth?" In *Revolution and Continuity,* 143–53. See Barker and Ariew.

———. "Levi ben Gerson's Preliminary Lunar Model." In *Theory and Observation.* See Goldstein. First published in *Centaurus* 18 (1974): 275–88.

———. "Medieval Observations of Solar and Lunar Eclipses." *Archives Internationales d'Histoire des Sciences* 29 (June–December 1979): 101–56.

———. "On the Theory of Trepidation according to Thabit b. Qurra and al-Zarqallu and Its Implications for Homocentric Planetary Theory." *Centaurus* 10 (1964): 232–47.

———. "The Status of Models in Ancient and Medieval Astronomy." *Centaurus* 24 (1980): 132–47.

———. *Theory and Observation in Ancient and Medieval Astronomy.* London: Variorum Reprints, 1985.

———. "Theory and Observation in Medieval Astronomy." *Isis* 63 (1972): 39–47.

Gossin, Pamela. "Poetic Resolutions of Scientific Revolutions: Astronomy and the Literary Imaginations of Donne, Swift, and Hardy." Ph.D. diss., University of Wisconsin–Madison, 1989.

Govi, G. "Galileo e i matematici del Collegio Romano nel 1611." *Atti del Accademia Nazionale dei Lincei,* ser. 2a, vol. 2 (1875): 230–40.

Grafton, Anthony, and Lisa Jardine. *From Humanism to the Humanities: Education and the Liberal Arts in Fifteenth- and Sixteenth-Century Europe.* Cambridge, Mass.: Harvard University Press, 1986.

Grant, Edward. "Celestial Incorruptibility in Medieval Cosmology, 1200–1687." In *Physics, Cosmology, and Astronomy, 1300–1700,* 101–27. See Unguru.

———. "Celestial Orbs in the Latin Middle Ages." *Isis* 78 (1987): 152–73.

———. "Cosmology." In *Science in the Middle Ages,* 265–302. See Lindberg.

————. "Eccentrics and Epicycles in Medieval Cosmology." In *Mathematics and Its Applications to Science and Natural Philosophy in the Middle Ages,* 189–214. See Grant and Murdoch.

————. "In Defense of the Earth's Centrality and Immobility: Scholastic Reaction to Copernicanism in the Seventeenth Century." *Transactions of the American Philosophical Society* 74, no. 4 (1984).

————. *Much Ado about Nothing: Theories of Space and Vacuum from the Middle Ages to the Scientific Revolution.* Cambridge: Cambridge University Press, 1981.

————. "A New Look at Medieval Cosmology, 1200–1687." *Proceedings of the American Philosophical Society* 129 (1985): 417–32.

————. "Were There Significant Differences between Medieval and Early Modern Scholastic Natural Philosophy? The Case for Cosmology." *Nous* 18 (1984): 5–14.

————, ed. *A Source Book in Medieval Science.* Cambridge, Mass.: Harvard University Press, 1974.

Grant, Edward, and John E. Murdoch, eds. *Mathematics and Its Applications to Science and Natural Philosophy in the Middle Ages: Essays in Honor of Marshall Clagett.* Cambridge: Cambridge University Press, 1987.

Grassi, Orazio. *De tribus cometis anni 1618 disputatio astronomica publice habita in Collegio Romano Societatis Iesu ab uno ex patribus eiusdem societatis.* Rome, 1619. Modern edition and translation in *The Controversy on the Comets of 1618.* See Galileo et al.

Grendler, Paul F. *Schooling in Renaissance Italy: Literacy and Learning, 1300–1600.* Baltimore: Johns Hopkins University Press, 1989.

Grienberger, Christoph. *Catalogus veteres affixarum Longitudines, ac Latitudines conferens cum novis. Imaginum caelestium prospectiva duplex.* Rome, 1612.

Hain, Ludovico. *Repertorium bibliographicum, in quo libri omnes ab arte typographica inventa usque ad annum MD.* Milan: G. G. Görlich, 1948.

Hamy, Alfred. *Essai sur l'iconographie de la Compagnie de Jésus.* Paris, 1875.

————. *Galerie illustree de la Compagnie de Jésus.* Paris, 1893.

Hargreave, D. "Reconstructing the Planetary Motions of the Eudoxean system." *Scripta Mathematica* 28, no. 4 (1970): 335–45.

Harris, Steven. "Jesuit Ideology and Jesuit Science: Religious Values and Scientific Activity in the Society of Jesus, 1540–1773." Ph.D. diss., University of Wisconsin–Madison, 1988.

————. "Transposing the Merton Thesis: Apostolic Spirituality and the Establishment of the Jesuit Scientific Tradition." *Science in Context* 3 (1989): 29–65.

Heilbron, J. L. *Elements of Early Modern Physics.* Berkeley and Los Angeles: University of California Press, 1982.

Hellman, C. Doris. *The Comet of 1577: Its Place in the History of Astronomy.* New York: Columbia University Press, 1944; Reprint, New York: AMS Press, 1971.

————. "Maurolyco's 'Lost' Essay on the New Star of 1572." *Isis* 51 (1960): 322–36.

Henderson, Janice Adrienne. *On the Distances between Sun, Moon, and Earth according to Ptolemy, Copernicus, and Reinhold.* Leiden: E. J. Brill, 1991.

Hirsch, Rudolf. *Printing, Selling, and Reading: 1450–1550.* Wiesbaden: Otto Harrassowitz, 1967.

The Holie Bible Faithfully Translated into English. 2 vols. Douai, 1609–10. Reprinted as vols. 265–66 of *English Recusant Literature, 1558–1640,* edited by D. M. Rogers. London: Scolar Press, 1975.

Homann, Frederick A. "Christopher Clavius and the Isoperimetric Problem." *Archivum Historicum Societatis Iesu* 49 (1980): 245–54.

———. "Christopher Clavius and the Renaissance of Euclidean Geometry." *Archivum Historicum Societatis Iesu* 52 (1983): 233–46.

Hoskin, M. A., and A. G. Molland. "Swineshead on Falling Bodies: An Example of Fourteenth-Century Physics." *British Journal for the History of Science* 3 (1966): 150–82.

Houzeau, Jean-Charles, and Albert Lancaster. *General Bibliography of Astronomy to the Year 1800.* Brussels, 1880–89. Reprinted with introduction and author index by D. W. Dewhirst. London: Holland Press, 1964.

Ignatius of Loyola. *The Constitutions of the Society of Jesus.* Translated by George E. Ganss. St. Louis: Institute of Jesuit Sources, 1970.

Jardine, Nicholas. *The Birth of History and Philosophy of Science: Kepler's "A Defence of Tycho against Ursus" with Essays on Its Provenance and Significance.* Cambridge: Cambridge University Press, 1984.

———. "The Forging of Modern Realism: Clavius and Kepler against the Sceptics." *Studies in History and Philosophy of Science* 10 (1979): 141–73.

———. "Scepticism in Renaissance Astronomy: A Preliminary Study." In *Scepticism from the Renaissance to the Enlightenment.* See Popkin and Schmitt.

———. "The Significance of the Copernican Orbs." *Journal for the History of Astronomy* 13 (1982): 168–94.

Jervis, Jane L. *Cometary Theory in Fifteenth-Century Europe.* Dordrecht: D. Reidel, 1985.

Johnson, Francis R. "Astronomical Text-books in the Sixteenth Century." In *Science, Medicine, and History* 1:285–302. See Underwood.

Kellner, Menachem. "On the Status of the Astronomy and Physics in Maimonides' *Mishneh Torah* and *Guide of the Perplexed:* A Chapter in the History of Science." *British Journal for the History of Science* 24 (1991): 453–63.

Kelly, Celsus, O.F.M. *Calendar of Documents: Spanish Voyages in the South Pacific from Alvaro de Menana to Alejandro Malaspina, 1567–1794, and the Franciscan Missionary Plans for the Peoples of the Austral Lands, 1617–1634.* Madrid: Franciscan Historical Studies, 1965.

Kelter, Irving Alan. "In Defense of Copernicus: Paolo Foscarini (ca. 1565–1616) and the Heliocentric System." Ph.D. diss., City University of New York, 1989.

———. "Paolo Foscarini's *Letter to Galileo:* The Search for Proofs of the Earth's Motion." *Modern Schoolman* 70 (1992): 31–44.

Kepler, Johannes. *Opera omnia*. Edited by C. Frisch. Frankfurt: Heyder & Zimmer, 1859.

———. *New Astronomy*. Translated by William H. Donahue. Cambridge: Cambridge University Press, 1992.

Knobloch, Eberhard. "Christoph Clavius—Ein Astronom zwischen Antike und Kopernikus." In *Vorträge des ersten Sympsions des Bamberger Arbeitskreises "Antike Naturwissenschaft und ihre Rezeption" (AKAN)*, 113–40. Edited by Klause Döring and Georg Wöhrle. Wiesbaden: Otto Harrassowitz, 1990.

———. "Christoph Clavius. Ein Namen- und Schriftenverzeichnis zu seinen *Opera mathematica*." *Bolletino di Storia delle Scienze Mathematiche* 10 (1990): 135–89.

———. "Masrin Mersennes Beiträge zur Kombinatorik." *Sudhoffs Archiv* 58 (1974): 356–79.

———. "Sur la vie et l'oeuvre de Christophore Clavius (1538–1612)." *Revue d'Histoire des Sciences* 41 (1988): 331–56.

Koyré, Alexandre. *The Astronomical Revolution*. Translated by R. E. W. Maddison. Ithaca, N.Y.: Cornell University Press, 1973.

Kren, Claudia. "Homocentric Astronomy in the Latin West: The *De reprobatione ecentricorum et epiciclorum* of Henry of Hesse." *Isis* 59 (1968): 269–81.

Kristeller, Paul Oskar. *Eight Philosophers of the Italian Renaissance*. Stanford, Calif.: Stanford University Press, 1964.

———. *Renaissance Thought: The Classic, Scholastic, and Humanistic Strains*. New York: Harper & Row, 1961.

Kuhn, Thomas S. *The Copernican Revolution: Planetary Astronomy in the Development of Western Thought*. Cambridge, Mass.: Harvard University Press, 1957.

———. *The Structure of Scientific Revolutions*. Chicago: University of Chicago Press, 1962.

Lamalle, Edmond. "L'Archivio generale di un grande ordine religioso: Quello della Compagnia di Gesù." *Estratte da 'Archiva Ecclesiae'* 24–25 (1981–82): 89–120.

———. *L'Histoire de la Compagnie de Jésus: Notes bibliographiques*. Brussels, 1930.

———. "Quattro secoli di attività scientifica tra i gesuiti." In *Quattro secoli: Nel IV centenario della morte di S. Ignazio di Loyola, fondatore della Compagnia di Gesù, 1556–1956*, 10–14. Pubblicazione periodica a cura del Comitato Nazionale Anno Ignaziano, 4–5.

Langford, Jerome J. *Galileo, Science, and the Church*. Ann Arbor: University of Michigan Press, 1971.

Lattis, James M. "Christopher Clavius and the *Sphere* of Sacrobosco: The Roots of Jesuit Astronomy on the Eve of the Copernican Revolution." Ph.D. diss., University of Wisconsin–Madison, 1989.

———. "Homocentrics, Eccentrics, and Clavius's Refutation of Fracastoro." *Physis* 28 (1991): 699–725.

Leibniz, Gottfried Wilhelm. *Philosophical Papers and Letters.* Dordrecht: D. Reidel, 1956.

Lemay, Richard. "The Teaching of Astronomy in Medieval Universities, Principally at Paris in the Fourteenth Century." *Manuscripta* 20 (1976): 197–217.

Lerner, Michel Pierre. "Le Problème de la matière céleste après 1550: Aspects de la bataille des cieux fluides." *Revue d'Histoire des Sciences* 42 (1989): 255–80.

———. *Tre saggi sulla cosmologia alle fine del Cinquecento.* Naples: Bibliopolis, 1992.

Lewis, C. S. *The Discarded Image: An Introduction to Medieval and Renaissance Literature.* Cambridge: Cambridge University Press, 1964.

Lindberg, David C. *The Beginnings of Western Science: The European Scientific Tradition in Philosophical, Religious, and Institutional Context, 600 B.C. to A.D. 1450.* Chicago: University of Chicago Press, 1992.

———. *A Catalogue of Medieval and Renaissance Optical Manuscripts.* Toronto: Pontifical Institute of Medieval Studies, 1975.

———. "Laying the Foundations of Geometrical Optics: Maurolyco, Kepler, and the Medieval Tradition." In *The Discourse of Light from the Middle Ages to the Enlightenment,* 3–65. See Lindberg and Cantor.

———. "On the Applicability of Mathematics to Nature: Roger Bacon and His Predecessors." *British Journal for the History of Science* 15 (1982): 3–25.

———, ed. *Science in the Middle Ages.* Chicago: University of Chicago Press, 1978.

Lindberg, David C., and Geoffrey Cantor, eds. *The Discourse of Light from the Middle Ages to the Enlightenment.* Los Angeles: William Andrews Clark Memorial Library, University of California, Los Angeles, 1985.

Lindberg, David C., and Ronald L. Numbers, eds. *God and Nature: Historical Essays on the Encounter between Christianity and Science.* Berkeley and Los Angeles: University of California Press, 1986.

Lindberg, David C., and Robert S. Westman, eds. *Reappraisals of the Scientific Revolution.* Cambridge: Cambridge University Press, 1990.

Litt, Thomas. *Les Corps célestes dans l'univers de Saint Thomas d'Aquin.* Louvain: Publications Universitaires, 1963.

Loria, Gino. *Storia delle matematiche.* Vol. 2, *I secoli XVI e XVII.* Turin: Società Tipografico Editrice Nazionale, 1931.

Lovejoy, Arthur O. *The Great Chain of Being: A Study of the History of an Idea.* Cambridge: Harvard University Press, 1936.

Lukács, Ladislaus, ed. *Monumenta paedagogica Societatis Iesu.* Vol. 1, 1540–1556. Monumenta Historica Societatis Iesu 92. Rome: Institutum Historicum Societatis Iesu, 1965.

———. *Monumenta paedagogica Societatis Iesu.* Vol. 2, 1557–1572. Monumenta Historica Societatis Iesu 107. Rome: Institutum Historicum Societatis Iesu, 1974.

———. *Ratio atque institutio studiorum Societatis Iesu (1586, 1591, 1599).* Mon-

umenta paedagogica Societatis Iesu. Vol. 5. Monumenta Historica Societatis Iesu 129. Rome: Institutum Historicum Societatis Iesu, 1986.

Maccagni, Carlo, ed. *Saggi su Galileo Galilei.* Vol. 2. *Pubblicazioni del Comitato Nazionale per le Manifestazioni Celebrative,* vol. 3. Florence: G. Barbèra Editore, 1972.

Machamer, Peter. "Galileo and the Causes." In *New Perspectives on Galileo,* 161–80. See Butts and Pitt.

Maelcote, Odo. *Astrolabium Aequinoctiale.* Brussels, 1607.

Maeyama, Y., and W. G. Saltzer, eds. *Prismata: Naturwissenschaftsgeschichtliche Studien.* Wiesbaden: Franz Steiner Verlag, 1977.

Mahoney, Edward P. "Philosophy and Science in Nicoletto Vernia and Agostino Nifo." In *Scienza e filosofia all'Università di Padova nel quattrocento,* 135–202. See Poppi.

Maierù, Luigi. "Il quinto postulato euclideo in Cristoforo Clavio." *Physis* 20 (1978): 191–212.

Mascart, Jean. *Clavius et l'astrolabe.* "Extrait du *Bulletin astronomique,* 1905." Paris: Imprimerie Gauthier-Villars.

Maurolico, Francesco. *Cosmographia: In tres dialogos distincta in quibus de forma, situ, numeroque tam coelorum quam elementorum, aliisque rebus ad astronomica rudimenta spectantibus satis differitur.* Venice, 1543.

———. The *"Photismi de Lumine"* of Maurolycus: A Chapter in Late Medieval Optics. Translated by Henry Crew. New York: Macmillan, 1940. Translation of the text edited for publication by Clavius.

McMenomy, Christe Ann. "The Discipline of Astronomy in the Middle Ages." Ph.D. diss. University of California, Los Angeles, 1984.

Monachino, Vincenzo. "La fortuna di Galileo in Oriente." In *Saggi su Galileo Galilei,* 812–30. See Maccagni.

Monumenta paedagogica Societatis Jesu quae primam Rationem studiorum anno 1586 editam praecessere. Monumenta Historica Societatis Iesu 19. Madrid, 1901.

Moore, James R. *The Post-Darwinian Controversies: A Study of the Protestant Struggle to Come to Terms with Darwin in Great Britain and America.* Cambridge: Cambridge University Press, 1979.

Moss, Jean Dietz. *Novelties in the Heavens: Rhetoric and Science in the Copernican Controversy.* Chicago: University of Chicago Press, 1993.

Moyer, Gordon. "Aloisius Lilius and the *Compendium Novae Rationis Restituendi Kalendarium.*" In *Gregorian Reform of the Calendar,* 171–88. See Coyne, Hoskin, and Pedersen.

Murr, Christoph Gottlieb von. *Merkwürdigkeiten der fürstbischöflichen Residenzstadt Bamberg.* Nuremberg, 1799.

Naux, Charles. "Le Père Christophore Clavius (1537–1612): Sa vie et son oeuvre." Pt. 1, "L'Homme et son temps." Pt. 2, Clavius astronome." Pt. 3, "Clavius mathématicien." *Revue des Questions Scientifiques* 154 (1983): 55–67, 181–93, 325–47.

Nifo, Agostino. *Preclara et admodum omnibus aliis in hac scientia resolutior Augustini Niphi Suessani in quattuor libros de celo et mundo et Aristote. et Avero. expositio.* Naples, 1517.

Norlind, Wilhelm. "A Hitherto Unpublished Letter from Tycho Brahe to Christopher Clavius." *Observatory* 74 (1954): 20–23.

Nuñez, Pedro. *De arte atque ratione navigandi libri duo. Eiusdem in theoricas planetarum Georgii Purbachii annotationes.* Coimbra, 1573.

———. *Obras.* Vol. 1. *Tratado da & Astronomici Introductorii de Spaera Epitome,* 245–67. Lisbon: Academia Das Ciêcias de Lisboa, 1940.

O'Dell, C. R., and A. Van Helden. "How Accurate Were Seventeenth-Century Measurements of Solar Diameter?" *Nature* 330 (1987): 629–31.

O'Malley, John W. *The First Jesuits.* Cambridge, Mass.: Harvard University Press, 1993.

Ong, Walter J. *Ramus: Method, and the Decay of Dialogue.* Cambridge, Mass.: Harvard University Press, 1958.

Orbaan, J. A. F., ed. *Documenti sul barocco in Roma.* Rome: R. Società Romana di Storia Patria, 1920.

Oresme, Nicole. *Le Livre du ciel et du monde.* Edited by Albert D. Menut and Alexander J. Denomy. Madison: University of Wisconsin Press, 1968.

Osler, Margaret J., ed. *Atoms, Pneuma, and Tranquillity: Epicurean and Stoic Themes in European Thought.* Cambridge: Cambridge University Press, 1991.

Osler, Margaret J., and Paul Lawrence Farber, eds. *Religion, Science, and Worldview: Essays in Honor of Richard S. Westfall.* Cambridge: Cambridge University Press, 1985.

Pachtler, G. M. *Ratio studiorum et institutiones scholasticae Societatis Jesu per Germaniam olim vigentes.* Edited by Karl Kehrbach. Tomus 1. *Monumenta germaniae paedagogica* 2. Berlin: A. Hofmann, 1887.

———. *Ratio studiorum et institutiones scholasticae Societatis Jesu per Germaniam olim vigentes collectae concinnatae dilucidatae.* Edited by Karl Kehrbach. Tomus 2. *Ratio studiorum ann. 1586, 1599, 1832. Monumenta Germaniae paedagogica* 5. Berlin: A. Hofmann, 1887.

———. *Ratio studiorum et institutiones scholasticae Societatis Jesu per Germaniam olim vigentes collectae concinnatae dilucidatae.* Edited by Karl Kehrbach. Volumen 3. *Monumenta Germaniae paedagogica* 9. Berlin: A. Hofmann, 1890.

Pannekoek, Anton. *A History of Astronomy.* London: George Allen & Unwin, 1961.

Pedersen, Olaf. "Astronomy." In *Science in the Middle Ages,* 303–37. See Lindberg.

———. "The Corpus Astronomicum and the Traditions of Medieval Latin Astronomy." *Colloquia Copernicana,* vol. 3. *Studia Copernicana* 13. Wroclaw: Polish Academy of Sciences Press, 1975.

———. "In Quest of Sacrobosco." *Journal for the History of Astronomy* 16 (1985): 175–221.

————. "The Origins of the 'Theorica Planetarum.'" *Journal for the History of Astronomy* 12 (1981): 113–23.

————. *A Survey of the Almagest*. Odense: University Press, 1974.

————. "The Theorica Planetarum Literature of the Middle Ages." *Cahiers de L'Institut du Moyen-Age Grec et Latin* 23 (1962): 225–32.

————. "The Theorica-planetarum Literature of the Middle Ages." *Proceedings of the Tenth International Congress of the History of Science* 1 (1962): 615–18.

Pedersen, Olaf, and Mogens Pihl. *Early Physics and Astronomy: A Historical Introduction*. New York: Science History Publications, 1974. Rev. ed. Cambridge: Cambridge University Press, 1993.

Pellegrini, F. "Un veronese precursore della riforma del calendario: Girolamo Fracastoro." *Studi Storici Veronese* 4 (1953): 135–44. Also published as *Raccolta Monografica di Studi Storici Veronese, 14*.

Peruzzi, Enrico. "Note e ricerche sugli 'Homocentrica' di Girolamo Fracastoro." *Rinascimento*, 2d ser., no. 25 (1985): 247–68.

Peurbach, Georg. *Theoricae novae planetarum*. Translated by E. J. Aiton. *Osiris*, 2d ser., no. 3 (1987): 5–44.

————. *Theoricae novae planetarum*. In *Opera collectanea*, 753–93. See Regiomontanus.

Phillips, Edward C. "The Correspondence of Father Christopher Clavius S.I. Preserved in the Archives of the Pontifical Gregorian University." *Archivum Historicum Societatis Iesu* 8 (1939): 193–222.

————. "The Proposals of Father Christopher Clavius, S.J., for Improving the Teaching of Mathematics." *Bulletin of the American Association of Jesuit Scientists, Eastern Section* 18, no. 4: 203–8.

Piccolomini, Alessandro. *Della Grandezza della Terra et dell'Acqua*. Venice, 1561.

————. *La Prima Parte delle Theoriche overo Speculationi de i Pianeti*. Venice, 1568.

————. *De la Sfera del Mondo & De le Stelle Fisse*. Venice, 1548.

Pifferi, Francesco (Sansovino). *Sfera di Gio. Sacrobosco Tradotta e Dichiarata*. Siena, 1604.

Popkin, Richard H. "Theories of Knowledge." In *Cambridge History of Renaissance Philosophy*, 668–84. See Schmitt.

Popkin, Richard H., and Charles B. Schmitt, eds. *Scepticism from the Renaissance to the Enlightenment*. Wiesbaden: Otto Harrassowitz, 1987.

Poppi, Antonio, ed. *Scienza e filosofia all'Università di Padova nel quattrocento*. Padua: Edizioni Lint, 1983.

Price, Derek J. de S. "Contra-Copernicus: A Critical Re-estimation of the Mathematical Planetary Theory of Ptolemy, Copernicus, and Kepler." In *Critical Problems in the History of Science*, edited by Marshall Clagett. Madison: University of Wisconsin Press, 1969.

Ptolemy. *Almagest*. Translated by G. J. Toomer. New York: Springer Verlag, 1984.

Rashdall, Hastings. *The Universities of Europe in the Middle Ages.* Vol. 2. Edited by F. M. Powicke and A. B. Emden. Reprint, London: Oxford University Press, 1936.

Regiomontanus, Johannes. *Opera collectanea: Faksimiledrucke von neun Schriften Regiomontans und einer von ihm gedruckten Schrift seines Lehrers Purbach.* Edited by Felix Schmeidler. Osnabrück: Otto Zeller Verlag, 1972.

Ribadeniera, Petro, Philippo Alegamba, and Nathanaele Sotvello. *Bibliotheca Scriptorum Societatis Iesu.* 1676. With an introduction by A. F. Allison. Reprint. Rome: Gregg International Publishers, 1969.

Ricci, Agostino. *De motu octave sphere: Opus mathematica atque philosophia plenum, ubi tam antiquorum quam iuniorum errores luce clarius demonstrantur.* Trino, 1513.

Riccioli, Giambattista. *Almagestum Novum.* Bologna, 1651.

Rose, Paul Lawrence. *The Italian Renaissance of Mathematics: Studies on Humanists and Mathematicians from Petrarch to Galileo.* Travaux d'Humanisme et Renaissance 140. Geneva: Librairie Droz, 1975.

Rose, Stewart. *St. Ignatius Loyola and the Early Jesuits.* London: Burns and Oates, 1891.

Rosen, Edward. "Copernicus' Spheres and Epicycles." *Archives Internationales d'Histoire des Sciences* 25 (1975): 82–92.

———. "The Dissolution of the Solid Celestial Spheres." *Journal of the History of Ideas* 46 (1985): 13–31.

———. "Francesco Patrizi and the Celestial Spheres." *Physis* 26 (1984): 305–24.

———. "Maurolico's Attitude toward Copernicus." *Proceedings of the American Philosophical Society* 101, no. 2 (April 1957): 177–94.

———. *The Naming of the Telescope.* New York: Henry Schuman, 1947.

Ross, Alexander. *The New Planet No Planet: or, the Earth No Wandring Star except in the Wandring Heads of Galileans.* London, 1646.

Rossi, Paolo. "Francesco Patrizi: Heavenly Spheres and Flocks of Cranes." In *Italian Studies in the Philosophy of Science,* 363–88. See Dalla Chiara.

———. *Immagini della scienza.* Rome: Editori Riuniti, 1977.

———. "La negazione delle sfere e l'astrobiologia di Francesco Patrizi." In *Il rinascimento nelle corti padane,* 401–37. See Rossi.

Rossi, Paolo, ed. *Il rinascimento nelle corti padane: Società e cultura.* Bari: De Donato editore, 1977.

———. "Sfere celesti e branchi di gru." In *Immagini della scienza,* 109–47. See Rossi.

Saitta, Giuseppe. *Il pensiero italiano nell'umanesimo e nel Rinascimento.* Vol. 2, *Il Rinascimento.* Florence: G. C. Sansoni Editore, 1961.

Salomone, Mario, trans. *Ratio atque institutio studiorum Societatis Jesu: L'ordinamento scolastico dei collegi dei gesuiti.* Milan: Feltrinelli Editore, 1979.

Sayili, Aydin. "An Early Seventeenth Century Persian Manuscript on the Tychonic System." *Anatolia* 3 (1958): 84–87.

Scaduto, Mario. *Catalogo dei gesuiti d'Italia: 1540–1565.* Rome: Institutum Historicum Societatis Iesu, 1968.

————. *L'Epoca di Giacomo Lainez: Il governo, 1556–1565.* Vol. 3 of *Storia della Compagnia di Gesù in Italia.* Rome: Edizioni Civiltà Cattolica, 1964.

————. "Galileo e i Gesuiti del Collegio Romano." In *Saggi su Galileo Galilei.* Florence: G. Barbèra Editore, 1967.

————. "Il matematico Francesco Maurolico e i gesuiti." *Archivum Historicum Societatis Iesu* 18 (1949): 126–41.

Scaglione, Aldo. *The Liberal Arts and the Jesuit College System.* Amsterdam: John Benjamins, 1986.

Scheibel, Johann Ephraim. *Astronomische Bibliographie.* [to 1590] Breslau, 1786.

————. *Astronomische Bibliographie.* [1591–1600] Breslau, 1787.

Scheiner, Christoph. *Rosa Ursina sive Sol.* Bracciani, 1630.

Schiaparelli, Giovanni. "Le sfere omocentriche di Eudosso, di Callippo, e di Aristotele." *Scritti sulla Storia della Astronomia antica.* Vol. 1, pt. 2. Bologna: Nicola Zanichelli Editore, 1926.

Schmitt, Charles B. *The Aristotelian Tradition and Renaissance Universities.* London: Variorum Reprints, 1984.

————. *Aristotle and the Renaissance.* Cambridge, Mass.: Harvard University Press, 1983.

————. *A Critical Survey and Bibliography of Studies on Renaissance Aristotelianism.* Padua: Editrice Antenore, 1971.

————. "Galilei and the Seventeenth-Century Text-book Tradition." In *Novità celesti e crisi del sapere,* 217–28. See Galluzzi.

————. "Toward a Reassessment of Renaissance Aristotelianism." *History of Science* 11 (1973): 159–93.

————, ed. *The Cambridge History of Renaissance Philosophy.* Cambridge: Cambridge University Press, 1988.

Schofield, Christine Jones. *Tychonic and Semi-Tychonic World Systems.* New York: Arno Press, 1981.

————. "The Tychonic and Semi-Tychonic World Systems." In *Planetary Astronomy from the Renaissance to the Rise of Astrophysics,* 33–44. See Taton and Wilson.

Schreiber, P. Johann. "P. Christoph Scheiner, S.J., und seine Sonnenbeobachtungen." *Natur und Offenbarung* 48 (1903): 1–20, 78–93, 145–58, 209–21.

Shank, Michael H. "The 'Notes on al-Biṭrūjī' Attributed to Regiomontanus: Second Thoughts." *Journal for the History of Astronomy* 23 (1992): 15–30.

Shapiro, Barbara J. *John Wilkins, 1614–1672: An Intellectual Biography.* Berkeley: University of California Press, 1969.

Shea, William R. "Galileo and the Church." In *God and Nature,* 114–35. See Lindberg and Numbers.

————. "Melchior Inchofer's 'Tractatus Syllepticus': A Consultor of the Holy Office Answers Galileo." *Novità celesti e crisi del sapere,* 283–92. See Galluzzi.

"The Solar Eclipse of 1567." *Nature* 15 (1877): 342.

Sommervogel, Carlos. *Bibliothèque de la Compagnie de Jésus.* Paris, 1891. Reprint, Paris, 1960.

————. *Dictionnaire des ouvrages anonymes et pseudonymes publiés par des religieux de la Compagnie de Jésus*. Paris: Librairie de la Société Bibliographique, 1884.

Swerdlow, Noel. "Aristotelian Planetary Theory in the Renaissance: Giovanni Battista Amico's Homocentric Spheres." *Journal for the History of Astronomy* 3 (1972): 36–48.

————. "On Copernicus' Theory of Precession." In *The Copernican Achievement*, 49–98. See Westman.

————. "Pseudodoxia Copernicana: or, Enquiries into Very Many Received Tenents and Commonly Presumed Truths, Mostly Concerning Spheres." *Archives Internationales d'Histoire des Sciences* 26 (1976): 108–58.

Swerdlow, Noel, and Otto Neugebauer. *Mathematical Astronomy in Copernicus's "De Revolutionibus."* New York: Springer-Verlag, 1984.

Taton, R., and C. Wilson, eds. *Planetary Astronomy from the Renaissance to the Rise of Astrophysics. Pt. A: Tycho Brahe to Newton.* Cambridge: Cambridge University Press, 1989.

Taub, Liba Chaia. *Ptolemy's Universe: The Natural Philosophical and Ethical Foundations of Ptolemy's Astronomy.* Chicago: Open Court Publishing, 1993.

Thomas Aquinas. *Exposition of Aristotle's Treatise "On the Heavens."* Translated by R. F. Larcher and Pierre H. Conway. Columbus, Ohio: College of St. Mary of the Springs, 1963.

Thoren, Victor E. "The Comet of 1577 and Tycho Brahe's System of the World." *Archives Internationales d'Histoire des Sciences* 29 (1979): 53–67.

————. *The Lord of Uraniborg: A Biography of Tycho Brahe.* Cambridge: Cambridge University Press, 1990.

Thorndike, Lynn. *A History of Magic and Experimental Science.* 8 vols. New York: Columbia University Press, 1923–41.

————. "Some Little Known Astronomical and Mathematical Manuscripts." *Osiris* 8 (1948): 41–72.

————. *The Sphere of Sacrobosco and Its Commentators.* Chicago: University of Chicago Press, 1949.

————, ed. *Latin Treatises on Comets between 1238 and 1368 A.D.* Chicago: University of Chicago Press, 1950.

Toomer, G. J. "A Survey of the Toledan Tables." *Osiris* 15 (1968): 5–174.

Twain, Mark. "Thirty Thousand Killed a Million." *Atlantic Monthly* 269, no. 4 (April 1992): 52–65.

Underwood, E. Ashworth, ed. *Science, Medicine, and History.* London: Oxford University Press, 1953.

Unguru, Sabetai, ed. *Physics, Cosmology, and Astronomy, 1300–1700: Tension and Accommodation.* Boston Studies in the Philosophy of Science 126. Dordrecht: Kluwer Academic Publishers, 1991.

Van Helden, Albert. "The 'Astronomical Telescope.' " *Annali dell'Istituto e Museo di Storia della Scienza di Firenze* 1 (1976): 13–36.

————. "Galileo on the Sizes and Distances of the Planets." *Annali dell'Istituto e Museo di Storia della Scienza di Firenze* 7 (1982): fascs. 265–86.

———. "The Invention of the Telescope." *Transactions of the American Philosophical Society* 67 (June 1977): 4.

———. *Measuring the Universe: Cosmic Dimensions from Aristarchus to Halley.* Chicago: University of Chicago Press, 1985.

———. "The Telescope and Cosmic Dimensions." In *Planetary Astronomy from the Renaissance to the Rise of Astrophysics,* 106–18. See Taton and Wilson.

Vasoli, Cesare. "Francesco Patrizi sull'infinità dell'universo." In *Filosofia e cultura: Per Eugenio Garin,* edited by Michele Ciliberto and Cesare Vasoli, 277–308. Rome: Editori Riuniti, 1991.

Villoslada, Riccardo G. *Storia del collegio romano dal suo inizio (1551) alla soppressione della Compagnia di Gesù (1773).* Rome: Gregorian University, 1954.

Wallace, William A. "Causes and Forces at the Collegio Romano." In *Prelude to Galileo,* 110–26. See Wallace.

———. "The Certitude of Science in Late Medieval and Renaissance Thought." *History of Philosophy Quarterly* 3 (July 1986): 281–91.

———. *Galileo and His Sources: The Heritage of the Collegio Romano in Galileo's Science.* Princeton: Princeton University Press, 1984.

———. "Galileo and Reasoning 'Ex Suppositione.' " In *Prelude to Galileo,* 129–59. See Wallace.

———. "Galileo Galilei and the *Doctores Parisienses.*" In *New Perspectives on Galileo,* 87–138. See Butts and Pitt.

———. "Galileo's Early Arguments for Geocentrism and His Later Rejection of Them." In *Novità celesti e crisi del sapere. Atti del Convegno Internazionale di Studi Galileiani,* 31–40. Edited by P. Galluzzi. Florence: G.Barbèra, 1984.

———. *Galileo's Early Notebooks: The Physical Questions. A Translation from the Latin, with Historical and Paleographical Commentary.* Notre Dame, Ind.: University of Notre Dame Press, 1977.

———. *Galileo's Logic of Discovery and Proof: The Background, Content, and Use of His Appropriated Treatises on Aristotle's "Posterior Analytics."* In Boston Studies in the Philosophy of Science 137. Dordrecht: Kluwer Academic Publishers, 1992.

———. *Galileo's Logical Treatises: A Translation, with Notes and Commentary, of His Appropriated Latin Questions on Aristotle's "Posterior Analytics."* Boston Studies in the Philosophy of Science 138. Dordrecht: Kluwer Academic Publishers, 1992.

———. *Prelude to Galileo: Essays on Medieval and Sixteenth-Century Sources of Galileo's Thought.* In Boston Studies in the Philosophy of Science 62D. Dordrecht: Reidel, 1981.

———. "The Problem of Apodictic Proof in Early Seventeenth-Century Mechanics: Galileo, Guevara, and the Jesuits." *Science in Context* 3 (1989): 67–87.

———, ed. *Reinterpreting Galileo.* Studies in Philosophy and the History of Philosophy 15. Washington, D.C.: Catholic University of America Press, 1986.

Westfall, Richard S. *Essays on the trial of Galileo*. Notre Dame, Ind.: University of Notre Dame Press, 1989.

———. "The Rise of Science and the Decline of Orthodox Christianity: A Study of Kepler, Descartes, and Newton." In *God and Nature*, 218–37. See Lindberg and Numbers.

———. "The Trial of Galileo: Bellarmino, Galileo, and the Clash of Two Worlds." *Journal for the History of Astronomy* 20 (1989): 1–23.

Westman, Robert. "The Astronomer's Role in the Sixteenth Century: A Preliminary Study." *History of Science* 18 (1980): 105–47.

———. "The Comet and the Cosmos: Kepler, Mastlin, and the Copernican Hypothesis." *Colloquia copernicana*, vol. 1. Studia Copernicana 5:57–30. Wroclaw: Polish Academy of Sciences Press, 1972.

———. "The Copernicans and the Churches." In *God and Nature*, 76–113. See Lindberg and Numbers.

———. "The Melanchthon Circle, Rheticus, and the Wittenberg Interpretation of the Copernican Theory." *Isis* 66 (1975): 165–93.

———. "Three Responses to the Copernican Theory: Johannes Praetorius, Tycho Brahe, and Michael Maestlin." In *The Copernican Achievement*, 285–345. See Westman.

———, ed. *The Copernican Achievement*. Berkeley: University of California Press, 1975.

Wilkins, John. *A Discourse Concerning a New Planet*. London, 1640.

———. *The Discovery of a World in the Moone*. London, 1638.

———. *Mathematical and Philosophical Works*. 2 vols. London: C. Whittingham, 1802.

Winkler, Mary G., and Albert Van Helden. "Representing the Heavens: Galileo and Visual Astronomy." *Isis* 83 (1992): 195–217.

Zaccagnini, Guido. *Bernardino Baldi nella vita e nella opere*. 2d. ed. Pistoia: Soc. An. Tipo-Litografica Toscana, 1908.

Ziggelaar, August. "The Papal Bull of 1582 Promulgating a Reform of the Calendar." In *Gregorian Reform of the Calendar*, 201–39. See Coyne, Hoskin, and Pedersen.

Zinner, Ernst. *Deutsche und Niederländische astronomische Instrumente des 11–18 Jahrhunderts*. Munich: C. H. Becksche Verlagsbuchhandlung, 1956.

———. *Entstehung und Ausbreitung der Coppernicanishen Lehre*. Erlangen, 1943.

———. *Geschichte und Bibliographie der astronomischen Literatur in Deutschland zur Zeit der Renaissance*. Edited by Anton Hiersemann. 1941. Reprint, Stuttgart, 1964.

———. *Leben und Wirken des Joh. Müller von Königsberg gennannt Regiomontanus*. Osnabrück: Otto Zeller, 1968.

Index

Accademia dei Lincei, 188, 202
Achillini, Alessandro, 87, 90, 94, 109, 111, 134; *De orbibus,* 90
Acosta, José de, 232n.14
Air, clarity of, 91, 92
Aiton, E. J., 233n.19
Albategnius, 75, 164. *See also* al-Battānī
Albertus Magnus, 79
Alfonsine astronomers, 115, 147, 164, 166, 171
Alfonsine Tables, 32, 41, 175
Alfonso X, 63, 231n.7
Almagest. See under Ptolemy
Alperio, Gaspar, 24
Alphraganus, 126. *See also* al-Farghānī
Ambrose, Saint, 151
Amico, Giovanni Battista, 87, 89, 90, 109, 111, 134, 137
Amico, Bartolomaeo, 122
Anaxagoras, 124
Anaximander, 124
Andalo di Negro, 95
Apian, Peter, 43, 147
Apogee, 55, 77
Apollonius, 55, 80
Apse, 55
Aquaviva, Claudio, 6
Archimedes, 45

Arcturus, 155
Aristarchus, 106, 117, 118, 126, 133
Aristotelianism, 6, 7, 30, 66, 93, 94, 97, 113, 116, 141, 142, 146, 151–54, 157–59, 214, 218; diversity of in Renaissance, 87
Aristotle, 4, 6, 28, 33, 42, 50, 55, 61, 65, 70, 71, 80, 81, 88, 90, 95, 108, 119, 128, 192, 195, 203; Logic, 130; on comets, 147; cosmology, 60; *De caelo,* 61, 66, 79, 85, 90; *Metaphysics,* 79; on homocentrics, 53, 54; order of planets, 53; *Physics,* 61, 66; *Posterior Analytics,* 16, 34, 36
Arneth, Konrad, 224n.39
ARSI, 12, 14, 26
Astrolabe, 16
Aufgebauer, Peter, 10, 250n.5
Aux, 55
Averroës, 61, 79, 88, 90, 92, 109, 111, 116, 126, 127, 134, 141, 219
Averroism, 87, 91, 113, 117, 127, 146
Averroists, 80, 110

Bacon, Roger, 68, 79, 89
Baldi, Bernardino, 12, 14–17, 22

Diurnal motion, 45, 47, 71–73, 94, 100, 103, 104, 111, 112, 158
Diurnal path, 47
Diversitas aspectus. See Parallax, diurnal
Diversity of aspects. *See* Parallax, diurnal
Donahue, William, 10, 181, 230n.2
Donne, John, 7, 8, 29
Drake, Stillman, 28, 184
Dreyer, J. L. E., 86, 148, 173, 230n.5
Duhem, Pierre, 139, 247n.65

Earth: centrality and stability of, 50, 65; centrality of, 81; mobility of, 81; motion of, 50, 201, 202, 212; rotation of, 96
Eccentric, 66, 79–81, 88, 116, 128, 132, 135; circle, 55; materialized, 60, 68–70, 73, 78, 80, 110; monster or portent, 142; simple, 69, 115; solar, 131, 134; virtual, 69, 80, 84, 115–17, 162
Ecclesiastes, 123
Eclipse, lunar, 26, 114, 119, 135, 141, 156, 207; solar, 5, 12, 15, 18, 75
Eclipse, solar, 15, 18
Eclipse diagram, 74
Ecliptic, 47
Egyptians, 76
Empyrean heaven, 72, 82, 85, 96, 179
Epicycle, 55–57, 66, 79, 80, 88, 116, 128, 132, 135, 209; materialized, 60, 68–70, 73, 78, 110; monster or portent, 142; solar, 131, 134
Equant, 55, 57, 58, 69, 70, 102, 111, 143; circle, 57; materialized, 69
Equinox, 47, 48, 54
Eudoxus, 51, 88, 137
Evans, James, 230n.48

Faber, Johann (Giovanni), 239n.47, 264n.108

Falsehood derived from truth, 133, 134
al-Farghānī, 77, 78. *See also* Alphraganus
Favaro, Antonio, 161, 186
Feingold, Mordechai, 222n.5, 256n.83
Filliucci, Vincenzo, 24
Finé, Oronce, 43, 83
Firmament, 46, 50, 67
Flamsteed, John, 42
Fluid heavens, 62, 83, 86, 106, 114, 130, 145, 180, 201, 207, 209, 210, 219
Fonseca, Pedro, 14
Foscarini, Paolo Antonio, 181, 202, 204
Fracastoro, Girolamo, 87–90, 93, 96, 106, 109, 111, 113, 115, 116, 127, 134, 137, 144, 147, 219; *Homocentrica,* 90, 94; sublunar sphere, 91, 92
Frischlin, Nicodemus, 134
Frisius, Gemma, 43
Froidmond, Libert, 7, 28
Fugger, Georg, 7

Gabbey, Alan, 253n.49
Galilei, Galileo, 1, 2, 4–7, 10, 12, 27, 35, 38, 45, 113, 122, 124, 153, 161, 214, 215, 218; *Dialogue,* 86; "Letter to the Grand Duchess," 28, 181; *Sidereus nuncius,* 11, 24, 26, 28, 180, 183, 198; telescopic discoveries, 5, 63, 108, 146, 160, 179, 181, 210, 212, 218
Gallanzoni, Gallanzone, 259n.46
Gassendi, Pierre, 5, 203
Gemma, Cornelius, 158
Generation and corruption, 128
Genesis, 215
Giannettasio, Nicola Partenio, 103
Giglio, Luigi, 20
Gingerich, Owen, 236n.61, 250n.3, 256n.88